THE PROSTAGLANDINS
VOLUME 3

THE PROSTAGLANDINS

VOLUME 3

Edited by Peter W. Ramwell

Department of Physiology and Biophysics
Georgetown University Medical Center
Washington, D. C.

PLENUM PRESS • NEW YORK AND LONDON

The Library of Congress cataloged the first volume of this title as follows:

Ramwell, Peter W
  The prostaglandins. Edited by Peter W. Ramwell. [1st ed.]
New York. Plenum Press. 1973-75
  v. 1-2 illus. 26 cm.
  Includes bibliographies.

  1. Prostaglandin. 1. Title. [DNLM: 1. Prostaglandins. QU90 R 184p]
QP801.P68R35   612'.405   72-76858

Library of Congress Catalog Card Number 72-76858
ISBN 0-306-37793-4

© 1977 Plenum Press, New York
A Division of Plenum Publishing Corporation
227 West 17th Street, New York, N.Y. 10011

Printed in the United States of America

Contributors

TOM P. BARDEN

Department of Obstetrics and Gynecology
University of Cincinnati Medical Center
Cincinnati, Ohio

MELVIN BLECHER

Department of Biochemistry
Georgetown University Medical Center
Washington, D.C.

K. E. COOPER

Division of Medical Physiology
Faculty of Medicine
The University of Calgary
Calgary, Alberta, Canada

WALTER FLAMENBAUM

Department of Nephrology
Walter Reed Army Institute of Research
Washington, D.C.
Current address:
Veterans Administration Hospital
Boston, Massachusetts

J. C. FRÖLICH

Departments of Medicine and Pharmacology
Vanderbilt University
Nashville, Tennessee

VIVIAN GOLDBERG

Department of Physiology
Georgetown University Medical Center
Washington, D.C.

BERNARD M. JAFFE

Department of Surgery
Washington University School of Medicine
St. Louis, Missouri

JAMES G. KENIMER

Department of Biochemistry
Georgetown University Medical Center
Washington, D.C.

JACK G. KLEINMAN

Department of Nephrology
Walter Reed Army Institute of Research
Washington, D.C.
Current address:
Veterans Administration Hospital
Milwaukee, Wisconsin

PETER A. KOT

Department of Physiology and Biophysics
Georgetown University Medical Center
Washington, D.C.

LAWRENCE LEVINE

Department of Biochemistry
Brandeis University
Waltham, Massachusetts

ALEKSANDER A. MATHÉ

Laboratory of Biogenic Amines and
Allergy, and Departments of Psychiatry
and Pharmacology
Boston University School of Medicine
Boston, Massachusetts

Q. J. PITTMAN

Division of Medical Physiology
Faculty of Medicine
The University of Calgary
Calgary, Alberta, Canada

SHENG-SHUNG PONG

Department of Biochemistry
Brandeis University
Waltham, Massachusetts

ANDRÉ ROBERT

Department of Experimental Biology
The Upjohn Company
Kalamazoo, Michigan

JOHN C. ROSE

Department of Physiology and Biophysics
Georgetown University Medical Center
Washington, D.C.

M. GABRIELLA SANTORO

Department of Surgery
Washington University School of Medicine
St. Louis, Missouri

W. L. VEALE

Division of Medical Physiology
Faculty of Medicine
The University of Calgary
Calgary, Alberta, Canada

Contents

Chapter 4

Induction of Labor with Prostaglandins

Tom P. Barden

Chapter 5

Cardiovascular Responses to the Prostaglandin Precursors

John C. Rose and Peter A. Kot

Chapter 6
Role of Prostaglandins in Fever and Temperature Regulation
W. L. Veale, K. E. Cooper, and Q. J. Pittman

Chapter 7
Prostaglandins and the Lung
Aleksander A. Mathé

Chapter 8
Prostaglandins and the Digestive System
André Robert

Gas Chromatography–Mass
Spectrometry of Prostaglandins

J. C. Frölich

Departments of Medicine and Pharmacology
Vanderbilt University
Nashville, Tennessee 37232

I. INTRODUCTION

The purpose of this chapter is to review the contributions mass spectrometry has made to the field of prostaglandin research and show how mass spectrometry can be used in the qualitative and quantitative analysis of prostaglandins, thromboxanes, and their metabolites. The rapid advancement of technology in the field of instrument design and electronics has greatly facilitated the application of mass spectrometry to biological problems. From the initial application of stable isotopes as tracers in biological systems (Hevesy, 1923), the application of mass spectrometry (Rittenberg *et al.*, 1939) has proven to be a dramatic advance since it can provide a very sensitive and accurate method for measurement of isotope ratios. Measurement of isotope ratios was utilized to trace the fate of ^{15}N-labeled animo acids (Schoenheimer *et al.*, 1939), ^{13}C-labeled CO_2 (Wood *et al.*, 1940), and ^{18}O-labeled sodium sulfate (Aten and Hevesy, 1938). As early as 1939 it was recognized that mass spectrometry was "highly sensitive and accurate and most impurities are without effect on the analysis. The procedure is rapid and applicable to routine" (Rittenberg *et al.*, 1939). Even though it took about 25 years for mass spectrometric analysis to become truly routine, the most important content of this quote—namely the ability of the mass spectrometer to provide specificity of analysis—indicates the major asset of this method. Initially, mass spectrometers were built by the investigators. When instruments became available commercially, they were designed predominantly for use by chemists and physicists and could be used for analysis of samples of biological origin only with difficulty. This was because sample introduction was accomplished via direct inlet or heated inlet, neither of which methods is suitable for many biological samples which contain impurities interfering with the analysis. Gas chromatography with its high separating power offers the necessary resolution for study of complex mixtures of biological origin. The interfacing of the gas chromatograph with the mass spectrometer accomplished by Gohlke (1959) and Ryhage (1964) provided instrumentation that has been applied very successfully to the qualitative and quantitative analysis of biological materials. Furthermore, the purification of samples prior to gas chromatography has been an area of considerable interest because a high degree of purification is a prerequisite for gas chromatographic–mass spectrometric analysis. In addition to silicic acid and reversed-phase partition chromatography (Bygdeman and Samuelsson, 1966; Norman, 1953), high-performance liquid chromatography (HPLC) has provided a powerful new instrument for this purpose (Carr *et al.*, 1976).

Mass spectrometry has been utilized extensively for the identification and quantification of substances in biological samples (for review, see McFadden, 1973). Applications of mass spectrometry were of great importance in the structure elucidation of the primary prostaglandins. This work formed the basis for the subsequent work on the biochemistry, physiology, pharmacology, and chemical syntheses of prostaglandins.

A. Identification of Prostaglandin Synthetase Dependent Substances

Initial attempts at structure elucidation of a biologically active material isolated from sheep prostate gland were carried out by Bergström and Sjövall (1960*a,b*). Elemental analysis and mass spectrometric analysis established that the two compounds isolated were unsaturated acids with molecular weights of 356 (PGF$_1$) and 354 (PGE$_1$) with the elemental compositions $C_{20}H_{36}O_5$ (PGF$_1$) and $C_{20}H_{34}O_5$ (PGE$_1$). Further work by Bergström *et al.* (1962*a*) produced mass spectra of PGE$_1$, PGF$_{2\alpha}$, and PGF$_{1\alpha}$ and the correct chemical structure was proposed by Bergström *et al.* (1962*b*, 1963). Mass spectrometry was used extensively to identify the side-chain degradation products obtained from these prostaglandins, as well as for the determination of the sites of attachment of the side chains to the hydroxycyclopentyl ring and the location of the keto group (PGE) within the ring. Two new prostaglandins, PGD_2 and $PGF_{2\alpha}$, were identified by similar methods by Bergström *et al.* (1962*d*), and Samuelsson (1963*b*) established the structure of PGE_3. Once the exact chemical nature of the prostaglandins had been established, it became much easier to show whether compounds with similar biological properties were or were not prostaglandins. In a study of various tissues from sheep, it was discovered by Bergström *et al.* (1962*c*) that the lung contained a smooth muscle stimulating substance. Extraction of 880 lb of swine lung yielded a few milligrams of this compound, quite enough to give a complete mass spectrum which identified this substance unequivocally as PGE_2. This was the first reported occurrence of prostaglandins outside of the reproductive system and showed in an exemplary way the approach to definitive identification of suspected prostaglandin compounds. Gas chromatography–mass spectrometry was utilized for the demonstration of $PGF_{3\alpha}$ in bovine lung (Samuelsson, 1964*a*), in bovine brain (Samuelsson, 1964*b*), and prostaglandins in human seminal plasma (Hamberg and Samuelsson, 1965, 1966), including the novel 19-hydroxyprostaglandins. Outside of the mammaliam organism, 15(*S*)-prostaglandin–PGA_2 has been identified mass spectron etrically in *Plexaura homamalla* (Bundy *et al.*, 1972; Schneider *et al.*, 1972). This material now is chemically modified for large-scale production of PGE_2, $PGF_{2\alpha}$, and PGA_2.

After it had been established by Bergström *et al.* (1964) and van Dorp *et al.* (1964) that arachidonic acid was the precursor of the prostaglandins E_2 and $F_{2\alpha}$, the origin of the oxygen atoms was studied with the help of $^{18}O_2$. Ryhage and Samuelsson (1965) and Nugteren and van Dorp (1965) arrived at the conclusion that both of the hydroxyl functions of PGE originated from molecular oxygen; however, the origin of the oxygen atom at C-9 remained obscure. In elegant experiments using a mixture of $^{18}O_2$ and $^{16}O_2$ in the biosynthesis of PGE_1, Samuelsson (1965) demonstrated that the oxygen atoms at C-9 and C-11 were either both ^{18}O or both ^{16}O, thus proving that the oxygen in these positions must have been derived from the same molecule. This finding led to the postulate of an endoperoxide intermediate, the structure of which was proposed for the first time in Samuelsson's communication. This endoperoxide was subsequently isolated by Nugteren and Hazelhoff (1973) and Hamberg and Samuelsson (1973) and shown to rearrange to 11-dehydro-$PGF_{2\alpha}$, PGE_2, and PGF_{2a}, all three of which were identified by gas chromatography–mass spectrometry (Hamberg and Samuelsson, 1973). The exact mechanism of conversion of the endoperoxide to PGE_2 and $PGF_{2\alpha}$ has been studied by Wlodawer and Samuelsson (1973) by mass spectrometric measurements of isotope ratios of the products formed from octadeuteroarachidonic acid. This work revealed that one endoperoxide is the common precursor for both PGE_2 and $PGF_{2\alpha}$ and that elimination of the hydrogen of C-9 is the rate-limiting step for the formation of PGE_2. In 1974 Hamberg and Samuelson described hitherto unknown transformation products of arachachidonic acid in platelets from human blood including 12-L-hydroxy-5,8,10-heptadecatrienoic acid (HHT) and thromboxane B_2. The structural assignment was largely accomplished by gas chromatographic–mass spectrometric analysis of the parent compounds and their fragments obtained by oxidative ozonolysis.

B. Metabolites of Prostaglandins

The first report on the metabolism of prostaglandins revealed that PGE_1 was converted by enzymes in lung tissue to 15-keto-PGE_1 and 15-keto-dihydro-PGE_1 (Änggård and Samuelsson, 1964) and that analogous transformations occurred with PGE_2, PGE_3, and $PGF_{2\alpha}$ (for overview, see Änggård and Samuelsson, 1967). The sequence of these steps in the lung was elucidated by Änggård and Larsson (1971) and in the guinea pig liver by Hamberg and Samuelsson (1971a). The pathway leading to the dinor and tetranor analogues of these metabolites of $PGF_{2\alpha}$ was discovered by Granström et al. (1965), Granström and Samuelsson (1969a, 1971), and Gréen (1971), and for PGE_1 and PGE_2 by Hamberg and Samuelsson (1971b).

Later work carried out with $PGF_{2\alpha}$ in the rat (Sun, 1974) and the monkey (Sun and Stafford, 1974) confirmed the principal metabolic pathways and indicated that β and ω oxidation could proceed further when the prostaglandin was given by constant infusion. A review of prostaglandin metabolism has appeared recently (Samuelsson et al., 1971). The biological activity of the metabolites is low or absent in comparison to the parent prostaglandins (Änggård and Samuelsson, 1966) and therefore the means of discovering and differentiating between them are limited. In all of the work on metabolism of prostaglandins mentioned above, gas chromatography–mass spectrometry played a crucial role in the structural assignment of the metabolites of PGE_2 and the 14 metabolites of $PGF_{2\alpha}$ isolated from primate urine. The importance of our understanding of the exact chemical pathways of metabolism is readily appreciated when one considers the impact this information had on the development of synthetic prostaglandin analogues [i.e., 16,16-dimethyl-PGE_2 and 15(S)-15-methyl-PGE_2] as well as on our ability to monitor prostaglandin synthesis by measuring prostaglandin metabolites in blood plasma and urine (Samuelsson et *al.,* 1975; Granström and Kindahl, 1976; Gréen *et al.,* 1973, 1974; Samuelsson and Gréen, 1974; Hamberg, 1972, 1973; Granström and Samuelsson, 1971).

C. Quantification

Gas chromatography–mass spectrometry has been successfully applied to the quantitative analysis of prostaglandins, prostaglandin metabolites, and thromboxanes. The principle of utilization of the mass spectrometer as a highly specific detector for a gas chromatograph was first demonstrated by Henneberg (1961), who employed the mass spectrometer to monitor the ion current profile of a single selected mass in the effluent from a gas chromatograph. Subsequently, Sweeley *et al.* (1966) developed an instrument that could monitor two ion current profiles, allowing the use of a stable isotope analogue of a biological substance as an internal standard. Such an internal standard should behave identically to its biological counterpart on extraction, purification, and derivatization and should be differentiated from the biological material only in the final step of analysis by the mass spectrometer. One of the main difficulties in the application of this methodology to prostaglandin analysis resulted from the poor gas chromatographic properties of the PGE and PGF derivatives at the very low levels encountered in biological samples. This problem was overcome by Samuelsson *et al.* (1970), who added a deuterated prostaglandin analogue in large excess to the sample prior to gas chromatographic analysis. This deuterated material thus served the dual purpose of internal standard and carrier. With this procedure, it was possible to analyze prostaglandins at the subnanogram level by a physical method for the first time. The deuterated internal standard in this case was synthesized by reacting PGE$_1$ with d$_3$-methoxylamine. A significant advancement in this method was made when PGE$_2$ and PGF$_{2\alpha}$ were synthesized with the deuterium incorporated into the 3,3,4,4-positions (Axen *et al.,* 1971). These deuterated prostaglandins now functioned as "ideal" internal standards in that they could be added directly to the biological sample and served as internal standards through all steps of the procedure including extraction, purification, and derivatization (Axen *et al.,* 1971). This method was extended to include the major circulation metabolites of PGE$_2$ and PGF$_{2\alpha}$, namely 15-keto-13,14-dihydro-PGE$_2$ and 15-keto-13,14-dihydro-PGF$_{2\alpha}$, after the appropriate deuterated internal standards had been synthesized (Gréen *et al.,* 1973; Samuelsson and Gréen, 1974). Measurement of these metabolites in peripheral plasma can reflect acute changes in prostaglandin biosynthesis in most cases more accurately than measurement of the primary prostaglandins since the primary prostaglandins are generated quite rapidly from platelets during the blood collection process whereas these metabolites are not generated by blood elements *in vitro*. The speed with which levels of these metabolites increase following an intervention is demonstrated in the experiments of Gréen *et al.* (1974) in which an allergen-provoked asthma produced increased levels of 15-keto-13,14-dihydro-PGF$_{2\alpha}$ within minutes after exposure to the allergen. Measurement of the major urinary metabolites of PGE (Hamberg, 1972) and PGF (Hamberg, 1973) has also been accomplished. These methods allowed determination of total body turnover rates for these prostaglandins in man and were useful to show the effect of pharmacological inhibition of prostaglandin synthesis on these rates. A recent review discusses these methods (Gréen *et al.*, 1976).

There is considerable discrepancy between the values reported for the primary prostaglandins and metabolites measured by different methods. This problem became apparent when a comparison was made between various laboratories that measured levels of PGF$_{2\alpha}$ from the same set of plasma samples by radioimmunoassay and by gas chromatography–mass spectrometry (Samuelsson, 1973*b*). There were also large discrepancies between the levels of 15-keto-dihydro-PGF$_{2\alpha}$ measured by radioimmunoassay (Levine, 1973) and gas chromatography–mass spectrometry, and a similar problem arose when blood plasma was analyzed for PGA$_2$ (for comparison of radioimmunoassay and gas chromatography–mass spectrometry methods, see Frölich *et al.*, 1975*c;* Steffenrud, 1976). In most cases, the gas chromatography–mass spectrometry method revealed values which were between ten- and a thousandfold lower than those obtained by radioimmunoassay. This would appear to indicate that the gas chromatographic–mass spectrometric method is more specific and therefore should be made the reference method. Furthermore, on the basis of the metabolic rate of clearance from plasma and rate of appearance of prostaglandin metabolites in urine, it has been calculated that the levels of prostaglandins and 15-keto-dihydroprostaglandins correspond only to those reported by gas chromatography–mass spectrometry (Samuelsson, 1973*a*). Thus measurement of prostaglandins and prostaglandin metabolites by radioimmunoassay can be accepted only if the assay has been compared to a completely independent and specific method under the same experimental conditions. The common practice of checking the assay specificity by measuring cross-reactivities with prostaglandins, thromboxanes, and their metabolites is limited in that many of the metabolites are not available and indeed are not even known at the time of this writing. In some cases, radioimmunoassays of prostaglandin metabolites have been compared to gas chromatography–mass spectrometry with favorable results (Granström and Kindahl, 1976; Axen *et al.*, 1973). A detailed discussion of this problem has appeared recently (Samuelsson *et al.*, 1975).

II. MASS SPECTROMETRY SYSTEMS FOR PROSTAGLANDIN ANALYSIS

A. Magnetic Focusing Instruments

Currently two types of mass spectrometers which operate on different physical principles are utilized for prostaglandin analysis. The first system is based on momentum separation of ions passing through a magnetic field (magnetic sector mass spectrometer) and the second system accomplishes mass separation by application of direct current and radiofrequency electrical fields to four parallel poles (quadrupole mass spectrometer). Up to the present time, time-of-flight mass spectrometers have not been utilized for prostaglandin analysis. In principle, mass spectrometers can be used for qualitative identification or for quantitative analysis. In the first case a complete mass spectrum is obtained, in the second case only selected ions are monitored (selected ion monitoring, multiple ion detection, mass fragmentography) to obtain a profile of ion abundance versus time, a tracing which is similar to a gas chromatogram. An alternate way of obtaining ion abundance profile is to scan either the entire mass spectrum or a brief portion thereof, storing the information and later reconstructing the desired profiles (repetitive scanning). For an application of this latter method to gas chromatographic–mass spectrometric analysis of prostaglandins, see Axen *et al.*, (1972, 1973). In gas chromatographic–mass spectrometric systems, the effluent of the gas chromatographic column is introduced into the mass spectrometer via a separator that removes most of the carrier gas (Fig. 1). In the ion source, ionization of the sample molecules occurs by bombardment with electrons of a certain energy generated from a rhenium filament and accelerated by a trap electrode. This bombardment with electrons will ionize and fragment the sample molecules. The percent ionization and extent of fragmentation depend, in part, on the constitution of the molecule, on the electron energy, and on the temperature of the ion source. The positively charged molecular ions and fragment ions resulting from ionization are accelerated into the mass analyzer. A portion of the ion beam (about 10%) is directed toward an electrode, and its output signal results in a tracing comparable to the FID output of a gas chromato-

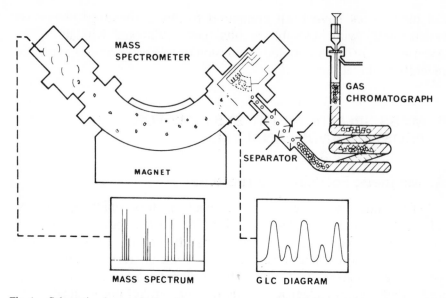

Fig. 1.   Schematic of magnetic focusing mass spectrometer showing introduction of sample into and separation of various compounds on gas chromatograph, removal of carrier gas in separator, and ionization chamber. Instrument is shown in the scanning mode producing a mass spectrum. Courtesy of LKB Instruments, Inc., Rockville, Md.

graph and is referred to as total ion current. The analyzer tube is curved (usually 60°) and has a magnet at its curvature (Fig. 1). In order to obtain a complete mass spectrum, the magnetic field is continuously increased while a constant amount of sample molecules is entering the ion source. This causes ions of increasing mass to impinge on the electron multiplier which is positioned behind an exit slit. The relationship between the mass of the ion (m), its charge (e) (usually 1), the magnetic field strength (H), and the accelerating voltage (V) is given by

$$m/e = R^2H^2/2V \tag{1}$$

where R is the radius of the curvature of the analyzer tube (a constant). If V is held constant, then at a particular value of H only one specific ion of m/e will pass through the exit slit to reach the electron multiplier. By continuously changing H, one ion after another is brought into focus and its abundance is registered. The resulting mass spectrum is specific for each compound and therefore has been called the fingerprint of the molecule. The high sensitivity of the mass spectrometer usually allows one to obtain a complete mass spectrum of a sample weighing between about 1 and 0.1 μg.

   Two examples may illustrate the application of gas chromatography–mass spectrometry to problems we had encountered in our work on renal and platelet prostaglandins. We had discovered material in urine of human females that behaved chromatographically like $PGF_{2\alpha}$ and gave a contractile response of the rat stomach strip that was antagonized by SC 19220 (Frölich *et al.*, 1973). Definitive identification of this material as $PGF_{2\alpha}$ was attempted by extensive purification of 1500 ml of urine from human females by four different sequential open-column chromatographic procedures. The material was analyzed by gas chromatography–mass spectrometry on an LKB-9000 instrument and a complete mass spectrum was obtained (Fig. 2). Comparison of the spectrum of the biological material with a mass spectrum of authentic PGF_2 showed identity in all major ions and thus provided conclusive evidence for the presence of $PGF_{2\alpha}$ in urine (Frölich *et al.*, 1973). In work with human blood platelets, we found that on incubation of washed platelets with arachidonic acid and purification by high-pressure liquid chromatography (HPLC, see later) radioactively labeled arachidonic acid was converted to material that behaved chromatographically identically to PGE_2. However, the amount of label appearing in the PGE_2 fraction was much greater than would have been anticipated on the basis of previously published data (Hamberg *et al.*, 1974). A methylester-tris-trimethylsilyl derivative of the material in the PGE fraction was prepared and a mass spectrum was obtained. Prominent ions at *m/e* 585 ($M - 15$), 529 ($M - 71$), 510 ($M - 90$), 439 ($M - 71 - 90$), and 420 ($M - 2 \times 90$) (Fig. 3) revealed that this compound was identical with thromboxane B_2 (formerly PHD) (Hamberg and Samuelsson, 1974). This finding provided a ready explanation for the excessive amounts of label found in the prostaglandin E fraction: a considerable amount of the labeled arachidonic acid was converted to thromboxane B_2, which cochromatographed on HPLC with prostaglandin E_2.

For quantitative analysis, at least two ions are usually monitored, one representing the biological compound and one an internal standard, which is frequently a deuterated analogue of the biological sample. In order to monitor these two ions with a single detector, the instrument must be focused for some time on one ion and then on the other ion. The time spent on each ion (dwell time) should be long enough to sample enough ions to obtain a detector response yet short enough to obtain an undistorted profile for each ion. A common dwell time is on the order of about 100 msec. In order to switch from one ion to another, it is theoretically possible to change H or V (see equation 1). However, because of hysteresis of the magnet it is more advantageous to change V as suggested by Sweeley *et al.* (1966). For this purpose, the instrument is focused on the lower mass unit and variable resistors (which can be connected to the accelerating voltage power

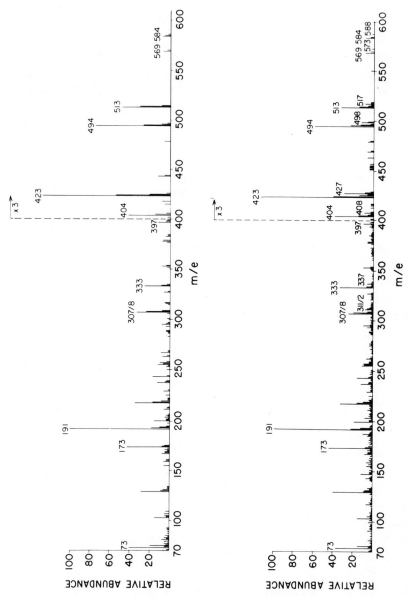

Fig. 2. Mass spectrum of PGF$_{2\alpha}$-methylester-tris-trimethylsilylether (top) and material isolated from human female urine to which 3,3,4,4-tetradeutero-PGF$_{2\alpha}$ had been added (bottom). The spectra are identical in all major ions (Frölich *et al.*, 1975e). LKB-9000 mass spectrometer.

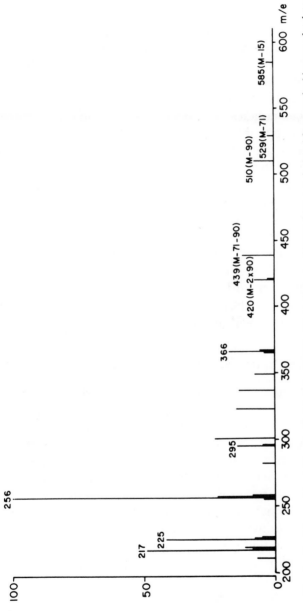

Fig. 3. Partial mass spectrum of material derivatized to the methylester-TMS, which was isolated from washed human platelets that were incubated with arachidonic acid:Thromboxane B_2 (PHD).

supply) are adjusted to focus other ions on the detector. The accelerating voltage is then rapidly switched between two (or more) settings, thus bringing each ion into focus on the detector. The identification of a compound depends now on the detection of one or more representative ions appearing at the appropriate retention time. Monitoring of the abundance of a particular ion can be accomplished conveniently with a strip-chart recorder. For an original ion tracing, see Fig. 16. Quantification depends on the determination of the ratio of the abundance of ions representing the internal standard and the biological material. Measurement of this ratio has been accomplished by measuring peak area, central peak area, or peak height. In order to increase specificity of detection, it is possible to monitor a different set of ions out of the mass spectrum of the biological compound and the internal standard after refocusing. However, one limitation in the application of this method using magnetic focusing instruments is the limited mass range. At significantly reduced accelerating voltage (>20%), the ion optics become suboptimal, resulting in a loss of sensitivity and distortion of the ion ratio. The operational conditions for the magnet at high masses are suboptimal because a strong magnetic field is needed. Under these conditions, the magnet warms up and tends to drift out of focus in some instruments. In order to reduce this problem, the magnet is usually allowed to equilibrate for several hours before use and frequent recalibration is performed throughout the operation.

B. Quadrupole Instruments

The quadrupole mass spectrometer was first described theoretically by Paul and Steinwedel in 1953 and a prototype was built and evaluated in 1955 (Paul and Raether, 1955). Ions are generated from the effluent of the gas chromatograph in a manner analogous to that of the magnetic focusing instrument. The mass filter operates on the following principle: four rods are arranged in parallel and connected in pairs (as shown in Fig. 4), and when an alternating voltage ($U_0 \cos \omega t$) is applied to the two pairs of rods an ion in the center between the four rods will be set into oscillating motion by the electrical field generated by the alternating voltage. Two situations emerge: the ion either increases its amplitude of oscillation and crashes into one of the rods (unstable flight path) or it oscillates with a constant amplitude and reaches the detector (stable flight path) (Fig. 4). From theoretical considerations (Paul and Raether, 1955), it can be shown that three different mass-region ions will have stable flight paths at a certain frequency ωt so that the filter has limited selectivity. However, when in addition to the alternating voltage a d.c. voltage (U_g) is applied, it is

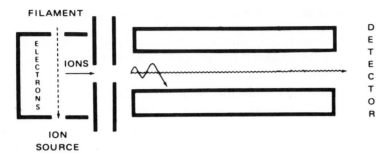

Fig. 4.   Schematic of quadrupole mass spectrometer showing the flight paths of two ions between the rods; one is stable and will reach the detector, one is unstable and will be terminated by contact with one of the rods.

possible to eliminate two of the three mass regions entirely and to narrow the first (highest) region by manipulation of U_g and $U_0 \cos \omega t$, so that only a specific mass (m/e) can pass through the filter. Transmission and resolution can be traded off readily by manipulation of the ratio U_g/U_0 (Paul and Raether, 1955). The resolution of the instrument is markedly affected by the numbers of oscillations an ion will undergo before it reaches the detector; therefore, the initial velocity of the ion at the time of entry into the mass filter and the length of the rods are of importance: the slower the velocity and the longer the rods, the better the resolution. In practice, resolutions of 500–700 are achieved in most quadrupole instruments, but resolution as high as 1000 or more has been achieved (Paul and von Zahn, 1961). There are several conditions that must be met in order to achieve good performance of the instrument. Apart from the necessity for accuracy and constancy of the a.c. and d.c. voltages, the precision of the rod configuration and alignment are critical. The rods have to be mounted in parallel with tolerances of a few micrometers. The quadrupole instrument has a number of attractive properties. The energy range of the ions entering the mass analyzer is not critical. Therefore, simple and highly efficient ion sources can be used. The transmission, i.e., the ratio of the number of ions of a specific mass entering the mass analyzer to the number of these ions leaving the analyzer, is high and this increases sensitivity. Finally, the mass analyzer can be manipulated very rapidly and fast scanning rates can be achieved. This is useful for repetitive scanning of a segment of the mass spectrum for characteristic ions in quantitative analysis and in the selected ion-monitoring mode. In contrast to the magnetic focusing instruments, any ion within the mass range of the instrument can be monitored. It is easy with modern quadrupole instruments to maintain exact focus on the selected ions for more than 18 hr and the instrument can be focused within minutes.

C. Choice of Internal Standard and Derivative

Gas chromatography–mass spectrometry affords the possibility of using an internal standard that has nearly "ideal" properties in that one can use a compound that is different from the biological compound only by virtue of its molecular weight; it may contain ^{13}C, ^{18}O, or deuterium. In prostaglandin research, deuterated analogues have been used most often. These analogues can be added directly to a biological sample and behave identically to the biological sample on extraction, on derivatization, and, in most instances, on chromatography.

Only in the final step of analysis will the difference between the internal standard and the biological material become measurable because the mass spectrometer can recognize the small mass difference between the biological material and the internal standard. Standard curves for PGA_2, PGE_2, and PGF_2 (Figs. 5, 6, and 7) illustrate this approach. To vessels containing a constant amount of 3,3,4,4-tetradeutero-PGA_2 (d_4-PGA_2) increasing amounts of PGA_2 are added. Monitoring ions m/e 349 (for PGA_2) and m/e 353 (for d_4-PGA_2), one can see that with increasing amounts of PGA_2 the ratio increases linearly. To get an impression of the sensitivity of the method, the actual amount of PGA_2 present in the aliquot injected into the gas chromatograph has

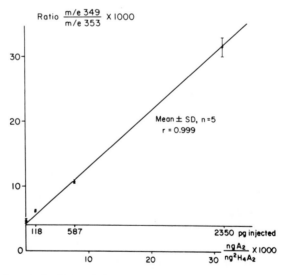

Fig. 5. Standard curve for PGA_2-methylester-trimethylsilylether. The lower (bottom) abscissa indicates the ratio PGA_2 and 3,3,4,4-tetradeutero-PGA_2 obtained by adding increasing amounts of PGA_2 to a fixed amount of 3,3,4,4-tetradeutero-PGA_2. The abscissa above indicates the actual amount of PGA_2 injected. The ordinate gives the ratio of ion m/e 349 (corresponding to PGA_2) and m/e 353 (corresponding to 3,3,4,4-tetradeutero-PGA_2) measured (LKB-9000 instrument). Nor correction for the "blank" has been made.

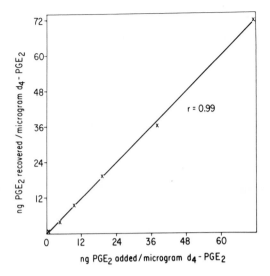

Fig. 6. Standard curve for PGE_2-methylester-methoxime-bisacetate. Increasing amounts of PGE_2 were added to a fixed amount of 3,3,4,4-tetradeutero-PGE_2 (d_4-PGE_2). The amount of PGE_2 recovered (ordinate) was calculated from the ratio of ions *m/e* 419 (corresponding to PGE_2) and *m/e* 423 (corresponding to d_4-PGE_2) measured on a Hewlett-Packard model 5982A quadrupole mass spectrometer. Data were corrected for blank.

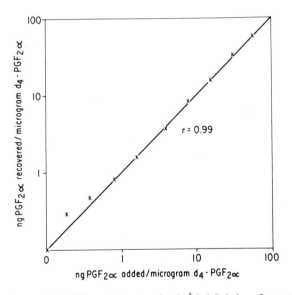

Fig. 7. Standard curve for $PGF_{2\alpha}$-methylester-tris-trimethylsilylether. Increasing amounts of $PGF_{2\alpha}$ were added to a fixed amount of 3,3,4,4-tetradeutero-$PGF_{2\alpha}$(d_4-$PGF_{2\alpha}$). The ratios of ions *m/e* 423 (corresponding to $PGF_{2\alpha}$) and *m/e* 427 (corresponding to d_4-$PGF_{2\alpha}$) were corrected for blank and used to calculate the amount of $PGF_{2\alpha}$ recovered (ordinate). Note logarithmic scale. Measurement on Hewlett-Packard model 5982A quadrupole mass spectrometer.

been indicated. It is important to note that the line connecting the data points does not go through the origin. This is due to the fact that on injection of the internal standard by itself a measurable response is obtained not only for ion m/e 353 but also for ion m/e 349. This response ("blank") usually is on the order of 2.5–5 units (arbitrary units of peak height) for ion m/e 349 when the response is 1000 units for ion m/e 353. It is important to recognize that the ability to accurately and reproducibly measure this blank critically determines the ultimate sensitivity and accuracy of a gas chromatographic–mass spectrometric method. The blank therefore has to be determined repeatedly in each run, and the ratios measured for generation of a standard curve or a biological sample have to be corrected for this value. Figures 6 and 7, for example, are shown corrected for this blank value.

We have made a comparison between the LKB-9000 and Hewlett-Packard 5982A mass spectrometers with respect to measurement of the blank of PGA_2-methylester-trimethylsilylether using d_4-PGA_2-methylester-trimethylsilylether (Table I). The blank on the former instrument is slightly larger and shows a greater standard deviation than on the latter instrument. Therefore, the quadrupole instrument will allow measurement of biological samples with levels close to the blank with an accuracy similar to that of the magnetic focusing instrument.

The choice of a derivative is of great importance in achieving favorable assay conditions. Several requirements have to be met: derivatization should be quantitative, and the derivative should be stable, have good gas chromatographic properties (i.e., show little adsorption and thermal decomposition), and have a fragmentation pattern that results in few ions of high mass. The advantages of such a fragmentation pattern are obvious. If the molecule fragments in only a few ions, the relative abundance of each will be high, resulting in increased sensitivity. If the fragments are of high mass, the specificity of the assay is enhanced since the likelihood of a measurable contribu-

Table 1. Blank Determination for PGA$_2$-Me-TMS (m/e 349) Using d$_4$-PGA$_2$-Me-TMS (m/e 353)

Instrument	n^a	Abundance of ion m/e 353	Abundance of ion m/e 349
LKB-9000	6	1000	$5.19 \pm 0.39\,(\pm 7.5\%)$
HP-5982A	6	1000	$3.29 \pm 0.10\,(\pm 3\%)$

^{a} Number of determinations.

tion of the biological background to the ion under observation is decreased. The generation of suitable derivatives provides a continual challenge for the organic chemist.

The use of a deuterated internal standard in prostaglandin analysis is of particular importance when the internal standard serves the dual purpose of internal standard and carrier (Samuelsson *et al.*, 1970), which is a requirement for low-level analysis when the gas chromatographic properties are poor. In the case of PGA_2-methylester-trimethylsilylether, one encounters very favorable gas chromatographic properties so that as little as 113 pg could be measured quantitatively without the use of a deuterated carrier (Sweetman *et al.*, 1973*a*). This property of the PGA derivatives was used to advantage when in experiments on the origin of PGA_2 it became necessary to use two internal standards simultaneously and PGA_1 was chosen as an internal standard for the measurement of PGA_2 (Frölich *et al.*, 1976*a*). Even though PGA_1 was separated completely from PGA_2 by the gas chromatagraph, a linear standard curve was obtained with the lowest point representing about 200 pg PGA_2.

III. METHODS OF EXTRACTION AND CHROMATOGRAPHY

An important prerequisite for gas chromatographic—mass spectrometric analysis of material of biological origin is extensive purification. A "dirty" sample will lead to buildup of deposits on the gas chromatographic column with resultant loss of resolution and peak broadening causing loss of sensitivity. In addition, the introduction of large amounts of organic material into the mass spectrometer may lead to deterioration of separator efficiency, high blank values, deterioration of the ion source, and possibly disturbed measurement of the ion ratios in case of quantitative analysis.

A. Extraction of PGs, Their Circulating Metabolites, and Thromboxane B_2 from Blood Plasma

Numerous methods for the extraction and purification of PGs and their metabolites have been described. These include extraction by adsorption to Amberlite XAD-2 (Gréen, 1971) and solvent extraction (Unger *et al.*, 1971). The advantage of the former method is that it is equally useful for the extraction of all the primary prostaglandins as well as for the major circulating and urinary metabolites of PGE_2 and $PGF_{2\alpha}$ from their respective matrices in man and animal. We have found solvent extraction (Fig. 8) useful for the extraction of PGA_2,

```
┌─────────────────────────────────────────────────────────────┐
│  Collect blood in 10% (v/v) of 3.2% aqueous (w/v) sodium citrate │
│                                                               │
│                        Centrifuge                             │
│                                                               │
│       Add internal and tritiated standard(s) to plasma        │
│                                                               │
│       Precipitate with −20°C acetone                          │
│                                                               │
│                        Centrifuge                             │
│                                                               │
│       Supernatant, wash with and discard petroleum ether      │
│                                                               │
│       Adjust aqueous phase pH 3,4 with formic acid            │
│                                                               │
│       Extract with chloroform                                 │
│                                                               │
│       Roto-evaporate and azeotrope with ethanol              │
│                                                               │
│       High performance liquid chromatography                  │
│                                                               │
│       Derivatization                                          │
│                                                               │
│       Gas chromatography-mass spectrometer analysis           │
└─────────────────────────────────────────────────────────────┘
```

Fig. 8. Extraction of prostaglandins and thromboxane B_2 from blood plasma.

PGE_2, and $PGF_{2\alpha}$, 15-keto-13,14-dihydro-PGE_2, 15-keto-13,14-dihydro-$PGF_{2\alpha}$, and thromboxane B_2 from blood plasma. This method exploits the weakly acidic properties of the prostanoids for separation from neutral lipids. In the initial step, the protein is precipitated with ice-cold acetone, which results in a very fine precipitate. Neutral lipids are removed by washing with petroleum ether. The petroleum ether also removes part of the acetone and reduces the volume of the aqueous phase. This is of advantage in the next step when the pH of the aqueous phase is adjusted with concentrated formic acid to 3.6 and extraction into $CHCl_3$ is performed. Rotevaporation removes the $CHCl_3$ containing the PGs and has to be continued until all traces of formic acid (smell!) are removed. This is best accomplished by repeated (3–4 times) addition of ethanol and evaporation of the azeotrope. At this point, the recovery should be >90%. This method of extraction results in a sample that is sufficiently clean for further purification on HPLC.

B. Extraction of Prostaglandins from Urine

We have shown that urine contains prostaglandins (Frölich *et al.*, 1973) and that these prostaglandins are predominantly of renal origin

(Frölich *et al.*, 1975*e*), thus providing a parameter for measuring renal prostaglandin synthesis. Much of the initial work was performed analyzing urine by competitive protein-binding assay (Frölich *et al.*, 1974). To obtain data on different populations and under various drug treatments, an assay with greater accuracy was sought and therefore a method for analysis of urinary prostaglandins by gas chromatography–mass spectrometry was developed. Urine (10–50 ml) is extracted into chloroform after addition of the internal standards and acidification to pH 3.4 with formic acid. After flash evaporation of the chloroform phase, the residue is dissolved in ethylacetate and is applied to and eluted from a 500-mg SiO_2 (Mallinckrodt, CC-4) column with ethylacetate which elutes PGE_2 and $PGF_{2\alpha}$ almost quantitatively. The dry residue of the ethylacetate eluate is dissolved in 0.1 ml of methanol and treated with diazomethane (Sweetman *et al.*, 1973*a*) to form the prostaglandin methylesters. Further purification is accomplished by exploiting the altered polarities of the PG methylesters. They are dissolved in $CHCl_3$ and applied to a 500-mg SiO_2 column packed in $CHCl_3$. The column is washed with 20 ml of 20% ethylacetate–80% $CHCl_3$ (v/v) and the prostaglandin methylesters are eluted with 90% ethylacetate–10% $CHCl_3$ (v/v). Similar methods were not applicable to 7α-hydroxy-5,11-diketotetranorprosta-1,16-dioic acid (PGE-M), the major urinary metabolite of PGE_1 and PGE_2 in man, since it is too polar to be extracted in good yield into $CHCl_3$ and extraction with more polar solvents led to unmanageable amounts of contaminants. For the isolation of PGE-M, therefore, extraction with Amberlite XAD-2 as described by Gréen (1971) is the preferred method. The Amberlite resin retains PGE-M (and prostaglandins) avidly even when washed with copious amounts of water for the removal of salts. PGE-M is subsequently eluted with methanol. Similar to the procedure described for isolation of prostaglandins from urine, high-pressure liquid chromatography is preceded by methylation and SiO_2 chromatography of the Amberlite eluate.

C. Methods of Chromatography of Prostaglandins and Prostaglandin Metabolites

A great number of methods for chromatography of prostaglandins have been reported (for review, see Shaw and Ramwell, 1969). Methods for class separation include thin-layer chromatography (Gréen and Samuelsson, 1964; Wickramasinghe and Shaw, 1973), silicic acid chromatography (Samuelsson, 1963*a*; Bygdeman and Samuelsson, 1966), and Sephadex LH-20 chromatography (Änggård and Bergkvist, 1970).

Separation of individual prostaglandins on the basis of their degree of unsaturation has been accomplished by argentation chromatography (Gréen and Samuelsson, 1964) and partition chromatography (Bergström and Sjövall, 1960*b*; Hamberg and Samuelsson, 1966), the principles of which were described by Howard and Martin (1950) and Norman (1953). The latter method has proven to be extremely useful for the isolation of individual prostaglandins and metabolites. Most of these chromatographic methods suffer from variable and often low recoveries. In the case of partition chromatography, the column support has to be prepared, the column is temperature sensitive, recovery is not quantitative, the retention volume is variable, a new column has to be packed for each sample, the column requires several hours to run, and the prostaglandins appear dissolved in a sizable volume of water which requires time-consuming evaporation. Recently, high-pressure liquid chromatography has found numerous applications and taken its place alongside thin-layer and gas chromatography (for review, see Heftmann, 1975). Modern column-packing methods and pumps result in highly efficient separations on columns packed with very small particles which are eluted under high pressure and flow. Early application of high-pressure liquid chromatography to prosta-glandins (Dunham and Anders, 1973; Anderson and Leavey, 1974) showed that group separation could easily be accomplished as well as separation of synthetically prepared prostaglandin isomers. We have investigated the possibility of using high-pressure liquid chromatography for sample purification prior to analysis by gas chromatography–mass spectrometry. For this work we used a column (4 by 300 mm) prepacked with 10-μm particles of silica (Microporasil). These columns are made of stainless steel tubing with the inner wall finished in such a manner that channeling of the solvent is prevented. Hence the solvent moves at a relatively uniform flow rate throughout the diameter of the column. The relatively large inner diameter (4 mm instead of the usual 2 mm) enhances column capacity so that samples of 10 mg and more can be handled. Also, separation is not so much affected on these larger columns when the samples are injected dissolved in a larger volume, which may be valuable in cases of poor solubility of the sample in the initial solvent. The column is eluted with the help of two pumps controlled by a gradient control device (all equipment from Waters Assoc., Milford, Mass., pump model 60000, injector model U6K, and solvent programmer model 660). The samples are injected dissolved in 100–200 μl of the initial solvent. Separation of PGA_2, PGE_2, and $PGF_{2\alpha}$ could be achieved readily using solvents C and D (solvent C, chloroform; solvent D, chloroform, methanol, and acetic acid in a ratio

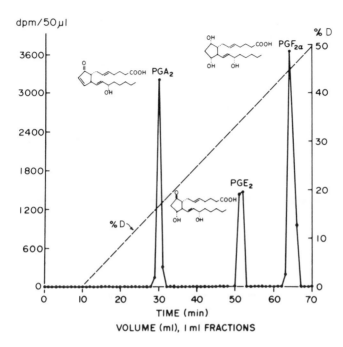

Fig. 9. High-pressure liquid chromatography of tritiated PGA_2, PGE_2, and $PGF_{2\alpha}$ isolated from 25 ml of human blood plasma. Note symmetrical peak shape. The dotted line indicates the percentage of solvent D.

of 500:50:11, v/v/v/) at a flow rate of 1 ml/min. Figure 9 shows a typical separation of these three prostaglandins added as tritiated tracers to and isolated from 25 ml of human blood plasma by the method described above using a linear gradient of 0–70% solvent D within 70 min preceded by 10 min of solvent C. The symmetrical peak shape shows that the amount of biological extract does not lead to an overloading of the column. Conversely, the amount of biological material is sufficient to prevent nonspecific adsorption, which may occur when only a few picograms of a very high specific activity prostaglandin are injected. The prostaglandins are eluted in only 2–3 ml of organic solvent. This is very convenient since it permits removal of the solvent under a stream of N_2 in a vessel that is small enough for derivatization. Reproducibility is very good. In seven consecutive samples of blood plasma the retention volumes (mean ± SD) were 31.1 ± 1.17 ml for PGA_2, 54 ± 0.76 ml for PGE_2, and 64 ± 0.6 ml for $PGF_2\alpha$. Recoveries from this column were greater than 90%. Human blood plasma of a normal volunteer pretreated with indomethacin (50 mg 3 times daily for 2 days) was obtained and spiked with increasing amounts of PGA_2, PGE_2, and $PGF_{2\alpha}$. 3,3,4,4-Tetradeutero-PGA_2, 3,3,-

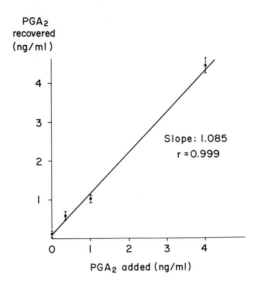

Fig. 10. Recovery of PGA$_2$ added to plasma of a normal subject pretreated with indomethacin (LKB-9000 instrument).

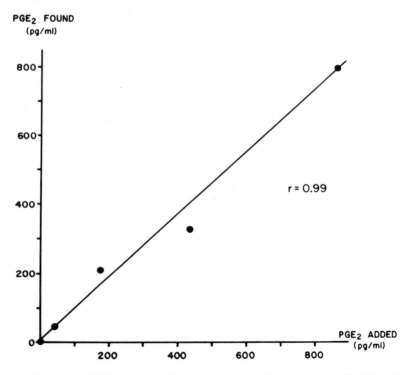

Fig. 11. Recovery of PGE$_2$ added to plasma of a normal subject pretreated with indomethacin (Hewlett-Packard 5982A instrument).

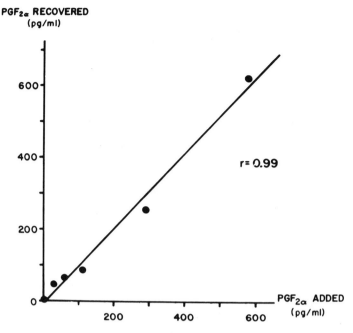

Fig. 12.   Recovery of PGF$_{2\alpha}$ added to plasma of a normal subject pretreated with indomethacin (Hewlett-Packard 5982A instrument).

4,4-tetradeutero-PGE$_2$, and 3,3,3,4-tetradeutero-PGF$_{2\alpha}$ were used as internal standards together with small amounts of tritiated analogues of these prostaglandins for peak localization. Figures 10, 11, and 12 show that the recovery of these spikes was quantitative.

The measurement of levels of 15-keto-dihydro-PGE$_2$ and 15-keto-dihydro-PGF$_{2\alpha}$, the major circulating metabolites of PGE$_2$ and PGF$_{2\alpha}$, has been utilized for monitoring prostaglandin-related events in plasma (Samuelsson and Gréen, 1974; Gréen *et al.*, 1974). The advantage of measuring these two parameters is that they are unaffected by the blood collection process that so significantly hampers determination of PGE$_2$ and PGF$_{2\alpha}$ because of the generation of these prostaglandins by blood platelets. High-pressure liquid chromatography is useful for conveniently analyzing PGE$_2$, PGF$_{2\alpha}$, and their major circulating metabolites in the same plasma sample. The appropriate four internal standards are added to blood plasma, and the plasma is extracted in the same manner as described for PGE$_2$. Utilizing the high-pressure liquid chromatography method for prostaglandins, one can achieve a satisfactory separation as indicated in Fig. 13. This method of chromatography is also quite useful for isolation of the tritiated metabolites from swine kidney incubations (Gréen *et al.*, 1973).

In the analysis of urinary prostaglandins, the purification and

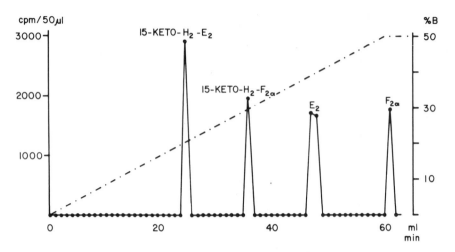

Fig. 13. Separation of tritiated PGE$_2$, PGF$_{2\alpha}$, 15-keto-13,14-dihydro-PGE$_2$, and 15-keto-13,14-dihydro-PGF$_{2\alpha}$ from human blood plasma. The dotted line indicates the percentage of second solvent.

separation of the methylesters of PGE$_2$ and PGF$_{2\alpha}$ were readily accomplished by high-pressure liquid chromatography (Fig. 14). The degree of purification was sufficient to measure levels of less than 100 pg/ml, which one encounters frequently in man after indomethacin therapy.

PGE-M can be purified by high-pressure liquid chromatography after its derivatization to the methylester-methoxime by isocratic elution with CHCl$_3$. Four isomers are obtained. The two major isomers represent about 70% of the total PGE-M. They can be pooled and further derivatized to the trimethylsilylether prior to gas chromatographic–mass spectrometric analysis.

In some applications it may be desirable to measure both the

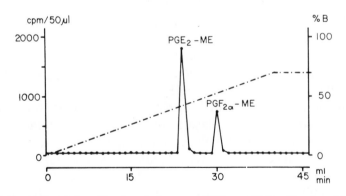

Fig. 14. Separation of the tritiated methylesters of PGE$_2$ and PGF$_{2\alpha}$. The dotted line indicates percentage of second solvent.

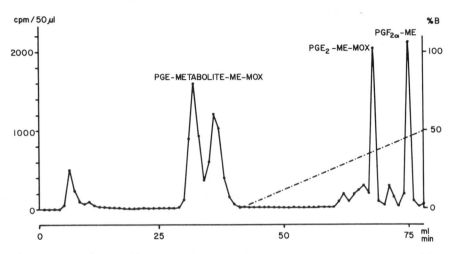

Fig. 15. Separation of tritiated 7α-hydroxy-9,11-diketotetranorprosta-1,16-dioic acid (urinary metabolite of PGE₁ and PGE₂), PGE₂, and PGF₂α recovered from human urine and treated with diazomethane and methoxyamine:HCl. The dotted line indicates the percentage of second solvent.

primary prostaglandins, i.e., PGE_2 and $PGF_{2\alpha}$ and PGE-M. For this purpose, one can conveniently extract the prostaglandins together with PGE-M from the same sample of urine using Amberlite XAD-2 and silicic acid open-column chromatography as described above. The mixture of all three compounds can then be derivatized to the methylester-methyloxime and chromatographed isocratically by high-pressure liquid chromatography with chloroform. This will elute the predominant two isomers of PGE-M-methylester-methoxime first. Subsequently, a linear program is started which will elute PGE_2-methylester-methoxime and $PGF_{2\alpha}$-methylester (Fig. 15). After each sample, the column is purified by washing it with 100% of the most polar solvent used for about 20 min at a flow rate of 1 ml/min. The column can be used many times (>200) without loss of efficiency.

The high separating power of this chromatographic system is further demonstrated in Fig. 16. Here a partial separation of tritiated and protonated PGE_2 is achieved. A mixture of 100,000 dpm of [³H]PGE_2 (specific activity 80 Ci/mmole, Amersham-Searle) and 10 μg PGE_2 was injected and eluted with a program used for separation of PGE_2 and $PGF_{2\alpha}$, collecting fractions of 0.5 ml. An aliquot of each fraction was analyzed for radioactivity by liquid scintillation counting and another aliquot was treated with methanolic KOH, thus converting PGE_2 to PGB_2. The absorbance was determined at 280 nm and was due nearly exclusively to the PGB_2 obtained from PGE_2 since the high

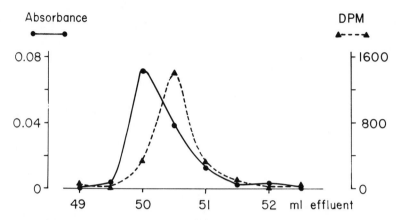

Fig. 16. Partial separation of [³H]-PGE₂ and d₄-PGE₂ on μ-Porasil. Solvents and program as Fig. 13. Fraction volume 0.5 ml. Analysis of d₄-PGE₂ by treatment with methanolic KOH (conversion to d₄-PGB₂) and measurement of [³H]-PGE₂ by liquid scintillation counting.

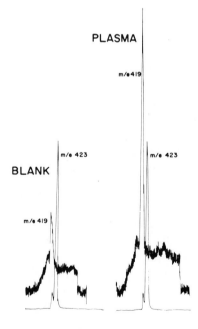

Fig. 17. Ion tracings of PGE₂ derivatized to the methylester-methoxime-bisacetate isolated from arterial blood plasma of dog in hemorrhagic shock. Blank: Ion profile obtained after injection of internal standard alone for blank determination. The bottom tracing (ion *m/e* 423) represents the internal standard; the upper tracing (ion *m/e* 419), recorded at 500-fold greater sensitivity, shows a small peak (blank) that appears to antecede the peak of ion *m/e* 423. This is due to displacement of the pens of the strip chart recorder used for data printout and is not the result of differences in retention time between PGE₂ derivative and the internal standard. Plasma: Ion profiles corresponding to internal standard (*m/e* 423, bottom tracing) and PGE₂ (ion *m/e* 419). The increase in the peak height of ion *m/e* 419 as compared to the blank is due to the presence of PGE₂ in this blood sample.

specific activity of the $[^3H]PGE_2$ precluded a significant contribution. It is important to recognize these marginal separations to avoid discarding most of the actual sample, i.e., those fractions that contain PGE_2 but little of the radioactive tracer. That in fact a partial separation of isotopes can be accomplished by chromatographic methods has been known for a long time. The ion profiles of glucose and d_7-glucose on analysis by gas chromatography–mass spectrometry (Sweeley *et al.*, 1966) showed a considerable difference in retention times of these two compounds.

IV. BIOLOGICAL APPLICATIONS

A. Circulating Prostaglandins

It is well recognized that the analysis of the primary prostaglandins in blood plasma may result in levels which do not reflect the *in vivo* situation since prostaglandins may be readily formed from blood platelets during blood sampling and handling (Samuelsson, 1973a; Samuelsson *et al.*, 1975). Therefore, measurement of the major metabolites in the circulation (15-keto-13,14-dihydro-PGE_2, and 15-keto-13,14-dihydro-$PGF_{2\alpha}$) has been suggested to gain insight into changes in the rate of prostaglandin synthesis as reflected in the circulation. However, measurement of these parameters allows only a statement as to the change in rate of prostaglandin synthesis to be made, while the site of synthesis remains obscure. For this reason, numerous efforts have been made to measure prostaglandin release from organs into their venous effluent or from the lung into the systemic circulation.

We became interested in the possibility that prostaglandins might be present in arterial blood following the report of Jakschik *et al.* (1974) indicating that the lung releases prostaglandins during hemorrhagic shock as determined by bioassay in chloralose-anesthetized dogs. We measured prostaglandin levels in blood plasma sampled from the right atrium and the carotid artery in pentobarbital-anesthetized and unanesthetized dogs. Whereas hemorrhage produced no increase in arterial PGE_2 levels in the former group, the latter group showed a striking increase. A typical ion profile representing PGE_2 is shown in Fig. 17 and demonstrates the favorable relation between background noise and actual prostaglandin profile. The ion profile of the arterial plasma sample shown was obtained during hemorrhage. A time course of PGE_2 release following hemorrhage is shown in Fig. 18. In this dog the

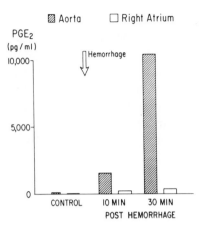

Fig. 18. Prostaglandin E_2 concentrations in the right atrium and aorta before and 10 and 30 min after hemorrhage.

maximal value for PGE_2 in the aorta was seen only at 30 min after hemorrhage. The low levels of PGE_2 in simultaneously collected blood samples from the right atrium exclude the possibility that the levels of PGE_2 seen in the aorta are artifacts generated during blood sampling. This is the first demonstration of a circulating prostaglandin by gas chromatography–mass spectrometry and it points out that under some well-controlled conditions measurement of the primary prostaglandins in the circulation can be rewarding.

Prostaglandin A_2 is another prostaglandin that has attracted considerable attention for its possible role as a classical, circulating hormone. PGA_2 was discovered and identified by mass spectrometry in kidney extracts (Lee *et al.*, 1965, 1967) as the result of efforts to identify a compound with antihypertensive properties known to be released from the renal medulla (Muirhead *et al.*, 1960). Speculations as to the possible antihypertensive role of PGA_2 were fueled by the observation that PGA_2 in pharmacological testing was a potent vasodilator and natriuretic (Lee *et al.*, 1967, 1971; Krakoff *et al.*, 1973; Spector *et al.*, 1974; Gross and Bartter, 1973). Furthermore, PGA_2, in contrast to PGE_2 and $PGF_{2\alpha}$, was reported to largely escape pulmonary metabolism (McGiff *et al.*, 1969), and thus might function as a circulating hormone. Early doubts as to the biosynthetic origin of PGA_2 in renal extracts (Crowshaw, 1973; Daniels *et al.*, 1967) were swept away by the finding of material in peripheral and renal venous plasma that reacted in a radioimmunoassay for PGA (Zusman *et al.*, 1972, 1973a; Attalah and Lee, 1973). The initial contention that PGA_2 might represent the natriuretic hormone (Lee, 1972) became untenable when the radioimmunnoassay indicated increasing levels during sodium deprivation (Zusman *et al.*, 1973*a*); however, the development of hypertension in a patient after removal of a tumor that appeared to secrete PGA_2 concurrently with a postoperative fall in radioimmunnoassayable PGA (Zusman *et al.*, 1973*b*) reinforced speculations as to a possible role for PGA in the control of blood pressure.

There was a remarkable discrepancy in the levels of PGA_2 reported for rabbit renal medulla (Lee *et al.*, 1967; Daniels *et al.*, 1967). Since PGE_2 is readily dehydrated to PGA_2, especially under the acidic conditions generally applied during extraction, we felt that part of the discrepancy might be due to variable amounts of PGE_2 dehydrated to PGA_2. In order to solve the question of a biosynthetic or *in vitro* origin of renal medullary PGA_2, gas chromatography–mass spectrometry was found to be uniquely useful. The experimental design is shown in Fig. 19. A fresh renal medullary homogenate was split into a fraction that was worked up immediately and another fraction that was incubated. The former fraction then would reflect the levels of prostaglandins in intact tissue and the latter could reflect the *in vitro* capacity to form prostaglandins. The internal standards used in both cases were 3,3,4,4-tetradeutero-PGE_2(d_4-PGE_2) and PGA_1. The d_4-PGE_2 served to quantitate PGE_2 and the PGA_1 to quantitate PGA_2 and d_4-PGA_2. Since no d_4-PGA_2 had been added, any amount of it would have to originate from dehydration of d_4-PGE_2 during the workup procedure and hence reflect the percent conversion of PGE_2 to PGA_2. The PGE fraction was derivatized to the methylester-methoxime-bisacetate and the PGA-fraction to the methylester-trimethylsilylether. Ions monitored were *m/e* 419 and 423 for PGE_2 and d_4-PGE_2 and *m/e* 349, 351, and 353 for PGA_2, PGA_1, and d_4-PGA_2, respectively. The

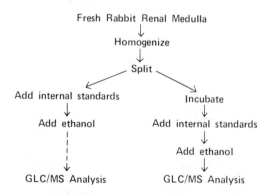

Fig. 19. Experimental design utilized to discriminate biosynthetic from chemical origin of PGA_2 in rabbit renal medulla.

calculations are given in Fig. 20. The factor K in equation II was the result of different ionic abundance of m/e 351 and m/e 349 in PGA_1 and PGA_2, respectively.

With the help of the internal standards, one can calculate the percentage conversion of d_4-PGE_2 to d_4-PGA_2 and therefore the percentage conversion of PGE_2 to PGA_2. In all of six samples of rabbit renal medulla, all the prostaglandins (PGE_2, PGA_2, and d_4-PGA_2) could be measured. On incubation, PGE_2 increased from 4430 to 14,950 ng/g tissue ($p < 0.01$). This indicated that the homogenate still could actively synthesize prostaglandins. That this increase in PGE_2 levels was due to enzymatic activity and not an autoxidative process was shown in separate experiments in which phenylbutazone produced a dose-dependent decrease in PGE_2 formation in such homogenates (Sweetman *et al.,* 1973*b*). PGA_2 levels were 35 ± 20 ng/g before and 150 ± 166 ng/g (p:Ns) after incubation. However, when the PGA_2 levels were corrected for the amount of PGA_2 generated during the workup procedure from PGE_2 by subtracting the amount of PGA_2 that corresponded to the percent conversion of d_4-PGE_2 to d_4-PGA_2, the levels of PGA_2 became indistinguishable from zero (Frölich *et al.,* 1975*b*).

The failure to detect PGA_2 in renal medulla made it highly unlikely that PGA_2 in the circulation originated from the kidney. To investigate the origin and possible function of circulating PGA_2, we measured PGA_2 in plasma by the methods described above. HPLC was of particular help in these investigations since the high degree of purification that could be accomplished by this method aided in the measurement of low levels. On the basis of previously published data obtained by immunoassay, PGA_2 levels in peripheral venous plasma were esti-

$$\text{I} \qquad d_4\text{-}PGA_2(ng) = PGA_1(ng) \times CA\, \frac{d_4\text{-}PGA_2(353)}{PGA_1(351)} \times K$$

$$\text{II} \qquad PGA_2(ng) = d_4\text{-}PGA_2(ng) \times CA\, \frac{PGA_2(349)}{d_4\text{-}PGA_2(353)}$$

Fig. 20. Formulas used to calculate the percent conversion of 3,3,4,4-tetradeutero-PGE_2 (d_4-PGE_2) to 3,3,4,4-tetradeutero-PGA_2 (d_4-PGA_2). In formula I, PGA_1 was known (internal standard) and the ratio of the central areas (CA) of ion m/e 353 representing d_4-PGA_2 and ion m/e 351 representing PGA_2 was measured. For explanation of K, see text. In formula II, d_4-PGA_2 was known (from formula I) and the ratio of the central areas of ion m/e 349 representing PGA_2 and ion m/e 353 representing d_4-PGA_2 was measured.

mated to be 1.6 ng/ml, increasing on sodium depletion to 2.1 ng/ml in parallel with the plasma renin activity (Zusman *et al.*, 1973*a*). Our gas chromatographic–mass spectrometric method therefore would quite readily allow measurement of these levels. In 12 healthy volunteers on random sodium intake, PGA_2 plasma levels were determined by collecting blood into 1 part of a 3.4 sodium citrate solution per 9 parts of blood. A small amount of indomethacin was added to the citrate in order to prevent formation of prostaglandins from blood platelets. In 23 samples of blood, the plasma levels of PGA_2 were 0.056 ± 0.134 ng/ml. Six of these volunteers were restudies, once on random sodium intake and once after acute sodium depletion with furosemide. Plasma renin activity showed the expected increase from 3.3 to 11 ng angiotensin $ml^{-1}\ hr^{-1}$; however, the plasma PGA_2 levels were 0.015 ± 0.166 ng/ml before and 0.025 ± 0.085 ng/ml after sodium depletion. None of these PGA_2 levels was different from each other or from zero (Frölich *et al.*, 1975*c*).

We also measured human seminal PGA_2 with the same method that we had applied to renal medulla. Even though PGA_2 could be detected in each of four samples, all of it resulted from *in vitro* generation. Using gas chromatographic–mass spectrometric analysis of PGA_2 in blood plasma, Steffenrud (1976) also could not find detectable levels of PGA_2 with a lower limit of detection of 6 pg/ml. Indeed, the probability of PGA_2 surviving as a hormone in the circulation is remote since Smith *et al.* (1975) showed that PGA_2 was rapidly converted by red blood cells into a biologically inactive compound.

B. Urinary Prostaglandins

The difficulties encountered in measuring primary prostaglandins in blood plasma became apparent when we wanted to study the levels of PGE_2 and $PGF_{2\alpha}$ in renal venous blood in man to gain information on renal prostaglandin synthesis. Levels of these prostaglandins were low (usually <100 pg/ml for either prostaglandin) but there was no difference between these levels and the levels in peripheral plasma (Whorton *et al.*, 1976). Thus this parameter is not a useful indicator of renal prostaglandin synthesis in man. The demonstration of prostaglandins in urine (Frölich *et al.*, 1973) and the indirect evidence produced for their intrarenal origin (Frölich *et al.*, 1975*e*) were followed by a study that provided direct evidence for an intrarenal origin. Stop-flow experiments in the dog showed that there was a net secretion of PGE between the proximal and distal tubule, with the loop of Henle remaining as the most probable site of entry (Williams *et al.*, 1976; Frölich *et al.*, 1976a). Measurement of urinary prostaglandin levels subsequently proved to be helpful in a number of studies that aimed at elucidating the relationship between plasma renin activity (PRA) and the renal prostaglandins.

C. Prostaglandins and Renin

In patients with postmalignant hypertension who were studied to evaluate the possible influence of endogenously synthesized prostaglandins on aldosterone, indomethacin lowered plasma aldosterone by 43% but, unexpectedly, it also lowered plasma renin activity by 58%, and sodium excretion was reduced (Frölich *et al.*, 1976b). This effect of indomethacin on plasma renin activity was subsequently studied in a group of normal volunteers, who again showed suppression of plasma renin activity in response to indomethacin and a complete blockade of the increase in plasma renin activity following intravenous furosemide (Frölich *et al.*, 1975a). In both groups, indomethacin lowered excretion of urinary PGE by more than 50%. The finding of a correlation between plasma renin activity and renal prostaglandin synthesis was extended into an experimental model in which sodium retention and changes in β-sympathetic tone could be excluded as contributing factors (Frölich *et al.*, 1975d, 1976b). Thus a regulatory influence of prostaglandins on plasma renin activity was likely and raised the possibility that in states of abnormal regulation of renin an abnormal regulation of prostaglandins might be causative. An abnormal regulation with excessive values of plasma renin activity characterizes Bartter's syndrome. The report of successful treatment with indomethacin of a patient with Bartter's syndrome raised the possibility that prostaglandin synthesis might be abnormal in this condition (Verberckmoes *et al.*, 1975, 1976). We measured urinary PGE_2 in four female patients with Bartter's syndrome by gas chromatography–mass spectrometry. Normal women have a mean excretion rate of 185 ng PGE_2/day. The patients with Bartter's syndrome excreted about 3 times as much PGE_2 (average of 640 ng PGE_2/day), indicating increased renal prostaglandin synthesis in the syndrome (Gill *et al.*, 1976). Treatment with indomethacin or ibuprofen lowered urinary PGE_2 excretion rate and plasma renin activity in parallel in these patients. This is another example of the close correlation between prostaglandin synthesis and plasma renin activity.

D. Assessment of Prostaglandin Synthesis

From the previous examples, one can see that measurement of the primary prostaglandins can provide useful information. In many cases, it will be preferable to measure prostaglandin metabolites because they may be present in higher concentrations than the primary prostaglandins, and they may not be subject to artifactual increase due to variable contributions by blood platelet prostaglandins (Samuelsson *et al.*, 1975; Gréen *et al.*, 1976). Measurement of the major urinary metabolite of PGE (Hamberg, 1972), for example, is much more easily accomplished than measurement of PGE_2 in urine because we found that its levels are 10 times higher than those of PGE_2 (Seyberth *et al.*, 1976). This metabolite can be measured interchangeably with PGE_2 in urine when one examines the effect of indomethacin on renal and total-body PGE biosynthesis since we have demonstrated a direct correlation between these two parameters (Seyberth *et al.*, 1976). However, this correlation is by no means predictable. Of the patients with Bartter's syndrome, only one had elevated PGE-M levels, whereas all had elevated urinary PGE_2 levels (Gill *et al.*, 1976).

The discovery of the inhibitory effect of aspirin-like drugs on prostaglandin synthesis (Vane, 1972) has led to widespread use of these agents in biological experiments to elucidate the involvement of prostaglandin synthetase dependent events. The effect of these drugs has been observed so consistently that many investigators do not feel the need to ascertain suppression of prostaglandin synthesis after administration of these drugs. This may lead to erroneous conclusions. We have studied a patient with medullary carcinoma of the thyroid in whom administration of indomethacin had no effect on the excretion of PGE-M (unpublished observation). Terragno *et al.* (1974) have shown recently that under the special circumstances of pregnancy inhibition of prostaglandin synthetase could be accomplished only at doses of indomethacin far in excess of those previously found effective. Progress in our understanding of prostaglandin and thromboxane mechanisms will be accelerated by the application of highly specific assay methods such as gas chromatography–mass spectrometry.

ACKNOWLEDGMENTS

This work was supported in part by the following grants: HL 16489, HD 05797, GM 15431, AM 18281, and 5MO 1-RR-00095. Drs. U. Axen and J. Pike, The Upjohn Company, Kalamazoo, Michigan, kindly supplied the prostaglandins used in these studies. The PGE-M was prepared by Drs. A. Rosegay and D. Taub, and supplied by Dr. W. J. A. Van den Heuvel, Merck, Sharp and Dohme, Rahway, New Jersey. I gratefully acknowledge the contributions of B. J. Sweetman, Ph.D., who routinely operated the mass spectrometer and provided the information used for Table I. R. A. Whorton, Ph.D., developed the method for chromatography of urinary prostaglandins and reviewed the manuscript. Invaluable, expert help was rendered by Mr. K. Carr, Ms. K. Davis, and Ms. S. Fulcher.

V. REFERENCES

Andersen, N. H., and Leavey, A. M. K., 1974, Identification and quantitative determination of prostaglandins by high pressure liquid chromatography, *Prostaglandins* 6:361.

Änggård, E., and Bergvist, H., 1970, Group separation of prostaglandins on Sephadex LH-20, *J. Chromatog.* 48:542.

Änggård, E., and Larsson, C., 1971, The sequence of the early steps in the metabolism of prostaglandin E_1, *Eur. J. Pharm.* 14:66.

Änggård, E., and Samuelsson, B., 1964, Prostaglandins and related factors, *J. Biol. Chem.* 239(12):4097.

Änggård, E., and Samuelsson, B., 1966, Metabolites of prostaglandins and their biological properties, in: *Endogenous Substances Affecting the Myometrium* (V. Pickles and R. Fitzpatrick, eds), pp. 107–117, Cambridge University Press, Cambridge.

Änggård, E., and Samuelsson, B., 1967, The metabolism of prostaglandins in lung tissue, in: *The Second Nobel Symposium* (S. Bergström and B. Samuelsson, eds), pp. 97–105, Almqvist and Wiksell, Stockholm.

Aten, A. H., and Hevesy, G., 1938, Fate of the sulphate radical in the animal body, *Nature (London)* 142:952.

Attalah, A. A., and Lee, J. B., 1973, Radioimmunoassay of prostaglandin A: Intrarenal PGA_2 as a factor mediating saline-induced diuresis, *Circ. Res.* 33:696.

Axen, U., Gréen, K., Hörlin, D., and Samuelsson, B., 1971, Mass spectrometric determination of picomole amounts of prostaglandins E_2 and $F_{2\alpha}$ using synthetic deuterium labeled carriers, *Biochim. Biophys. Acta* 90:207.

Axen, U., Baczynskyj, L., Duchamp, D. J., and Zierserl, J. F., 1972, Gas chromatography–mass spectrometry assay for prostaglandins, *J. Reprod. Med.* 9(6):372.

Axen, U., Baczynskyj, L., Duchamp, D., Kirton, K., and Zieserl, J., 1973, Differentiation between endogenous and exogenous (administered) prostaglandins in biological fluids, in: *Advances in the Biosciences,* Vol. 9, (S. Bergström and S. Bernhard, eds.), pp. 109–116, Pergamon Press-Vieweg, Oxford.

Bergström, S., and Sjövall, J., 1960a, The isolation of prostaglandin E from sheep prostate glands, *Acta. Chem. Scand.* 14(8):1701.

Bergström, S., and Sjövall, J., 1960b, The isolation of prostaglandin F from sheep prostate glands, *Acta. Chem. Scand.* 14(8):1693.

Bergström, S., Krabisch, L., Samuelsson, B., and Sjövall, J., 1962a, Preparation of prostaglandin F from prostaglandin E, *Acta. Chem. Scand.* 16(4):969.

Bergström, S., Ryhage, R., Samuelsson, B., and Sjövall, J., 1962b, The structure of prostaglandin E, F_1, and F_2, *Acta Chem. Scand.* 16(2):501.

Bergström, S., Dressler, F., Krabisch, L., Ryhage, R., and Sjövall, J., 1962c, The isolation and structure of a smooth muscle stimulating factor in normal sheep and pig lungs, *Ark. Kemi* 20(6):63.

Bergström, S., Dressler, F., Ryhage, R., Samuelsson, B., and Sjövall, J., 1962*d,* The isolation of two further prostaglandins from sheep prostate glands, *Ark. Kemi* **19**(42):563.

Bergström, S., Ryhage, R., Samuelsson, B., and Sjövall, J., 1963, Prostaglandins and related factors: The structures of prostaglandin E_1, $F_{1\alpha}$, and $F_{1\beta}$, *J. Biol. Chem.* **238**(11):3555.

Bergström, S., Danielsson, H., and Samuelsson, B., 1964, The enzymatic formation of prostaglandin E_2 from arachidonic acid: Prostaglandins and related factors 32, *Biochim. Biophys. Acta* **90**:207.

Bygdeman, M., and Samuelsson, B., 1966, Analysis of prostaglandins in human semen, *Clin. Chim. Acta* **13**:465.

Bundy, G. L., Daniels, E. L., Lincoln, F. H., and Pike, J. E., 1972, Isolation of a new naturally occurring prostaglandin, 5-*trans*-PGA_2. Synthesis of 5-*trans*-PGE_2 and 5-*trans*-$PGF_{2\alpha}$, *J. Am. Chem. Soc.* **94**:2124.

Carr, K., Sweetman, B. J., and Frölich, J. C., 1976, High performance liquid chromatography of prostaglandins: Biological applications, *Prostaglandins* **11**:3.

Crowshaw, K., 1973, The incorporation of (1-14C) arachidonic acid into the lipids of rabbit renal slices and conversion to prostaglandins E_2 and $F_{2\alpha}$, *Prostaglandins* **3**:607.

Daniels, D. G., Hinman, J. W., Leach, B. E., and Muirhead, E. E., 1967, Identification of prostaglandin E_2 as the major vasodepressor lipid of rabbit renal medulla, *Nature (London)* **215**:1298.

Dunham, E., and Anders, M., 1973, High speed liquid chromatographic analysis of prostaglandins in rat kidney, *Prostaglandins* **4**:85.

Frölich, J. C., Sweetman, B. J., Carr, K., Hollifield, J. W., and Oates, J. A., 1973, Occurrence of prostaglandins in human urine, in: *Advances in the Biosciences,* Vol. 9 (S. Bergström and S. Bernard, eds.), International Conference on Prostaglandins, Vienna, 1972, pp. 321–330, Pergamon Press-Vieweg, Oxford.

Frölich, J. C., Wilson, T. W., Smigel, M., and Oates, J. A., 1974, A competitive protein binding assay specific for prostaglandin E, *Biochim. Biophys. Acta* **348**:241.

Frölich, J. C., Hollifield, J., Wilkinson, B. R., and Oates, J. A., 1975*a,* Effect of indomethacin on furosemide stimulated renin and sodium excretion, *Circulation* **52**:99 (Suppl. II).

Frölich, J. C., Sweetman, B. J., Carr, K., and Oates, J. A., 1975*b,* Prostaglandin synthesis in rabbit renal medulla, *Life Sci.* **17**:1105.

Frölich, J. C., Sweetman, B. J., Carr, K., Hollifield, J. W., and Oates, J. A., 1975*c,* Assessment of the levels of PGA_2 in human plasma by gas chromatography–mass spectrometry, *Prostaglandins* **10**:185.

Frölich, J. C., Hollifield, J. W., Wilson, J. P., Sweetman, B. J., Seyberth, H. W., and Oates, J. A., 1975*d,* Suppression of plasma renin activity in man by indomethacin: Independence of sodium retention, *Clin. Res.* **24**:271.

Frölich, J. C., Wilson, T. W., Sweetman, B. J., Nies, A. S., Carr, K., Watson, J. T., and Oates, J. A., 1975*e,* Urinary prostaglandins: Identification and origin, *J. Clin. Invest.* **55**:763.

Frölich, J. C., Williams, W. M., Sweetman, B. J., Smigel, M., Carr, K., Hollifield, J. W., Fleischer, W., Nies, A. S., Frisk-Holmberg, M., and Oates, J. A., 1976*a,* Analysis of renal prostaglandin synthesis by competitive protein binding assay and gas chromatography–mass spectrometry, in: *Progress in Prostaglandin and Thromboxane Research* (B. Samuelsson and R. Paoletti, eds.), pp. 65–80, Raven Press, New York.

Frölich, J. C., Hollifield, J. W., Dormois, J. C., Seyberth, H. J., Michelakis, A. M., and Oates, J. A., 1976*b,* Suppression of plasma renin activity by indomethacin in man, *Circ. Res.* (in press).

Gill, J. R., Frölich, J. C., Bowden, R. E., Taylor, A. A., Keiser, H. R., Seyberth, H. W., Oates, J. A., and Bartter, F. C., 1976, Bartter's syndrome: A disorder characterized by high urinary prostaglandins and a dependence of hyperreninemia on prostaglandin synthesis, *Am. J. Med.* **61**:43–51.

Gohlke, R. S., 1959, Time-of-flight mass spectrometry and liquid-gas partition chromatography, *Anal. Chem.* **31**(4):535.

Granström, E., and Kindahl, H., 1976, Radioimmunoassays for prostaglandin metabolites, in:

Advances in Prostaglandin and Thromboxane Research I (B. Samuelsson and R. Paoletti, eds.), pp. 81–92, Raven Press, New York.

Granström, E., and Samuelsson, B., 1969a, The structure of the main urinary metabolite of prostaglandin $F_{2\alpha}$ in the guinea pig, *Eur. J. Biochem.* **10**:411.

Granström, E., and Samuelsson, B., 1969b, The structure of a urinary metabolite of prostaglandin $F_{2\alpha}$ in man, *J. Am. Chem. Soc.* **91(12)**:3398.

Granström, E., and Samuelsson, B., 1971, On the metabolism of prostaglandin $F_{2\alpha}$ in female subjects, *J. Biol. Chem.* **245**:5254.

Granström, E., Inger, U., and Samuelsson, B., 1965, The structure of a urinary metabolite of prostaglandin $F_{1\alpha}$ in the rat, *J. Biol. Chem.* **240(1)**:457.

Gréen, K., 1971, The metabolism of prostaglandin $F_{2\alpha}$ in the rat, *Biochim. Biophys. Acta* **231**:419.

Gréen, K., and Samuelsson, B., 1964, Prostaglandins and related factors. XIX. Thin-layer chromatography of prostaglandins, *J. Lipid Res.* **5**:117.

Gréen, K., Granström, E., Samuelsson, B., and Axen, U., 1973, Methods for quantitative analysis of $PGF_{2\alpha}$, $PGE_{2\alpha}$, 9α, 11α, 15-keto-prost-5-enoic acid and 9α, 11α, 15-trihydroxy-prost-5-enoic acid from body fluids using deuterated carriers and gas chromatography-mass spectrometry, *Anal. Biochem.* **54**:434.

Gréen, K., Hedqvist, P., and Swanborg, N., 1974, Increased plasma levels of 15-keto-13,14 dihydro-prostaglandin F_2 after allergen provoked asthma in man. *Lancet* **2(2)**:1419.

Gréen, K., Hamberg, M., and Samuelsson, B., 1976, Quantitative analysis of prostaglandins and thromboxanes by mass spectrometric methods, in: *Advances in Prostaglandin and Thromboxane Research,* Vol. 1 (B. Samuelsson and R. Paoletti, eds), pp. 47–58, Raven Press, New York.

Gross, J. B., and Bartter, F. C., 1973, Effects of prostaglandin E_1, A_1, and $F_{2\alpha}$ on renal handling of salt and water, *Am. J. Physiol.* **225**:218.

Hamberg, M., 1972, Inhibition of prostaglandin synthesis in man, *Biochem. Biophys. Res. Commun.* **49(3)**:720.

Hamberg, M., 1973, Quantitative studies on prostaglandin synthesis in man. II. Determination of the major urinary metabolite of prostaglandins $F_{1\alpha}$ and $F_{2\alpha}$, *Anal. Biochem.* **55**:368.

Hamberg, M., and Samuelsson, B., 1965, Isolation and structure of a new prostaglandin from human seminal plasma: Prostaglandins and related factors 43, *Biochim. Biophys. Acta* **106**:215.

Hamberg, M., and Samuelsson, B., 1966, Prostaglandins in human seminal plasma, *J. Biol. Chem.* **241(2)**:257.

Hamberg, M., and Samuelsson, B., 1971a, Metabolism of prostaglandin E_2 in guinea pig liver: Pathways in the formation of the major metabolites, *J. Biol. Chem.* **246(22)**:6713.

Hamberg, M., and Samuelsson, B., 1971b, On the metabolism of prostaglandins E_1 and E_2 in man, *J. Biol. Chem.* **246(22)**:6713.

Hamberg, M., and Samuelsson, B., 1973, Detection and isolation of an endoperoxide intermediate in prostaglandin biosynthesis, *Proc. Natl. Acad. Sci. U.S.A.* **70(3)**:899.

Hamberg, M., and Samuelsson, B., 1974, Prostaglandin endoperoxides: Novel transformations of archidonic acid in human platelets, *Proc. Natl. Acad. Sci. U.S.A.* **71(9)**:3400.

Hamberg, M., Svenson, J., and Samuelsson, B., 1974, Prostaglandin endoperoxides: A new concept concerning the mode of action and release of prostaglandins, *Pro. Natl. Acad. Sci. U.S.A.* **71**:3824.

Heftman, E., 1975, *Chromatography,* Van Nostrand-Reinhold, New York.

Henneberg, D., 1961, Eine Kombination von Gaschromatograph und Massen Spektrometer zur Analyse Organischer Stoffgemische, *Anal. Chem.* **183**:12.

Hevesy, G., 1923, The absorption and translocation of lead by plants: A contribution to the application of the method of radioactive indicators in the investigation of the change of substance in plants, *Biochem. J.* **17**:439.

Howard, G. A., and Martin, A. J. B., 1950, The separation of the C_{12}-C_{13} fatty acids by reversed-phase-partition chromatography, *Biochem. J. (London)* **4**:532.

Jakschik, B., Marshall, G., Kourik, J., and Needleman, P., 1974, Profile of circulating vasoactive substances in hemorrhagic shock and their pharmacologic manipulation, *J. Clin. Invest.* **54(4)**:842.

Krakoff, L. R., Deguia, R. D., Deguia, D., Vlachakis, N., Stricker, J., and Goldstein, M., 1973, Effect of sodium balance on arterial blood pressure and renal response to prostaglandin A_1 in man, *Circ. Res.* **33**:539.

Lee, J. B., 1972, Natriuretic hormone and the renal prostaglandins, *Prostaglandins* **1**:55.

Lee, J. B., Covino, B. J., Takman, B. H., and Smith, E. R., 1965, Renomedullary vasodepressor substance medullin: Isolation, chemical characterization, and physiological properties, *Circ. Res.* **17**:57.

Lee, J. B., Crowshaw, K., Takman, B. H., Atrep, K. A., and Gougoutas, J. Z., 1967, The identification of prostaglandin E_2, $F_{2\alpha}$, and A_2 from rabbit kidney medulla, *Biochem. J.* **105**:1251.

Lee, J. B., McGiff, J. C., Kannegiesser, H., Aykent, Y., Mudd, J. G., and Frawley, T. F., 1971, Prostaglandin A_1, antihypertensive and renal effects: Studies in patients with essential hypertension, *Ann. Intern. Med.* **74**:703.

Lee, S. J., Johnson, J. G., Smith, C. J., and Hatch, F. E., 1971, Renal effects of prostaglandin A_1 in patients with essential hypertension, *Kidney Int.* **1**:254.

Levine, L., Gutierrez-Cernosek, R., and Vunakis, H., 1973, Specific antibodies: Reagents for quantitative analysis of prostaglandins, in: *Advances in the Biosciences,* Vol. 9, (S. Raspe and S. Bernhard, eds.), pp. 71–82, Pergamon Press-Vieweg, Oxford.

McFadden, W. H., 1973, *Techniques in Combined Gas Chromatography–Mass Spectrometry: Applications in Organic Analysis,* Wiley, New York.

McGiff, J. C., Terragno, N. A., Strand, J. C., Lee, J. C., Lonigro, A. J., and Ng, K. K. F., 1969, Selective passage of prostaglandins across the lung, *Nature (London)* **223**:743.

Muirhead, E. E., Jones, F., and Stirman, J. A., 1960, Antihypertensive property in renoprival hypertension of extract from renal medulla, *J. Lab. Clin. Med.* **56**:167.

Norman, A., 1953, Separation of conjugated bile acids by partition chromatography-bile acids and steroids 6, *Acta Chem. Scand.* **7(10)**:1413.

Nugteren, D. H., and Hazelhof, E., 1973, Isolation and properties of intermediates in prostaglandin biosynthesis, *Biochim. Biophys. Acta* **326**:448.

Nugteren, D. H., and van Dorp, D. A., 1965, The participation of molecular oxygen in biosynthesis of prostaglandins, *Biochim. Biophys. Acta* **98**:654.

Paul, W., and Raether, M., 1955, Das elektrische Massenfilter, *Z. Phys.* **140**:262.

Paul, W., and Steinwedel, H., 1953, Ein neues Massenspektrometer ohne Magnetfeld, *Z. Naturforsch.* **8(a)**:448.

Paul, W., and von Zahn, U., 1961, Präzisionsmessungen mit dem elektrischen Massenfilter, *Phys. Verhdlg.* **12**:222.

Rittenberg, D., Keston, A. S. Rosebury, F., and Schoenheimer, R., 1939, Studies in protein metabolism. II. The determination of nitrogen isotopes in organic compounds, *J. Biol. Chem.* **36**:759.

Ryhage, R., 1964, Use of a mass spectrometer as a detector and analyzer for effluents emerging from high temperature gas liquid chromatography columns, *Anal. Chem.* **36**:759.

Ryhage, R., and Samuelsson, B., 1965, The origin of oxygen incorporated during the biosynthesis of prostaglandin E_1, *Biochem. Biophys. Res. Commun.* **19**:279.

Samuelsson, B., 1963*a,* Isolation and identification of prostaglandins from human seminal plasma, *J. Biol. Chem.* **238(10)**:3229.

Samuelsson, B., 1963*b,* Prostaglandins and related factors, 17: The structure of prostaglandin E_3, *J. Am. Chem. Soc.* **85**:1878.

Samuelsson, B., 1964*a,* Identification of prostaglandin $F_{3\alpha}$ in bovine lung prostaglandins and related factors, 26, *Biochim. Biophys. Acta* **84**:707.

Samuelsson, B., 1964*b,* Identification of a smooth muscle stimulating factor in bovine brain: Prostaglandins and related factors, 25, *Biochim. Biophys. Acta* **84**:218.

Samuelsson, B., 1965, On the incorporation of oxygen in the conversion of 8,11,14-eicosatrienoic acid to prostaglandin E_1, *J. Am. Chem. Soc.* **87**:3011.

Samuelsson, B., 1973*a*, Quantitative aspects of PG synthesis in man, in: *Advances in the Biosciences*, Vol. 9, (S. Bergström and S. Bernhard, eds.), pp. 7–14, Pergamon-Vieweg, Oxford.

Samuelsson, B., 1973*b*, Roundtable discussion of analytical methods, in: *Advances in the Biosciences*, Vol. 9, (S. Bergström and S. Bernhard, eds.), pp. 121–123, Pergamon-Vieweg, Oxford.

Samuelsson, B., and Gréen, K., 1974, Endogenous levels of 15-keto-dihydro prostaglandins in human plasma: Parameters for monitoring prostaglandin synthesis, *Biochem. Med.* **11**:298.

Samuelsson, B., Hamberg, M., and Sweeley, C., 1970, Quantitative gas chromatography of prostaglandin E_1 at the nonogram level: Use of deuterated carrier and multiple-ion analyzer, *Anal. Biochem.* **38**:301.

Samuelsson, B., Granström, E., Gréen, K., and Hamberg, M., 1971, Metabolism of prostaglandins, *N.Y. Acad. Sci.* **180**:138.

Samuelsson, B., Granström, E., Gréen, K., Hamberg, M., and Hammarström, S., 1975, Prostaglandins, *Annu. Rev. Biochem.* **44**:669.

Schneider, W. P., Hamilton, R. D., and Rhuland, L. E., 1972, Occurrence of esters of (15*S*)-prostaglandin A_2 and E_2 in coral, *J. Am. Chem. Soc.* **94(6)**:2122.

Schoenheimer, R., Ratner, S., and Rittenberg, D., 1939, Studies in protein metabolism. X. The metabolic activity of body proteins investigated with L($-$)-leucine containing two isotopes, *J. Biol. Chem.* **130**:703.

Seyberth, H. W., Sweetman, B. J., Frölich, J. C., and Oates, J. A., 1976, Assessment of the quantitative analysis of the major urinary metabolite of prostaglandin E in man, *Prostaglandins* **11**:381–397.

Shaw, J., and Ramwell, P. W., 1969, Separation, identification, and estimation of prostaglandins, *Methods Biochem. Anal.* **17**:328.

Smith, J. B., Silver, M. J., Ingerman, C. M., and Kocsis, J. J., 1975, Uptake and inactivation of A-type prostaglandins by human red cells, *Prostaglandins* **9(1)**:135.

Spector, D., Zusman, R. M., Caldwell, B. V., and Speroff, L., 1974, The distribution of prostaglandins A, E, and F in the human kidney, *Prostaglandins* **6**:263.

Steffenrud, S., 1976, Methods for gas chromatographic mass spectrometric quantification of PGA_2, in: *Advances in Prostaglandin and Thromboxane Research*, Vol. 2, (B. Samuelsson and R. Paoletti, eds.), p. 866, Raven Press, New York.

Sun, F., 1974, Metabolism of prostaglandin $F_{2\alpha}$ in the rat, *Biochim. Biophys. Acta* **348**:249.

Sun, F., and Strafford, J. E., 1974, Metabolism of prostaglandin $F_{2\alpha}$ in rhesus monkeys, *Biochim. Biophys. Acta* **369**:95.

Sweeley, C., Elliott, W., Fries, I., and Ryhage, R., 1966, Mass spectrometric determination of unresolved components in gas chromatographic effluents, *Anal. Chem.* **38(11)**:1249.

Sweetman, B. J., Frölich, J. C., and Watson, J. T., 1973*a*, Quantitative determination of prostaglandins A, B, or E in the sub-nanogram range, *Prostaglandins* **3**:75.

Sweetman, B. J., Frölich, J. C., Carr, K., Danon, A., Oates, J. A., and Watson, J. T., 1973*b*, Quantitative analysis of renal prostaglandins by selected ion monitoring (SIM) using GLC-MC computer system, Proceedings of the NIGMS Meeting, May 1973, Washington, D.C., pp. 211–215.

Terragno, N. A., Terragno, D. A., Pacholczyk, D., and McGiff, J. C., 1974, Prostaglandins and the regulation of uterine blood flow in pregnancy, *Nature (London)* **249**:57.

Unger, W., Stamford, I., and Bennett, A., 1971, Extractions of prostaglandins from human blood, *Nature (London)* **233**:336.

van Dorp, D. A., Beerthuis, R. K., Nugteren, D. H., and Vonkeman, H., 1964, The biosynthesis of prostaglandins, *Biochim. Biophys. Acta* **90**:204.

Vane, J. R., 1972, Inhibition of prostaglandin synthesis as a mechanism of action for aspirin-like drugs, *Nature (London) New Biol.* **231**:232.

Verberckmoes, R., Clement, J., Michielsen, P., and van Damme, B., 1975, Bartter's syndrome with hyperplasia of renomedullary interstitial cells: Successful treatment with indomethacin, VI International Congress of Nephrology, Florence, Italy, Abs. No. 558.

Verberckmoes, R. van Damme, B., Clement, J., Amery, A., and Michielsen, P., 1976, Bartter's syndrome with hyperplasia of renomedullary cells: Successful treatment with indomethacin, *Kidney Int.* **9**:302.

Whorton, A. R., Frisk-Holmberg, M., and Frölich, J. C., 1976, Analysis of renal venous prostaglandin in man by gas chromatography—mass spectrometry. Technical note, submitted.

Wlodawer, P., and Samuelsson, B., 1973, On the organization and mechanism of prostaglandin synthetase, *J. Biol. Chem.* **248(16)**:5673.

Wickramasinghe, J. A. F., and Shaw, S. R., 1973, Separation of prostaglandins A, B, and C by thin-layer chromatography on ferric chloride impregnated silica gel, *Prostaglandins* **4(6)**:903.

Williams, W. M., Frölich, J. C., Nies, A. S., and Oates, J. A., 1976, Urinary prostaglandins: Site of entry into renal tubular fluid, *Kidney International* (in press).

Wood, H. G., Werkman, C. H., Hemingway, A., and Nier, A., 1940, Heavy carbon as a tracer in bacterial fixation of carbon dioxide, *J. Biol. Chem.* **135**:789.

Zusman, R. M., Caldwell, B. V., Speroff, L., and Behrman, H. R., 1972, Radioimmunoassay of the A prostaglandins, *Prostaglandins* **2**:41.

Zusman, R. M., Spector, D., Caldwell, B. V., Speroff, L., Schneider, G., and Mulrow, P. J., 1973*a*, The effect of chronic sodium loading and sodium restriction on plasma prostaglandin A, E, and F concentrations in normal humans, *J. Clin. Invest.* **52**:1093.

Zusman, R. M., Schneider, J., Cline, A., Caldwell, B., and Speroff, L., 1973*b*, Antihypertensive function of a renal-cell carcinoma, *New Engl. J. Med.* **290**:843.

NOTE ADDED IN PROOF

The initial investment necessary to obtain a mass spectrometer useful for prostaglandin analysis used to be in excess of $120,000, making it affordable for only a few laboratories. More recently, quadrupole instruments have become available which cost only a fraction of this amount. An instrument capable of quantitative analysis of prostaglandins at the nanogram level can be purchased for less than $50,000. While sample work-up was quite time-consuming in the past, high-performance liquid chromatography has greatly aided in reducing this factor. As long as other methods for the analysis of prostaglandins have not been compared to gas chromatography—mass spectrometry, they must be viewed with greatest caution. The greatly reduced initial cost and the relatively large number of samples that can be processed, coupled with greater ease of operation of the instrument, make the quadrupole mass spectrometer a desirable tool for analysis of prostaglandins and their metabolites that is within the reach of many laboratories.

Prostaglandin Biosynthesis and Metabolism as Measured by Radioimmunoassay

Sheng-Shung Pong and Lawrence Levine

Department of Biochemistry
Brandeis University
Waltham, Massachusetts 02154

I. INTRODUCTION

The extent to which unlabeled ligand competes with radioactive ligand for a limited number of receptor sites in a macromolecule serves as a basis for quantitation in competitive protein binding assays. Often, picogram amounts of a particular compound can be estimated. Once the receptor molecule and labeled ligand are available, the assays are relatively simple to perform: mixtures containing labeled ligand, receptor molecule, and sample are incubated; free labeled ligand is separated from receptor-bound labeled ligand, and the extent of binding is determined. Although natural receptors have been employed in several competitive binding assays, their use is limited by their availability and stability. Thus antibodies have been produced to develop radioimmunoassays.

Antibodies are proteins found in the globulin fractions of blood that are produced by vertebrates in response to the presence of an antigen, i.e., a substance that is recognized by the host to be foreign. Antibodies can be produced against specific low-molecular-weight compounds if these are covalently linked to macromolecules. They show a remarkable ability to selectively bind the antigen that stimu-

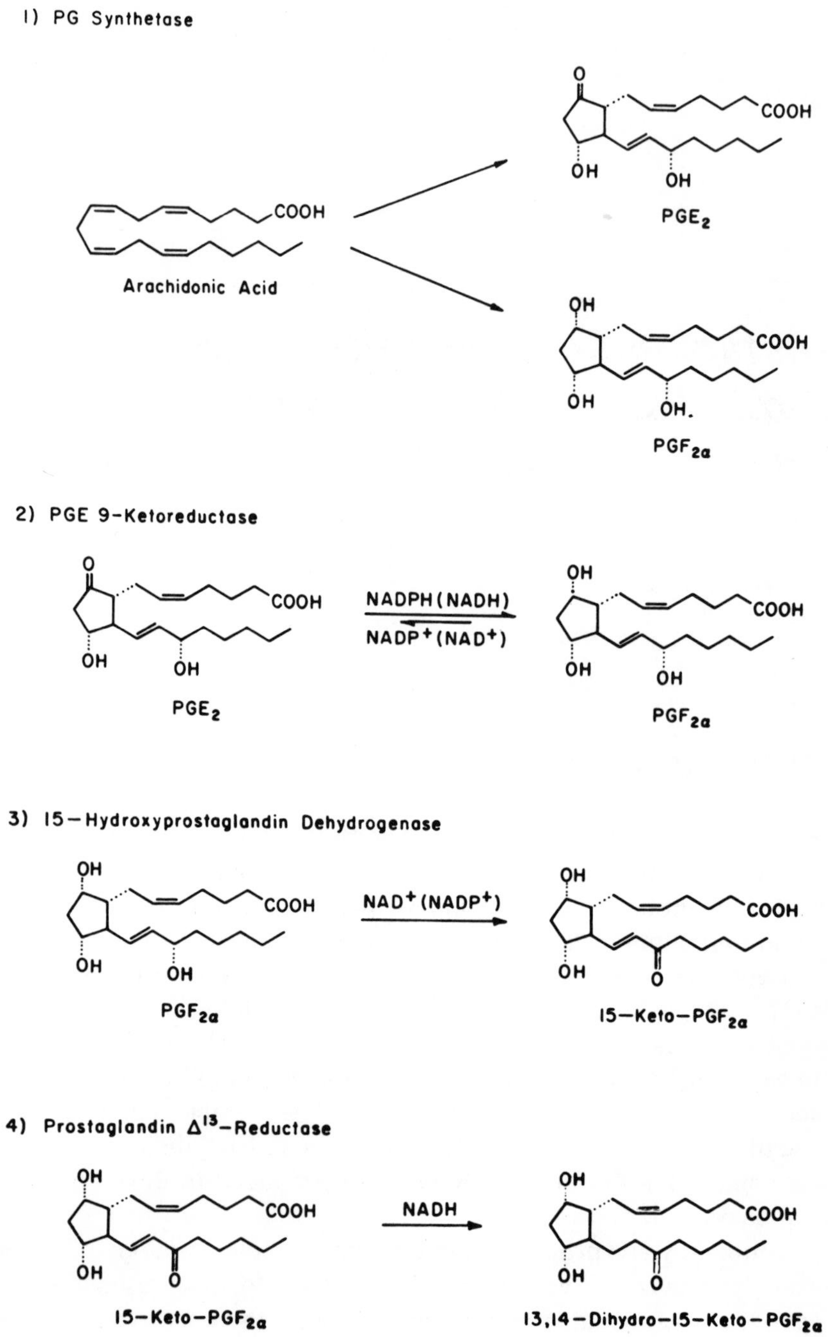

1) PG Synthetase

2) PGE 9-Ketoreductase

3) 15-Hydroxyprostaglandin Dehydrogenase

4) Prostaglandin Δ^{13}-Reductase

Fig. 1. Some enzyme reactions in the biosynthesis, metabolism, and regulation of the prostaglandins.

5) Dehydrase, PGA Isomerase and PGC Isomerase

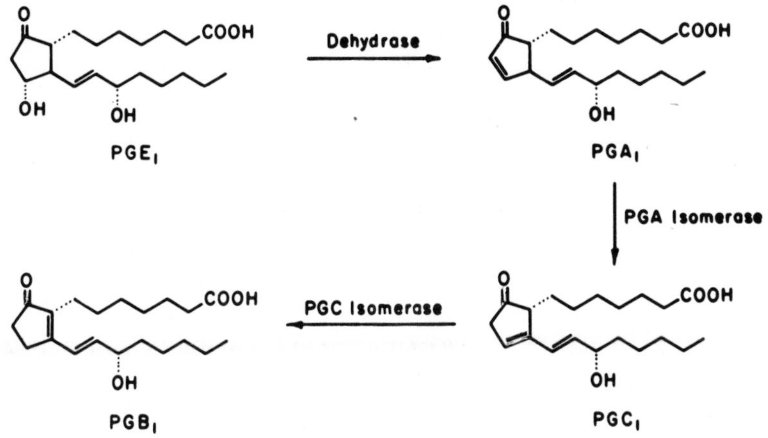

Fig. 1 (cont.)

lates their production. Their specificity may be regarded as comparable to that of an enzyme for substrate. The binding constants between antibody and antigen are often of the order of $10^7 - 10^9$ liters/mole. This ability of antibodies to discriminate between the homologous antigen and the other compounds of widely diverse structure that are found in biological fluids, e.g., serum, urine, or tissue extracts, is of fundamental importance in their use as analytical tools.

Radioimmunoassay is a simple, sensitive, and specific analytical method that has been used to study the biosynthesis and metabolism of the prostaglandins. Its principle was described by Berson and Yalow (1960). Radioimmunoassays have been developed for many different classes of compounds (Yalow, 1973; Butler and Beiser, 1973). Macromolecules such as proteins, nucleic acids, and polysaccharides usually can elicit an immune response when they are injected into the experimental animal directly or in the form of an electrostatic complex. However, low-molecular-weight compounds ordinarily cannot elicit an immune response unless they are bound covalently to an antigenic macromolecule (e.g., protein or polypeptide). Such low-molecular-weight compounds are called haptens. Although fatty acids are not good haptens, the more rigid structure of prostaglandins imparted by the cyclopentane ring increases their antigenicity. The first successful production of antibodies to certain prostaglandins was reported by Levine and Van Vunakis (1970). Shortly thereafter, several laboratories published serological methods that could be used for measuring prostaglandins (Jaffe *et al.*, 1971; Kirton *et al.*, 1972; Zusman *et al.*, 1972;

Yu and Burk, 1972; Caldwell *et al.*, 1971; Granstrom and Samuelsson, 1972; Bauminger *et al.*, 1973).

Radioimmunoassay offers distinct advantages over other analytical procedures. Chromatographic methods including thin-layer and gas–liquid chromatography are among the most widely used analytical tools to study the metabolism of the prostaglandins, especially when labeled compounds are available to aid in the identification and quantitation of the parent compound and its metabolites. Gas–liquid chromatography is particularly effective when combined with mass spectrometry (Gréen, 1973). With such systems, sensitivity down to the picomole level is not uncommon. These chromatographic methods normally require that samples be extracted and separated from substances that might interfere with the assays.

The sensitivity of radioimmunoassay equals or surpasses that of the available chromatographic and spectral methods. The added dimension of antigen–antibody specificity allows radioimmunoassays to be used for the direct analysis of compounds in enzyme reaction mixtures, which ordinarily must be processed before the other techniques can be applied.

We have used radioimmunoassays to detect enzymes that synthesize, metabolize, and/or regulate prostaglandins. In Fig. 1 are shown five enzymatic reactions: (1) synthesis of PGE_2 and $PGF_{2\alpha}$ from arachidonic acid, (2) enzymatic reduction of the C-9 keto group to form the C-9 hydroxyl group (conversion of PGE_2 to $PGF_{2\alpha}$), (3) enzymatic dehydrogenation of the C-15 hydroxyl group to form the C-15 keto group, (4) enzymatic reduction of the C-13,14 double bond, and (5) dehydration of PGE_2 to form PGA_2 and isomerization of PGA_2 to form PGC_2 and PGB_2.

The serological specificities relevant to the assay of each enzymatic reaction will be shown. Radioimmunoassays were developed to measure both the substrate and the product so that catalysis could be followed in two ways, loss of substrate and concomitant generation of product.

II. SEROLOGICAL SPECIFICITIES OF THE PROSTAGLANDIN IMMUNE SYSTEMS

A. Prostaglandin Synthetase

For measurement of synthetase activity, the antibodies should be directed toward the products PGE_2 and $PGF_{2\alpha}$ and should not react significantly with the substrate, arachidonic acid, or the metabolites of the products, the 15-keto- and 13,14-dihydro-15-ketoprostaglandins. In addition, the antibodies should distinguish between the products, PGE_2 and $PGF_{2\alpha}$. The specificities of both the anti-PGE_2 and -$PGF_{2\alpha}$ immune systems fulfill these requirements.

The serological specificity of the $PGF_{2\alpha}$–anti-$PGF_{2\alpha}$ reaction in terms of weights of ligand required for 50% inhibition of [^{3}H]$PGF_{2\alpha}$– anti-$PGF_{2\alpha}$ binding is shown in Table I. Greater than 2×10^6 times more substrate (arachidonic acid) than product ($PGF_{2\alpha}$) is required for equivalent activity with anti-$PGF_{2\alpha}$. PGE_2, another product of the synthetase reaction, cross-reacts between 1 and 2%. The isomer of $PGF_{2\alpha}$, $PGF_{2\beta}$, cross-reacts between 2 and 3%. Metabolites of the products react very poorly. The cyclopentane ring, the C-9 α-hydroxyl group, the C-15 keto group, and the $\Delta^{13,14}$ double bond all contribute to the binding energy of the $PGF_{2\alpha}$–anti-$PGF_{2\alpha}$ reaction.

The serological specificity of the PGE_2–anti-PGE_2 reaction is shown in Table II. Again the substrate (arachidonic acid) reacts ineffectively with the antibodies; 10^6 times more molecules are required for inhibition equivalent to that of PGE_2. $PGF_{2\alpha}$ reacts poorly, about 0.5%. The metabolites of PGE_2 also react poorly; only 0.5–1%. The C-9 keto group is immunodominant, but the degree and positions of the unsaturated bonds within the cyclopentane ring are not recognized. These antibodies which are binding to [^{3}H]PGE_2 do not distinguish PGE_2 from the dehydrated (PGA_2) and isomerized (PGB_2) products.

B. Prostaglandin E 9-Ketoreductase

The serological specificities of the immune systems required for measurement of both the substrate and product of the PGE 9-ketoreductase reaction, PGE_2–anti-PGE_2, and $PGF_{2\alpha}$–anti-$PGF_{2\alpha}$ have already been described (Tables I and II). PGE_2 reacts with anti-$PGF_{2\alpha}$

Table I. Serological Specificity of $PGF_{2\alpha}$ Immune System[a]

Ligand	ng required for 50% inhibition[a]	Ligand	ng required for 50% inhibition[a]
$PGF_{2\alpha}$	0.09	PGE_2	7.6
$PGF_{2\beta}$	3.7	PGA_2	180.0
15-Keto-$PGF_{2\alpha}$	23.0	PGB_2	>1,000.0
13,14-Dihydro-15-keto-$PGF_{2\alpha}$	90.0	Arachidonic acid	20,000.0
13,14-Dihydro-$PGF_{2\alpha}$	20.0		

[a]Inhibition of [^{3}H]$PGF_{2\alpha}$–anti-$PGF_{2\alpha}$ binding.

Table II. Serological Specificity of PGE$_2$ Immune System[a]

Ligand	ng required for 50% inhibition[a]	Ligand	ng required for 50% inhibition[a]
PGE$_2$	0.13	13,14-Dihydro-15-keto-PGE$_2$	14.0
PGE$_1$	0.13	PGF$_{2\alpha}$	36.0
PGE$_3$	0.65	15-Keto-PGF$_{2\alpha}$	>1,000.0
PGA$_2$	0.12	13,14-Dihydro-15-keto-PGF$_{2\alpha}$	>1,000.0
PGB$_2$	0.11	13,14-Dihydro-PGF$_{2\alpha}$	>1,000.0
15-Keto-PGE$_2$	20.0	Arachidonic acid	12,000.0

[a]Inhibition of [^{3}H]PGE$_2$–anti-PGE$_2$ binding.

to the extent of 1–2%. To detect stereospecific conversion of PGE$_2$ to PGF$_{2\alpha}$, antisera directed toward PGF$_{2\beta}$ and PGF$_{2\alpha}$ were used simultaneously for measurement of product (Levine *et al.*, 1975). Both antisera distinguish the homologous from the heterologous ligands. The specificities of such antisera are shown in Fig. 2. For these analyses, the substrate PGE$_2$ is converted to PGB$_2$ by alkaline treatment to minimize the serological interference by PGE$_2$.

C. 15-Hydroxyprostaglandin Dehydrogenase

Oxidation of the 15-hydroxyl groups of PGE$_2$ and PGF$_{2\alpha}$ is measured by the appearance of serological activity when assayed with antibodies directed toward 15-keto-PGF$_{2\alpha}$ (Levine, 1973; Lee and Levine, 1975*b*). The serological specificity of the monkey antiserum is shown in Table III. The most effective inhibitor is 15-keto-PGF$_{2\alpha}$. PGF$_{2\alpha}$ inhibits, but approximately 1000 times more PGF$_{2\alpha}$ than 15-

Table III. Serological Specificity of 15-Keto-PGF$_{2\alpha}$ Immune System[a]

Ligand	ng required for 50% inhibition
15-Keto-PGF$_{2\alpha}$	1.1
15-Keto-PGE$_2$	1.6
13,14-Dihydro-15-keto-PGF$_{2\alpha}$	30.0
PGF$_{2\alpha}$	>1000.0
15-Keto-PGE$_2$ (NaOH)	900.0
15-Keto-PGF$_{2\alpha}$ (NaOH)	1.1

[a]Inhibition of [^{3}H]15-keto-PGF$_{2\alpha}$ monkey anti-15-keto-PGF$_{2\alpha}$ binding (Levine and Gutierrez-Cernosek, 1972).

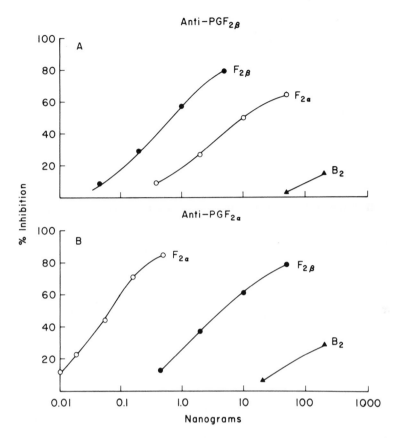

Fig. 2. Inhibition of [³H]PGF$_{2\beta}$–anti-PGF$_{2\beta}$ binding (A) and [³H]PGF$_{2\alpha}$–anti-PGF$_{2\alpha}$ binding (B) by increments of PGF$_{2\alpha}$, PGF$_{2\beta}$, and alkaline-treated PGE$_2$ (B$_2$). From Levine *et al.* (1975).

keto-PGF$_{2\alpha}$ is required for 50% inhibition. The immunodominant role of the C-13,14 double bond can also be seen by the 25-fold decrease in inhibiting effectiveness when the double bond in 15-keto-PGF$_{2\alpha}$ is reduced. Recognition of the substituent on the C-9 group by these antibodies is not striking. 15-keto-PGE$_2$, which has a keto group in the C-9 position, is almost as effective a competitor as 15-keto-PGF$_{2\alpha}$, which has a hydroxyl group in the C-9 position. However, the serological activity of 15-keto-PGE$_2$ is labile to NaOH treatment while that of 15-keto-PGF$_{2\alpha}$ is not, so that these 15-keto-PG classes can easily be differentiated.

Serological detection of 15-hydroxyprostaglandin dehydrogenase activity is both very sensitive and specific. The substrates (PGF$_{2\alpha}$ and PGE$_2$) can be used at the 10 and 100 ng level, respectively, without significant reaction with the antibodies, and as little as 500 pg of the products (15-keto-PGF$_{2\alpha}$ or 15-keto-PGE$_2$) can be measured quantitatively. Oxidation of the 15-OH group can be measured also by loss of serological activity of the substrates, PGE_2 or $PGF_{2\alpha}$. Both 15-keto metabolites react with their respective antisera less than 1% as effectively as the homologous ligands (Tables I and II).

D. Prostaglandin Δ^{13}-Reductase

Reduction of the Δ^{13} double bond of 15-keto-$PGF_{2\alpha}$ is measured both by disappearance of serological activity when assayed with monkey antibodies directed toward 15-keto-$PGF_{2\alpha}$ (Table III) and by appearance of serological activity when measured with antibodies directed toward 13,14-dihydro-15-keto-$PGF_{2\alpha}$. Antibodies that recognize the reduced Δ^{13} double bond, anti-13,14-dihydro-15-keto-$PGF_{2\alpha}$, are used to identify the product of this reaction. The serological specificity of the anti-13,14-dihydro-15-keto-$PGF_{2\alpha}$ is shown in Table IV.

E. PGE Dehydrase and PGA and PGC Isomerases

To study the dehydration of PGE to PGA and the subsequent isomerization of PGA to PGB, an antiserum with specificity toward PGB_1 has been used (Levine et al., 1971). Antibodies produced in rabbits, guinea pigs, and monkeys to PGE or PGA usually are directed toward PGB. The serological specificity of one such rabbit antiserum is shown in Table V. This antiserum is much more sensitive to inhibition by PGB_1 than by PGA_1, which differs from PGB_1 only by the location of the unsaturated bond in the cyclopentane ring. Approximately 15 times more PGA_1 is needed to yield equivalent inhibition. The presence of a hydroxyl group on the cyclopentane ring of PGE_1 renders this prostaglandin still less effective as an inhibitor of the antigen–antibody reaction; 22 ng was required to inhibit 50% compared to 70 pg of PGB_1.

Table IV. Serological Specificity of the 13,14-Dihydro-15-Keto-$PGF_{2\alpha}$ Immune System[a]

Ligand	ng required for 50% inhibition
13,14-Dihydro-15-keto-$PGF_{2\alpha}$	0.2
15-Keto-$PGF_{2\alpha}$	3.1
13,14-Dihydro-15-keto-PGE_2	19.0
$PGF_{2\alpha}$	320.0

[a]Inhibition of [^{3}H]13,14-dihydro-15-keto-$PGF_{2\alpha}$ anti-13,14-dihydro-15-keto-$PGF_{2\alpha}$ binding (Lee and Levine, 1974).

Table V. Serological Specificity of PGB$_1$ Immune System[a]

Ligand	ng required for 50% inhibition	Ligand	ng required for 50% inhibition
PGB$_1$	0.070	PGE$_1$	22.0
PGB$_2$	0.260	PGE$_2$	230.0
PGA$_1$	1.0	PGF$_{1\alpha}$	>1000
PGA$_2$	5.0	PGF$_{2\alpha}$	>1000

[a] Inhibition of [^{3}H]PGB$_1$-anti-PGB$_1$ binding.

The F type of prostaglandin, which contains two hydroxyl groups in a saturated cyclopentane ring, is a poor inhibitor of the reaction.

The fact that PGB$_2$, which contains a second unsaturated double bond between C-5 and C-6 of the aliphatic chain, is a weaker inhibitor than PGB$_1$, indicates that the antibodies in this antiserum also recognize the aliphatic side chains. The other bis-unsaturated compounds, i.e., PGA$_2$ and PGE$_2$, are also less inhibitory than the corresponding monounsaturated prostaglandins (PGA$_1$ and PGE$_1$).

This antibody does not distinguish PGC$_1$ from PGA$_1$. Conversion of PGA$_1$ or PGC$_1$ to PGB$_1$ results in increased inhibition of the binding of [^{3}H]PGB$_1$ to the antibody. The inhibition by mixtures of PGA$_1$ plus PGB$_1$ of various compositions, but at a constant amount of total prostaglandins, is shown in Fig. 3. The presence of rabbit albumin increases the inhibition by PGA$_1$ but not PGB$_1$. These calibration curves can be used to determine the amount of PGB$_1$ in the presence of PGA$_1$ and/or PGC$_1$. The method is sensitive and accurate enough to allow for the measurement of initial rates of formation of product by PGA isomerase. If more than 30–40% PGB$_1$ is present, the method becomes less sensitive.

The conversion of PGE$_1$ to PGA$_1$ is also followed with this PGB$_1$ antiserum. The inhibition by 20 ng of PGE$_1$ is 5%, but 1 ng of PGA$_1$ inhibits 20%. Thus a conversion of 5% of PGE$_1$ to PGA$_1$ could easily be detected. If the enzymatically generated PGA$_1$ (or PGC$_1$) were further converted to PGB$_1$ by isomerases, the increased inhibition would reflect a mixture of PGA$_1$ and PGB$_1$ of unknown quantity and composition, and only qualitative information about the conversion of PGE$_1$ to PGA$_1$ could be obtained with this antiserum. The disappearance of PGE can be measured also by chemically reducing PGE to PGF, which could then be measured with antibodies directed toward PGF. After chemical reduction, PGA does not react to any measurable extent to anti-PGF. Thus the amount of PGF$_{2\alpha}$ present could be obtained directly from the inhibition curves.

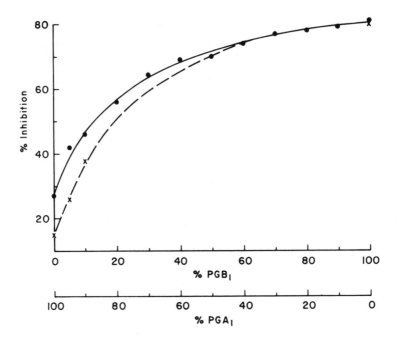

Fig. 3.   Inhibition of the binding of [³H]PGB₁ to rabbit anti-PGB₁ by mixtures of unlabeled PGA₁ and PGB₁. With rabbit albumin (●); without rabbit albumin (×). From Polet and Levine (1975a).

III. PROSTAGLANDIN SYNTHETASE

Prostaglandin synthetase is a multiple enzyme complex catalyzing the conversion of certain unsaturated fatty acids into a wide array of prostaglandins and prostaglandin-like materials. The reaction is initiated by the incorporation of two oxygen molecules at C-11 of the unsaturated eicosanoic acid, followed by the oxygenation at C-15 and concomitant cyclization of the carbon chain at C-8 and C-12 (Samuelsson, 1965, 1972). The unstable product has been isolated and referred as the 15-hydroperoxy-PGR or PGG (Nugteren and Hazelhof, 1973; Hamberg *et al.*, 1974). The hydroperoxy group at C-15 is then converted to a hydroxy group; this product also has been isolated and has been referred to as PGR or PGH (Nugteren and Hazelhof, 1973; Hamberg *et al.*, 1974). Finally reduction or isomerization of PGR or PGH gives rise to PGE or PGF.

Prostaglandin synthetase activity was first demonstrated in sheep seminal vesicle glands and subsequently in a variety of tissues, but tissue activities were low compared to that of seminal vesicle glands (van Dorp *et al.*, 1964; Bergström *et al.*, 1964; Christ and van Dorp, 1972). Most of the assays have used radioactive arachidonic acid as substrate and the products of the reaction have been identified and estimated by thin-layer chromatography. The Zimmerman reaction also has been used to measure the formation of PGE_2. Alternatively, oxygen electrodes have been employed to detect oxygenation during the course of the reaction. Because of the limitation of sensitivity of these assays, most of the studies on prostaglandin biosynthesis have used bovine or ovine seminal vesicle glands as the source of enzyme. A recent review has summarized the studies of this enzyme using these methods (Sih and Takeguchi, 1973).

We have studied the prostaglandin synthetase systems of various rabbit tissues by measuring the formation of either PGE_2 or $PGF_{2\alpha}$ by radioimmunoassay. The wide range of capacities for the prostaglandin synthesis by the microsomes of these tissues is shown in Table VI. The renal medulla is most active, followed by microsomes of the renal cortex, lung, brain, spleen, uterus, and heart. The synthetase of the heart microsomal preparation was 40 times less active than the renal medullary preparation.

A. Cofactor Requirements

The microsomes from all seven rabbit tissues require arachidonic acid, hydroquinone, and glutathione for enzymatic activity (Table VI). The optimal concentrations of arachidonic acid, hydroquinone, and

Table VI. Requirements of Substrate and Cofactors for the Formation of $PGF_{2\alpha}$ by Microsomes of Rabbit Tissues[a]

Source of microsomes	$PGF_{2\alpha}$ formed (ng)[b]			
	Complete system	Minus arachidonate	Minus glutathione	Minus hydroquinone
Brain	9.4	1.45	1.62	0.88
Heart	1.26	0.06	0.02	0.18
Lung	17.80	0.1	2.24	2.88
Renal cortex	37.2	2.8	4.3	5.6
Renal medulla	48.2[c]	1.7	1.4	2.8
Spleen	4.86	0.80	0.36	1.38
Uterus	2.28	0.10	0.10	0.38

[a] From Pong and Levine (1975).
[b] The complete reaction mixture (0.1 ml) contained 150 μM arachidonic acid, 2 mM reduced glutathione, 0.5 mM hydroquinone, and 0.1 M Tris-HC1 (pH 8.0). Microsomes prepared from 10.0 mg rabbit tissues were used. Incubation was at 37°C for 10 min.
[c] Renal medulla can produce five- to tenfold more $PGF_{2\alpha}$ if appropriate amounts of medullary cytoplasmic fractions are present.

glutathione for synthesis of $PGF_{2\alpha}$ are 0.15 mM, 1 mM, and 0.5 mM, respectively. For the synthesis of PGE_2, the optimal concentrations of arachidonic acid and glutathione are also 0.15 mM and 1 mM, respectively, but the hydroquinone concentration is 0.2 mM. At 0.5 mM, hydroquinone inhibits the production of PGE_2 by 50% (Pong and Levine, 1976*a*). Addition of glutathione to seminal vesicle microsomal preparations greatly enhances the yield of PGE_2 but not $PGF_{2\alpha}$ (Lands *et al.*, 1971). In contrast, with rabbit tissue microsomes, glutathione and hydroquinone enhance the synthesis of both PGE_2 and $PGF_{2\alpha}$. Hydroquinone can be replaced by serotonin, dopamine, norepinephrine, and epinephrine (Takeguchi *et al.*, 1971; Pong and Levine, 1976*a*).

B. Activation and Inhibition of Microsomal Prostaglandin Biosynthesis by Cytoplasmic Fractions

The prostaglandin synthetase activity is associated with the microsomal fraction. Samuelsson (1967) and van Dorp (1967) independently showed that boiled cytoplasmic fractions stimulated the synthetase activity of seminal vesicle microsomes. This stimulation could be replaced by addition of glutathione and a source of reducing equivalent, e.g., hydroquinone or epinephrine. Stimulation by the supernatant fluids could not be replaced by addition of NADH or NADPH. With washed rabbit renal medulla microsomes, small amounts of cytoplasmic fraction enhanced the synthetase activity (Pong and Levine, 1976*a*) even in the presence of hydroquinone and glutathione. An excess of cytoplasmic fraction inhibited the synthetase activity (Fig. 4). The stimulating activity of the cytoplasmic fraction is separable into two components by Sephadex G-100 chromatography; one component is a heat-labile nonheme protein and the other is relatively heat stable and is associated with hemoglobin. The heat-labile cytoplasmic factor did not stimulate the synthetic activity of microsomes prepared from other rabbit tissues, nor could hemoglobin stimulate microsomes of other rabbit tissues except those from small intestines. Hemoglobin has been reported to stimulate synthetic activity of bovine seminal vesicle microsomes (Yoshimoto *et al.*, 1970; Miyamoto *et al.*, 1974), but other investigators could not confirm this (Sih and Takeguchi, 1973). The differences in tissue specificity with respect to stimulation by cytoplasmic factors suggest that some of the components of the synthetase system are unique to a particular tissue.

The differential effects of the cytoplasmic fraction on the synthetase activities from cortical and medullary regions of rabbit kidney are

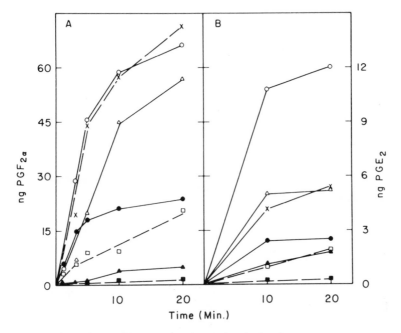

Fig. 4. Effect of addition of increments of cytoplasmic fraction to a constant amount of medullary microsomes on prostaglandin biosynthesis. All of the reaction mixtures contained $16 \mu g$ microsomes, substrate, and cofactors. (A) $PGF_{2\alpha}$; (B) PGE_2. Cytoplasmic protein ($76 \mu g$) alone (minus microsomes) (■); microsomes alone (▲) and in the presence of $7.4 \mu g$ cytoplasmic protein (△); $38 \mu g$ cytoplasmic protein (○); $76 \mu g$ cytoplasmic protein (×); $185 \mu g$ cytoplasmic protein (●); $380 \mu g$ cytoplasmic protein (□). From Pong and Levine (1976a).

shown in Fig 5. PGE_2 was actively synthesized by medullary homogenate but not by cortical homogenate; on the other hand, the washed microsomal fraction from each region was almost equally active in the synthesis of PGE_2.

C. Solubilization and Resolution of a Prostaglandin Synthetase System

The prostaglandin synthetase system of bovine seminal vesicle glands has been solubilized by treatment with detergent Tween 20 in the presence of ethylene glycol and resolved into oxygenase and isomerase components after DEAE-cellulose chromatography (Miyamoto *et al.*, 1974). The oxygenase fraction catalyzed the conversion of 8,11,14-eicosatrienoic acid into PGR_1 or PGH_1, which was subsequently transformed into PGE_1, $PGF_{1\alpha}$, and PGD_1 by the isomerase fraction. Both hemoglobin and tryptophan were necessary for oxygen-

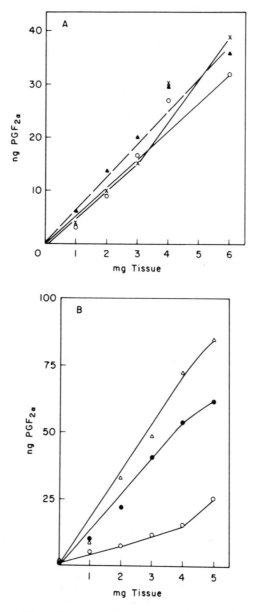

Fig. 5. Effect of crystalline rabbit hemoglobin and the partially purified, heat-labile medullary cytoplasmic factor on the synthetase activity of the cortical (A) and medullary (B) microsomes. The amount of microsomes is expressed as wet weight of tissue from which the microsomes were prepared. (A) Control (○); 6.6 μg of rabbit hemoglobin (×); 0.6 μg of medullary cytoplasmic factor (▲). (B) Control (○); 6.6 μg of hemoglobin (●); 0.6 μg of medullary cytoplasmic factor (Δ). From Pong and Levine (1976a).

ase activities while glutathione was required for the activity of the isomerase fraction. This is the first direct evidence for the multiple-subunit nature of prostaglandin synthetase molecule.

D. The pH Optima of Prostaglandin Synthetase Systems

The pH optimum for the prostaglandin synthetase system of bovine seminal vesicles is between 7.8 and 8.2 (Flower and Vane, 1974). The synthetase activity of rabbit renal medulla microsomes is optimum between pH 8.0 and 8.5 (Pong and Levine, 1976a). A very narrow pH optimum at 7.5 has been reported for the microsomes from whole rabbit kidney (Blackwell *et al.*, 1975). Rose and Collins (1974) have shown that microsomes from fresh rabbit medulla homogenate gave two pH optima at 7.0 and 9.0, while freeze-dried microsomes had a single peak of activity at pH 8.5. The differences in pH optima for kidney synthetase may be due to differences in the assay as well as in the preparation of enzyme from these studies.

E. Inhibition of Prostaglandin Synthetase from Different Tissues of the Same Species by Nonsteroidal Anti-inflammatory Drugs

Nonsteroidal anti-inflammatory drugs have been shown to inhibit the prostaglandin synthetase activity of many tissues (Vane, 1971; Flower, 1974). The drugs are diverse chemically, yet they all share to some extent the antipyretic, anti-inflammatory, and analgesic activities of aspirin (Shen *et al.*, 1974). It has been postulated by Vane and his co-workers that the broad-spectra pharmacological action of nonsteroidal anti-inflammatory drugs involves the inhibition of prostaglandin synthetase system (Vane, 1971; Ferreira and Vane, 1974). Indomethacin has been shown to inhibit the activity of the oxygenase component but not the isomerase component of the prostaglandin synthetase system (Miyamoto *et al.*, 1974).

Prostaglandins are synthesized by most mammalian tissues (Christ and van Dorp, 1972). Do these prostaglandin synthetase systems from all tissues of the same organism respond to the nonsteroidal anti-inflammatory drugs in an identical manner? Variation of inhibition of synthetase activities from different tissues by indomethacin has been reported (Flower and Vane, 1974; Bhattacherjee and Eakins, 1974). The effectiveness of these drugs depends on the concentration of substrate, arachidonic acid (Fig. 6). Similar findings have been reported in a study on the effect of indomethacin on prostaglandin synthetase activities of ovine or bovine seminal vesicle systems (Ham

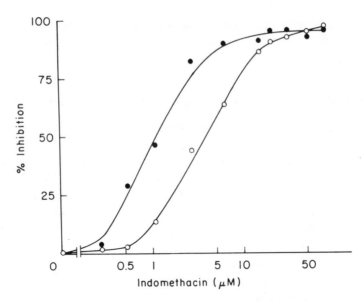

Fig. 6. Inhibition by indomethacin of PGF$_{2\alpha}$ biosynthesis by rabbit brain microsomes in the presence of 25 and 75 μM arachidonic acid. Rabbit brain microsomes (from 10 mg of tissue) were used. Arachidonic acid was 25 μM (•) or 75 μM (○). Control reactions in the absence of indomethacin produced 6.7 ng and 10.6 mg PGF$_{2\alpha}$ with the arachidonic acid at 25 μM and 75 μM, respectively. From Pong and Levine (1976c).

et al., 1972; Flower *et al.*, 1973). When assayed under identical substrate concentrations, the prostaglandin synthetase activities from seven rabbit tissues are inhibited identically by indomethacin or flufenamic acid or aspirin (Table VII). Thus several physiological functions might be disturbed by the administration of therapeutic doses of these

Table VII. Inhibition of PGF$_{2\alpha}$ Production by Indomethacin, Flufenamic Acid, and Aspirin[a]

Tissue	I_{50}[b] Indomethacin (μM)	Flufenamic acid (μM)	Aspirin (mM)
Brain	1.3	0.8	4.4
Heart	2.2	1.6	3.0
Lung	1.3	1.1	3.8
Renal cortex	1.8	1.7	5.4
Renal medulla	2.5	2.1	3.7
Spleen	2.3	1.2	1.9
Uterus	2.4	1.3	2.4

[a]From Pong and Levine (1976c).
[b]I_{50} was determined at 25 μM arachidonic acid.

drugs. These data suggest that prostaglandin synthetase may not exist in multiple forms within a species as postulated by Flower (1974).

IV. PROSTAGLANDIN E 9-KETOREDUCTASES

The biological activities of PGE and PGF are often diverse, and it is possible that their physiological effects can be controlled by enzyme(s) that interconvert them. It was initially believed that conversion of PGE to PGF did not occur in biological systems (Horton, 1972). However, small amounts of 9-α-hydroxy derivative had been detected in particle-free fractions of guinea pig liver after incubation with PGE_2 (Hamberg and Isrealsson, 1970) and reduction of the C-9 keto group to a 9-β-hydroxy derivative had been observed *in vivo* in the guinea pig (Samuelsson, 1972).

A. Cofactor Requirements and Tissue Distribution of PGE 9-Ketoreductase

Leslie and Levine (1973) demonstrated the enzymatic reduction of PGE_2 to $PGF_{2\alpha}$ in rat tissues. NADH was the only cofactor studied, and heart homogenates had the highest activity, followed by homogenates of kidney, brain, and liver. Subsequently, with NADPH as well as NADH as cofactors, the subcellular localization of the PGE 9-ketoreductase in various tissues was reported (Lee and Levine, 1974a). In the cytoplasmic fractions of brain, liver, blood, kidney, spleen, heart, lung, and uterus from monkey or pigeon, a 9-ketoreductase that used NADPH more effectively than NADH was detected (Table VIII). This type of enzyme has been purified from chicken heart and from human erythrocytes and has been characterized extensively (Lee and Levine, 1975a; Kaplan *et al.*, 1975). A second type of PGE 9-ketoreductase that used NADH more effectively than NADPH was found in the monkey liver microsomes (Lee and Levine, 1974a). A third type of PGE 9-ketoreductase that uses NADH as a cofactor was found in the cytoplasmic fraction of rat renal cortex and medulla (Katzen *et al.*, 1975). Unless precautions such as dialysis and addition of an NADPH- or NADH-generating system have been taken, it may be difficult to detect any enzymatic activities, especially in the crude preparation. Enzymatic reduction of PGE_2 to $PGF_{2\alpha}$ has been detected in the crude cellular fractions of blood from sheep but not from other animals (Henby, 1974). This reductase does not require either NADH or NADPH for activity.

The PGE 9-ketoreductases catalyze the reduction of the 9-keto group to a hydroxy group with an α configuration. With the use of both

Table VIII. NADPH- and NADH-Dependent Prostaglandin E 9-Ketoreductase Activity in Different Monkey and Pigeon Tissues[a]

Tissue	NADPH (units/mg protein)	NADH (units/mg protein)
Monkey		
Brain	1000	46
Liver cytoplasmic fraction	960	100
Liver microsomal fraction	52	450
Blood, formed elements	770	44
Kidney	480	34
Spleen	305	41
Heart	163	15
Uterus	140	33
Lung	98	21
Pigeon		
Brain	550	33
Heart	231	44
Liver	198	70
Lung	132	18
Blood, formed elements	110	32

[a]Reaction mixture contained 2.5 μg of PGE_2 and 3.3 mM NADPH or NADH and varying amounts of protein. Activity units are expressed as nonograms of $PGF_{2\alpha}$ produced per 10 min. With the exception of the microsomal fraction of monkey liver, the cytoplasmic fractions from all tissues were used. From Lee and Levine (1974).

anti-$PGF_{2\alpha}$ and anti-$PGF_{2\beta}$, the product of PGE_2 reduction by the 9-ketoreductase purified from chicken heart was identified as $PGF_{2\alpha}$. Neither guinea pig liver nor kidney homogenates reduced PGE_2 to $PGF_{2\beta}$ (Levine et al., 1975). Thus the β-reduction product in guinea pig urine found by Hamberg and Isrealsson (1970) may have been produced by other metabolic pathways.

B. Purification and Properties

Purification of the NADPH-dependent enzyme from chicken heart is summarized in Table IX. Two peaks of the activity were resolved during the phosphocellulose chromatographic step (Fig. 7). Both peaks were stimulated twentyfold by the fraction that was not bound to the phosphocellulose column. The stimulatory factor was heat stable, was not dialyzable, and was resistant to treatment with pronase, ribonuclease, and deoxyribonuclease. Sodium pyrophosphate also enhanced the activities of the phosphocellulose-purified enzymes. Angiotensin I, a precursor of the powerful pressor agent angiotensin II, stimulated both peaks of enzyme activity; angiotensin II did not. Angiotensin I

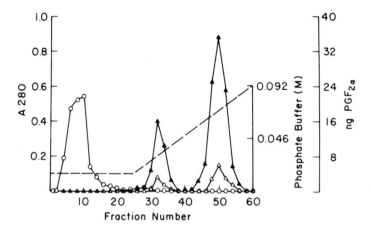

Fig. 7. Phosphocellulose column chromatography of prostaglandin E 9-ketoreductase from chicken heart. A_{280} (o) prostaglandin E 9-ketoreductase (Δ); prostaglandin E 9-ketoreductase in the presence of 10 mM sodium pyrophosphate (▲); phosphate buffer gradient (---). From Lee and Levine (1975*a*).

may have a role other than simply a precursor of angiotensin II (Itskovitz and Odya, 1971). In the presence of 1 mM 3′,5′-cyclic AMP, AMP, or several other ribonucleotides, the enhancing effects of the natural stimulatory substance or of sodium pyrophosphate or angiotensin I were inhibited.

Table IX. Purification of Chicken Heart Prostaglandin E 9-Ketoreductase[a]

Fraction	Activity (units)[b]	Total protein (mg)	Specific activity (units/mg protein)	Yield (%)
10,000*g* supernatant fluid	131,400	60,480	2.2	100
78,000*g* supernatant fluid	199,600	47,160	4.2	152
Ammonium sulfate (30–60% fraction)	164,600	18,540	8.9	125
DEAE-Sephadex chromatography	151,200	9,000	16.8	115
Hydroxylapatite chromatography	44,000	950	46.3	34
Phosphocellulose chromatography	17,200[c]	0.3[c]	57,300[c]	13[c]

[a] From Lee and Levine (1975*a*).
[b] Expressed as nanomoles of product generated per 10 min at 37°C.
[c] Values represent the sum of peaks I and II, and the activities are expressed without enhancement.

In crude enzyme preparations, both PGE_1 and PGE_2 were converted to the F series of prostaglandins in the presence of NADPH and NADH. 15-Keto-PGE_2 and 13,14-dihydro-15-keto-PGE_2 were converted less effectively to the corresponding $PGF_{2\alpha}$ metabolites (Lee and Levine, 1974*a*). Both the 9-keto group and the 15-keto group of 15-keto-PGE and the 15-keto group of 15-keto-$PGF_{2\alpha}$ can be converted to the corresponding hydroxy group by the PGE 9-ketoreductase purified from chicken heart (Lee and Levine, 1975*b*). The 15-keto group was a better substrate than the 9-keto group. The 11-keto group of PGD could not be converted to the 11-hydroxy group of $PGF_{2\alpha}$. The backward reaction, conversion of PGF to PGE by PGE 9-ketoreductase, was measurable with purified chicken heart and monkey liver microsome fractions (Lee and Levine, 1974*a*). Pace-Asciak (1975*a*) has described a 9-hydroxydehydrogenase activity which converts 15-keto-13,14-dihydro-$PGF_{2\alpha}$ to 15-keto-13,14-dihydro-PGE_2 in the adult rat kidney.

The purified chicken heart 9-ketoreductase has a broad pH range, 7.4–8.8. The molecular weight of the PGE 9-ketoreductase was estimated to be about 45,000–55,000 (Lee and Levine, 1975*b*).

V. 15-HYDROXYPROSTAGLANDIN DEHYDROGENASES

15-Hydroxyprostaglandin dehydrogenase (PGDH) is probably the important enzyme for biological inactivation of the prostaglandins. The enzyme oxidizes the C-15 alcoholic group into the 15-oxoderivatives, which are considerably less active pharmacologically.

A. Detection and Tissue Distribution of NAD$^+$-Dependent and NADP$^+$-Dependent 15-Hydroxyprostaglandin Dehydrogenases

PGDH activity was first detected in the pig lung (Änggård and Samuelsson, 1966). Subsequently, the enzyme was isolated from swine and beef lung (Nakano *et al.*, 1969; Tai *et al.*, 1974; Marrazzi and Anderson, 1974) and recently the PGDH from human placenta has been purified to homogeneity (Braithwaite and Jarabak, 1975). These enzyme preparations used NAD$^+$ as a cofactor, but not NADP$^+$. An NADP$^+$-dependent PGDH has been detected in monkey brain and red blood cells (Lee and Levine, 1975*b*). In kidneys of many species, both NAD$^+$- and NADP$^+$-dependent PGDH activities have been found (Lee *et al.*, 1975; Katzen *et al.*, 1975). The NAD$^+$-dependent and NADP$^+$-dependent PGDHs are best distinguished by their cofactor re-

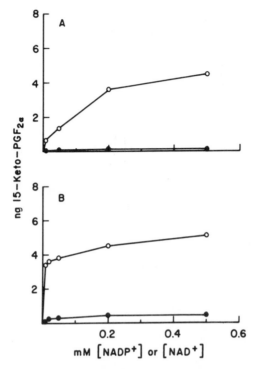

Fig. 8. Cofactor specificity of 15-hydroxyprostaglandin dehydrogenases. (A) Chicken heart 15-hydroxyprostaglandin dehydrogenase in the presence of NAD$^+$ (○) and NADP$^+$ (●). (B) Monkey brain 15-hydroxyprostaglandin dehydrogenase in the presence of NADP$^+$ (○) and NAD$^+$ (●). In the absence of cofactor, no 15-hydroxyprostaglandin dehydrogenase activity could be detected. From Lee and Levine (1975*b*).

quirements (Lee and Levine, 1975*b;* Lee *et al.,* 1975; Kaplan *et al.,* 1975). The chicken heart PGDH uses NAD$^+$ as cofactor much more effectively than NADP$^+$, whereas the monkey brain enzyme uses NADP$^+$ more effectively than NAD$^+$ (Fig. 8). Moreover, the NAD$^+$-dependent enzyme is strongly inhibited by NADH but not by NADPH, and the NADP$^+$-dependent enzyme is inhibited by NADPH but not by NADH. This inhibition is not due to the reversible reduction of the 15-ketoprostaglandins to the pharmacologically active prostaglandins because 15-keto-PGF$_{2\alpha}$ is not converted to PGF$_{2\alpha}$ in the presence of either NADH or NADPH.

B. Purification and Properties of the NAD$^+$- and NADP$^+$-Dependent Dehydrogenases

All PGDH activities have been found in the cytoplasmic fractions (100,000*g* supernatant fluids) of various tissue homogenates. Both enzymes can be separated during DEAE-Sephadex chromatography (Fig. 9). Under specified experimental conditions, the NADP$^+$-depen-

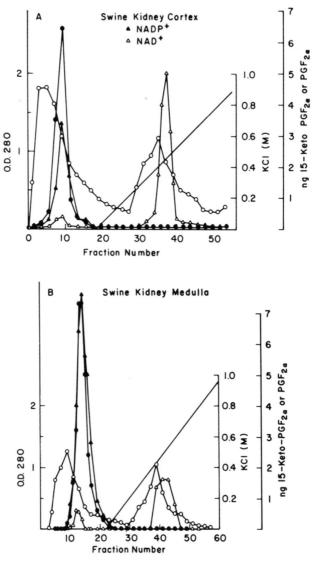

Fig. 9. DEAE-Sephadex chromatography of swine kidney cortex (A) and medulla (B). The fractions were assayed for PGDH activity for the generation of 15-keto-PGF$_{2\alpha}$ in a reaction mixture containing 1 μg PGF$_{2\alpha}$ and 1 mM NADP$^+$ (▲) or 1 mM NAD$^+$ (Δ). The activity of PGE 9-ketoreductase was assayed for the production of PGF$_{2\alpha}$ in a reaction mixture containing 1 mM NADPH and PGE$_2$ (●). OD$_{280}$ (○); KCl gradient (—). From Lee *et al.* (1975).

dent dehydrogenase passes through the column, but the NAD$^+$-dependent enzyme is absorbed and can only be eluted by higher salt concentrations. The NAD$^+$-dependent PGDH from chicken heart and the NADP$^+$-dependent enzyme from monkey brain have been purified (Table X) by ammonium sulfate fractionation, DEAE-Sephadex, and

Table X. Purification of Prostaglandin Dehydrogenase from Chicken Heart and Monkey Brain[a]

Fraction	Activity[b] (units)		Total protein (mg)		Specific activity (units/mg protein)		Yield (%)	
	Chicken heart	Monkey brain	Chicken heart	Monkey brain	Chicken heart	Monkey brain	Chicken heart	Monkey brain
10,000g supernatant fluid	1398.6	484.8	3024	1584	0.46	0.31	100	100
78,000g supernatant fluid	923.4	466.2	2358	1090	0.39	0.43	66	96
Ammonium sulfate (30–60%)	507.6	640.2	927	434	0.55	1.48	36	132
DEAE-Sephadex	496.8	840.6	203	84	2.45	10.01	35	225
Hydroxylapatite	421.2	1044.6	101	46	4.17	22.71	30	280

[a]From Lee and Levine (1975b).

[b]Enzymatic activity = nanomoles of product generated per 10 min at 37°C. The $PGF_{2\alpha}$ concentration is 10 μg/ml (approximately 28 μM).

hydroxylapatite chromatography (Lee and Levine, 1975b). Partial purification of the $NADP^+$-dependent PGDH from human erythrocytes also has been accomplished, but all attempts to separate PGE 9-ketoreductase activity from $NADP^+$-dependent PGDH activity have been unsuccessful (Lee and Levine, 1975b; Kaplan et al., 1975). The specific activities of these purified enzymes from chicken heart, monkey brain, and human erythrocytes were 0.4, 2.3, and 0.006 nmol 15-keto-PGF_2 formed/min/mg protein, respectively. The NAD^+-dependent dehydrogenase from chicken heart and the NAD^+-dependent enzyme from monkey liver have broad pH spectra; no differences in activities were observed over a pH range of 7.2–8.5. The molecular weight of these enzymes was estimated to be 60,000–70,000 (Lee and Levine, 1974b).

Human placenta is particularly rich in NAD^+-dependent dehydrogenase activity. Specific activities of 1765–649 nmol/min/mg protein for the purified enzymes have been reported (Braithwaite and Jarabak, 1975; Schlegel and Greep, 1975). The molecular weight of the enzyme was estimated to be 42,000–51,500 and no evidence was obtained for the existence in the enzyme of multiple subunits. The molecular weights of the NAD^+-dependent dehydrogenase from beef and pig lung have been reported to be 60,000–70,000 and 20,000 (Änggård, 1971; Marrazzi and Matschinsky, 1972).

Chicken heart NAD^+-dependent PGDH oxidizes the C-15 hydroxy group of both PGE_2 and PGF_2 with equal efficiency; K_m values for PGE_2 and PGF_2 were 14 and 25 μM, respectively (Lee and Levine, 1975b). The $NADP^+$-dependent PGDH of monkey brain converts $PGF_{2\alpha}$ to 15-keto-$PGF_{2\alpha}$ more effectively than it does PGE_2 to 15-keto-PGE_2; the apparent K_m values of PGE_2 and $PGF_{2\alpha}$ are 200 μM and 25 μM, respectively. These results suggest that *in vivo* the NAD^+-dependent dehydrogenase may metabolize both PGE_2 and PGF_2 equally, while the $NADP^+$-dependent enzymes metabolize $PGF_{2\alpha}$ preferentially. PGB_2 was not a substrate for these two types of dehydrogenase and was found to inhibit the chicken heart NAD^+-dependent PGDH but not the $NADP^+$-dependent enzymes from monkey brain and human erythrocytes (Lee and Levine, 1975b; Kaplan et al., 1975).

NAD^+-dependent dehydrogenase from human placenta is reported to have a K_m value of 1 μM (Schlegel and Greep, 1975). In the same study, the equilibrium constant with respect to PGE_2 was measured and found to be 18 μM. In other studies, however, a substantially lower value, 6.5×10^{-8} M, has been reported for the equilibrium constant (Braithwaite and Jarabak, 1975).

Thyroid hormones, thyroxine, 3,3′,5-triiodothyronine, and tetraiodothyroacetic acid inhibit the NAD^+-dependent dehydrogenase (Tai et al., 1974; Lee and Levine, 1975b). These hormones have little or no effect on $NADP^+$-dependent dehydrogenase activity from monkey brain and human erythrocytes (Lee and Levine, 1975*b*; Kaplan *et al.*, 1975). Inhibition of the NAD^+-dependent dehydrogenase from human placenta by estrogen and progesterone has been reported (Schlegel *et al.*, 1974).

VI. PROSTAGLANDIN Δ^{13}-REDUCTASE

Reduction of the Δ^{13} double bond during the formation of the 13,14-dihydro-15-keto-PGE_2 derivative must be preceded by the oxidation of the allylic alcoholic group (Hamberg and Samuelsson, 1971). Therefore, prostaglandin Δ^{13}-reductase, which reduces the C-13,14 double bond, is the enzyme immediately following the PGDH in the prostaglandin metabolic sequence. The Δ^{13}-reductase activity was first detected in guinea pig lungs (Änggård and Samuelsson, 1964) and later in pig adipose tissue, spleen, kidney, liver, and small intestine (Samuelsson *et al.*, 1971). The product of the reaction, 13,14-dihydro-15-keto-PGE or 13,14-dihydro-15-keto-$PGF_{2\alpha}$, retains considerable biological activity (Änggård, 1966).

Like the PGDHs, the Δ^{13}-reductase is located in the soluble fraction of cell preparations. With the use of serological methods for assaying catalytic activity, the Δ^{13}-reductase from chicken heart has been purified and characterized (Lee and Levine, 1974*b*). The enzyme was purified 7000-fold (Table XI) and contained no PGDH or PGE 9-

Table XI. Purification of Prostaglandin Δ^{13}-Reductase from Chicken Heart[a]

Fraction	Activity (units)[b]	Total protein (mg)	Specific activity (units/ mg protein) $\times 10^3$	Yield (%)
10,000g supernatant fluid	43.8	60,480	0.7	100
78,000g supernatant fluid	41.6	47,160	0.9	95
Ammonium sulfate precipitate (30–60%)	39.7	18,540	2.1	91
DEAE-Sephadex	34.9	9,000	3.9	80
Hydroxylapatite	20.9	1,250	16.7	48
Phosphocellulose	12.3	1.8	6833.3	28

[a] From Lee and Levine (1974*b*).
[b] Expressed as nanomoles of substrate lost per 10 min at 37°C.

ketoreductase activities. The enzyme activity was optimum at pHs ranging from 7.4 to 8.5 and the molecular weight of the enzyme was estimated to be 70,000–80,000.

The purified enzyme from chicken heart used NADPH as a cofactor much more effectively than NADH. The PG Δ^{13}-reductase activity in homogenates of human placenta, dog lung, swine lung, kidney, and liver, and monkey heart, lung, liver, spleen, kidney, and uterus also used NADPH more effectively than NADH. The enzyme is not inhibited by oxidized pyridine nucleotides. It specifically reduces 15-keto-prostaglandins but not 15-hydroxyprostaglandins. A Δ^{13}-reductase recently has been purified from human placenta (Westbrook and Jarabak, 1975). The enzyme, purified 800-fold, utilized NADH as a cofactor but not NADPH.

VII. PGE DEHYDRASE, PGA ISOMERASE, AND PGC ISOMERASE

Three PGE-metabolizing enzymes, a dehydrase that converts PGE to PGA (Cammock, 1972; Russell *et al.*, 1973; Polet and Levine,

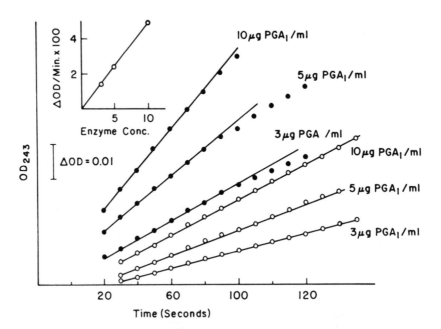

Fig. 10. Spectroscopic measurement of the initial rates of formation of PGC_1 from PGA_1 by PGA isomerase in the presence and absence of rabbit serum albumin. The reaction was initiated by the addition of 25 μl of PGA isomerase (purified fiftyfold) to a PGA_1 solution (and albumin) in buffer with 5% ethanol in a 1-ml cuvette. Without albumin (○); with 0.5 mg albumin per milliliter (●). The insert shows the initial rates as a function of the enzyme concentration in arbitrary units. From Polet and Levine (1975*b*).

Table XII. Influence of Serum Albumins and Other Proteins on the Conversion of PGC₁ to PGB₁[a]

	Protein (mg/ml)	Conversion of PGA_1 to PGB_1 (%)
Experiment 1	None	0
Rabbit serum albumin	0.1	15
	0.2	35
	0.4	60
	0.8	80
Bovine serum albumin	0.1	11
	0.2	20
	0.4	25
	0.8	30
Experiment 2	None	0
Human serum albumin	0.4	60
Human mercaptalbumin	0.4	58
Rabbit γ-globulin	0.4	0
Ovalbumin	0.4	0
β-Lactoglobulin	0.4	0
Ribonuclease	0.4	0

[a]PGA isomerase purified fiftyfold was incubated with PGA_1 (4 ng/ml) in the presence of various proteins for 1 hr at 37°C. From Polet and Levine (1975a).

1975a), a PGA isomerase that converts PGA to PGC (Jones, 1970; Polet and Levine, 1971), and a PGC isomerase that converts PGC to PGB (Polet and Levine, 1975a), have been described. All three activities are inhibited by sulfhydryl blocking reagents. Exhaustive dialysis does not alter their enzymatic activities. Heating for 20 min at 70°C completely inactivates the dehydrase and the PGA isomerase, whereas the PGC isomerase of rabbit albumin is only partially inactivated. In human serum, dehydrase activity is low (Polet and Levine, 1975a). At a concentration of 20 μg PGE_1 per milliliter of undiluted human serum, only 30% of the PGE_1 was converted to PGA in 4 hr of incubation at 37°C. Conversion of PGE to a less polar compound (possibly PGA) in human plasma has been reported by McDonald-Gibson et al., (1972). No dehydrase activity was detected in rabbit serum (Polet and Levine, 1975a).

Prostaglandin A isomerases have been partially purified (Jones and Cammock, 1973; Polet and Levine, 1975b). The initial rates of product formation by the purified isomerase from rabbit serum are linear and they are proportional to the enzyme concentration (Fig. 10). These initial rates increased two- to threefold in the presence of rabbit albumin, bovine albumin, or β-lactoglobulin, probably because of a solubilizing effect of the proteins on PGA. A K_m of 4×10^{-5} M in the absence of rabbit albumin and a K_m of 5×10^{-5} M in its presence were obtained for the partially purified enzyme. This value is in agreement with the K_m of 2.5×10^{-5} M reported by Jones *et al.* (1972) for cat plasma isomerase. The molecular weight of the enzyme from rabbit serum was estimated to be 110,000 (Polet and Levine, 1975*b*). This enzyme may be a mixture of enzymes as judged by the spread of activities over a relatively large pH range during isoelectric focusing. The enzyme appears to have a strict requirement for the 13,14 double bond since 13,14-dihydro-PGA$_1$ is not a substrate (Jones *et al.*, 1972).

In rabbit serum, PGA$_1$ is rapidly converted to PGB$_1$. The PGA isomerase converts PGA to PGC$_1$ (Jones and Cammock, 1973; Polet and Levine, 1975*a*) and the PGC is rapidly converted to PGB during incubation in rabbit serum (Polet and Levine, 1975*a*). Thus rabbit serum catalyzes not only the conversion of PGA$_1$ to PGC$_1$ but also the second isomerization of PGC$_1$ to PGB$_1$. This PGC isomerase activity cannot be separated from albumin. In fact, the activity found in human

Table XIII. Enzyme Activities in the Sera of Various Species[a]

	Isomerase activities				
	Dilution that converts 15% PGA$_1$ to PGB$_1$		PGA isomerase activity[b]	PGC isomerase activity[c]	Dehydrase activity[d]
Species	No RSA	+ RSA	(%)		
Rabbit	600	10,000	100	+	−
Rat	1500	4,000	40	+	−
Mouse	300	1,000	10	+	n.d.[e]
Pig	1000	1,000	10	+	+
Dog	300	500	5	+	+
Cat	100	400	5	+	n.d.
Sheep	10	10	0.1	+	+
Guinea pig	10	10	0.1	+	−
Horse			<0.05	n.d.	−
Ox			<0.05	n.d.	−
Chicken			<0.05	n.d.	−
Man			∼0.01	+	+
Monkey			n.d.	n.d.	+

[a] From Polet and Levine (1975*a*).
[b] Sera were incubated with PGA$_1$, in the presence or absence of 0.4 mg of rabbit albumin per milliliter for 1 hr (except for man), at 37°C.
[c] The presence of PGC isomerase was deduced from the conversion of PGA to PGB in these sera without additional PGC isomerase (rabbit albumin).
[d] Sera were diluted 1:10 and incubated with 20 μg of PGE$_2$ per milliliter for 20 hr at 37°C. The remaining PGE$_2$ was reduced to PGF and determined with PGF$_{2\alpha}$ antiserum.
[e] Not determined.

albumin quantitatively accounts for all the activity detected in human serum. Even human mercaptalbumin has PGC isomerase activity. In the presence of other proteins such as ovalbumin, β-lactoglobulin, or ribonuclease, no PGB was formed (Table XII).

The data shown in Table XIII describe the activities of these three enzymes in the sera of different species. In human serum, only PGC isomerase is very active whereas the two other enzymes were much less active. In rabbit serum, both isomerases were very active but the dehydrase could not be detected.

VIII. REGULATION OF PGE AND PGF LEVELS

Although mammalian cells have the capacity to synthesize prostaglandins, there is little evidence that the prostaglandins accumulate intracellularly, except in the seminal vesicles. Body fluids, cells, and tissues probably contain only trace amounts. Therefore, biological activity is associated mainly with newly synthesized prostaglandins rather than with their release.

Prostaglandins have been implicated in numerous diverse physiological functions, and many external stimuli such as hormonal and neural stimulation and trauma have been found to change the prostaglandin levels synthesized by the cells. Modulation of these levels may occur during synthesis or during metabolism. As described earlier, the prostaglandin synthetase system is localized in the particulate microsomal fraction and is probably a membrane-bound enzyme. On the other hand, most of metabolic enzymes, e.g., PGE 9-ketoreductases, PGDH, and PG Δ^{13}-reductase, are found in the cellular cytoplasmic fractions. One exception is a microsomal NADH-dependent PGE 9-ketoreductase found in monkey liver (Lee and Levine, 1974*a*). This intracellular compartmentation for biosynthesis and metabolism may have important implications for the function and the regulation of the prostaglandins. For example, $PGF_{2\alpha}$ has luteolytic properties in several species. The receptor for $PGF_{2\alpha}$ is associated with the particulate fraction of the corpora lutea (Powell *et al.*, 1975; Rao, 1974). There are reports that the prostaglandins may interact with adenylate cyclase, a membrane-bound enzyme. Possibly the sites of biosynthesis and function are physically related.

Exogenous arachidonic acid increases the levels of prostaglandins found in tissues. Perfusion of guinea pig lung or frog intestine with phospholipase A results in the release of large amounts of prostaglandins into the perfusate (Vogt *et al.*, 1969; Bartels *et al.*, 1970). Probably it is the release of arachidonic acid from membrane phospholipids by phospholipase A that is the rate-limiting step in the formation of prostaglandins and hence an important step in the regulation of biosynthesis (Kunze and Vogt, 1971). Acid phospholipase A_2 has been found in lysozymes of several tissues, and the release of this enzyme from decidual cells may be involved in mechanisms leading to abortion and labor (Gustavii, 1974; Akesson and Gustavii, 1975). Thyroid-stimulating hormone has been shown to stimulate arachidonic acid release from phosphatidylinositol in thyroid homogenates (Haye *et al.*, 1973). Trauma of cells by various mechanisms has been found to increase the synthesis of prostaglandins, possibly via phospholipase activity (Gryglewski and Vane, 1972).

Based on the quantitative differences among synthetases from different tissues to inhibition by indomethacin, Flower and Vane (1974) have proposed that the prostaglandin synthetase exists in multiple molecular forms within the organism. However, no differences in sensitivity to inhibition by indomethacin, flufenamic acid, and aspirin among the synthetases of seven rabbit tissues were found by Pong and Levine (1976c). Since a cytoplasmic heat-labile protein from renal medulla stimulates the prostaglandin synthetase of renal medullary microsomes but not the synthetases from other tissues (Pong and Levine, 1976a) and hemoglobin and heme compounds stimulate the synthetase system of bovine seminal vesicles, renal medulla, and small intestine but not other tissues, the prostaglandin synthetase system may have different regulatory subunit(s) in the tissues from the same organism (Pong and Levine, 1976a).

Hemoglobin and the renal medulla cytoplasmic factor appear to stimulate the renal medulla synthetase system at different steps of the synthesis. Whether hemoglobin is a natural activator under physiological conditions has been questioned, although heme compounds may be directly involved in the activation (Yoshimoto *et al.*, 1970). Inhibitory effects of cytoplasmic fractions also have been demonstrated. It is conceivable that these stimulatory and inhibitory activities in cytoplasmic fractions are modulating prostaglandin synthetase activity and regulating the levels of prostaglandins synthesized *in vivo* .

Still another type of regulation for prostaglandin synthetase activity is the negative feedback mechanism of the oxygenase reaction as described by Smith and Lands (1971, 1972) and Lands *et al.* (1971, 1974). This hypothesis was based on the observation of rapid termination of the reaction by bovine or ovine seminal vesicle prostaglandin synthetase systems. A mechanism of self-catalyzed destruction of enzyme was proposed. We have observed, however, that the reaction by homogenates and some microsomes prepared from rabbit tissues proceeds for much longer periods of time, e.g., 20–30 min (Pong and Levine, 1976c). The possibility that an unidentified cofactor in microsomes or cytoplasm is required for the reaction to continue should be considered.

The various prostaglandins often have different biological activities. A variety of compounds may interact with the prostaglandin synthetase system and preferentially direct the formation of PGE or PGF from the endoperoxide intermediates. In the presence of both glutathione and hydroquinone, the yield of PGE_2 was greatly stimulated and that of the other prostaglandins decreased to a negligible level (van Dorp, 1967). In ovine or bovine seminal vesicle synthetase systems, copper–dithiol complex and L-epinephrine selectively enhance the formation of $PGF_{2\alpha}$ at the expense of other prostaglandins (Lands *et al.*, 1971; Chan *et al.*, 1975). Rabbit renal medulla microsomes require both hydroquinone and glutathione for the synthesis of $PGF_{2\alpha}$ and PGE_2 (Pong and Levine, 1976*a*). However, little is known of the conditions at or near the site of *in vivo* biosynthesis. The activities of PGE 9-ketoreductases, which interconvert the E and F series of prostaglandins, may be significant in this respect. Three types of PGE 9-ketoreductases now have been found. Each can be distinguished by its specific cofactor requirement and subcellular localization. External and internal stimuli, e.g., angiotensin I, cyclic AMP, AMP, or natural stimulatory factors, affect the activity of these enzymes, and thus can alter the ratio of PGE and PGF_{α} (Lee and Levine, 1975*a*).

PGDH is the key enzyme in the biological inactivation of prostaglandins (Änggård, 1966; Änggård and Samuelsson, 1967; Samuelsson *et al.*, 1971). Two types of PGDH have been described; one requires NAD^+ as a cofactor and the other $NADP^+$ (Lee and Levine, 1975*b*). Most tissues have only one type of dehydrogenase, but some tissues have both. In the kidney, two types of PGDH activity have been found (Lee *et al.*, 1975; Katzen *et al.*, 1975). These PGDHs are sensitive to inhibition by the reduced form of their cofactors. They also show distinct substrate specificities: NAD^+-dependent dehydrogenase metabolizes PGE_2 and $PGF_{2\alpha}$ equally well, but the $NADP^+$-dependent enzyme metabolizes $PGF_{2\alpha}$ much more effectively than PGE_2. Thus a coupled reaction of NADPH-dependent PGE 9-ketoreductase and $NADP^+$-dependent PGDH, both of which depend on the relative concentrations of $NADP^+$ and NADPH, could regulate not only the ratio of PGE_2 to $PGF_{2\alpha}$ but also the duration of their activities in some tissues. In monkey brain and human erythrocyte, the PGE 9-ketoreductase was not able to be separated from the $NADP^+$-dependent dehydrogenase (Lee and Levine, 1975*b*; Kaplan *et al.*, 1975). Inactivation of the prostaglandins may be linked to the glycolytic activity of the cell which determines the amount of NAD^+, $NADP^+$, NADH, or NADPH present for PGDH and 9-ketoreductase activities.

It has been suggested that the prostaglandins synthesized in the renal medulla are transported to the renal cortex, where they regulate blood flow (Larsson and Änggård, 1973; Herbaczynska-Cedro and Vane, 1973). The hypothesis originally was based on the findings that highly active prostaglandin synthetase was present in the rabbit renal medulla, but only low levels, if any, were in the cortex. Also, the NAD$^+$-dependent PGDH was associated mainly with renal cortex in the rabbit. However, the renal medulla as well as the medulla of several species has metabolic activity equal to that of the cortex if both the NADP$^+$- and NAD$^+$-dependent dehydrogenase activities are estimated (Lee *et al.*, 1975; Katzen *et al.*, 1975), and the rabbit renal cortex does have considerable capacity to synthesize prostaglandins (Pong and Levine, 1976*b*). Thus transport of prostaglandins in the kidney from the medulla to the cortex may not be obligatory. Most likely, the prostaglandin levels in the renal cortex and medulla are regulated independently.

An important approach that may lead to our understanding of the regulation of biosynthesis and metabolism of prostaglandins is the changing enzymatic activities during development. It is clear that the lung and kidney of fetal and early newborn rats have different metabolic profiles than adult rats (Pace-Asciak and Miller, 1973, 1974; Pace-Asciak, 1975*b*). In the rat kidney, PGDH and PG Δ^{13}-reductase activities rise sharply after birth to a maximal value at 19 days and then by day 40 decrease to the low values seen in adults. This result suggests a possible relationship between prostaglandin-catabolizing activities and nephrogenesis.

IX. REFERENCES

Akesson, B., and Gustavii, B., 1975, Occurrence of phospholipase A$_1$ and A$_2$ in human decidua, *Prostaglandins* **9**:667.

Änggård, E., 1966, The biological activities of three metabolites of prostaglandin E$_1$, *Acta Physiol. Scad.* **66**:509.

Änggård, E., 1971, Studies on the analysis and metabolism of prostaglandins *Ann. N.Y. Acad. Sci.* **180**:200.

Änggård, E., and Samuelsson, B., 1964, Metabolism of prostaglandin E$_1$ in guinea pig lung: The structure of two metabolites, *J. Biol. Chem.* **239**:4097.

Änggård, E., and Samuelsson, B., 1966, Purification and properties of a 15-hydroxyprostaglandin dehydrogenase from swine lung, *Ark. Kemi.* **25**:293.

Änggård, E., and Samuelsson, B., 1967, The metabolism of prostaglandins in lung tissue, *Nobel Symposium 2, Prostaglandins* (S. Bergstrom and B. Samuelsson, eds.), pp. 97–105, Almqvist and Wiksell, Stockholm.

Bartels, J., Kunze, H., Vogt, W., and Willie, G., 1970, Prostaglandin: Liberation from and formation in perfused frog intestine, *Naunyn-Schmiedebergs Arch. Pharmakol.* **266**:199.

Bauminger, S., Zor, U., and Lindner, H. R., 1973, Radioimmunological assay of prostaglandin synthetase activity, *Prostaglandins* **4**:313.

Bergström, S., Danielsson, H., Klenberg, D., and Samuelsson, B., 1964, The enzymatic conversion of essential fatty acid into prostaglandins, *J. Biol. Chem.* **239**:PC4006.

Berson, S. W., and Yalow, R. S., 1960, Quantitative aspects of the reaction between insulin and insulin-binding antibody, *J. Clin. Invest.* **39**:1157.

Bhattacherjee, P., and Eakins, K. E., 1974, Inhibition of the prostaglandin synthetase systems in occular tissues by indomethacin, *Br. J. Pharmacol.* **50**:227.

Blackwell, G. J., Flower, R. J., and Vane, J. R., 1975, Some characterization of the prostaglandin synthesizing system in rabbit kidney microsomes, *Biochim. Biophys. Acta* **398**:178.

Braithwaite, S. S., and Jarabak, J., 1975, Studies on a 15-hydroxyprostaglandin dehydrogenase from human placenta: Purification and partial characterization, *J. Biol. Chem.* **250**:2315.

Butler, V. P., Jr., and Beiser, S. M., 1973, Antibodies to small molecules: Biological and clinical applications, *Adv. Immunol.* **17**:255.

Caldwell, B. V., Burstein, S., Brock, W. A., and Speroff, L., 1971, Radioimmunoassay of the F prostaglandins, *J. Clin. Endocrinol. Metab.* **33**:171.

Cammock, S., 1972, Conversion of PGE_1 to PGA_1-like compound by rat kidney homogenates, *Adv. Biosci. Suppl.* **9**:10.

Chan, J. A., Nagasawa, M., Takeguchi, C., and Sih, C. J., 1975, On agents favoring prostaglandin F formation during biosynthesis, *Biochemistry* **14**:2987.

Christ, E. J., and van Dorp, D. A., 1972, Comparative aspect of prostaglandin biosynthesis in animal tissues, *Biochim. Biophys. Acta* **270**:537.

Ferreira, S. H., and Vane, J., 1974, New aspects of the mode of action of nonsteroidal anti-inflammatory drugs, *Annu. Rev. Pharmacol.* **14**:57.

Flower, R. J., 1974, Drugs which inhibit prostaglandin biosynthesis, *Pharmacol. Rev.* **26**:33.

Flower, R. J., and Vane, J. R., 1974, Some pharmacologic and biochemical aspects of prostaglandin biosynthesis and its inhibition, in: *Prostaglandin Synthetase Inhibitors* (H. J. Robinson and J. R. Vane, eds.), pp. 9–18, Raven Press, New York.

Flower, R. J., Cheung, H. S., and Cushman, D. W., 1973, Quantitative determination of prostaglandins and malondialdehyde formed by the arachidonate oxygenase (prostaglandin synthetase) system of bovine seminal vesicle, *Prostaglandins* **4**:325.

Granström, E., and Samuelsson, B., 1972, Development and mass-spectrometric evaluation of a radioimmunoassay for $9\alpha,11\alpha$-dihydroxy-15-ketoprost-5-enoic acid, *FEBS Lett.* **26**:211.

Gréen, K., 1973, Methods for quantitative determination of prostaglandins using gas chromatography–mass spectrometry, in: *Prostaglandins 1973*, pp. 113–132, Inserm, Paris.

Gryglewski, R., and Vane, J., 1972, The release of prostaglandins and rabbit aorta contracting substance (RCS) from rabbit spleen and its antagonism by anti-inflammatory drugs, *Br. J. Pharmacol.* **45**:37.

Gustavii, B., 1974, Studies on the mode of action of intra-amniotically and extra-amniotically injected hypertonic saline in therapeutic abortion, *Acta. Obstet. Gynecol. Scand. Suppl.* **25**:1.

Ham, E. A., Cirillo, V. J., Zanetti, M., Shen, T. Y., and Kuehl, F. A., Jr., 1972, Studies on the mode of action of nonsteroidal anti-inflammatory agents, in: *Prostaglandins in Cellular Biology* (P. W. Ramwell and B. B. Pharriss, eds.), pp. 345–352, Plenum Press, New York.

Hamberg, M., and Isrealsson, U., 1970, Metabolism of prostaglandin E_2 in guinea pig liver. I. Identification of seven metabolites, *J. Biol. Chem.* **245**:5107.

Hamberg, M., and Samuelsson, B., 1971, Metabolism of prostaglandin E_2 in guinea pig liver, *J. Biol. Chem.* **246**:1073.

Hamberg, M., Svensson, J., Wakabayashi, T., and Samuelsson, B., 1974, Isolation and structure of two prostaglandin endoperoxides that cause platelet aggregation, *Proc. Natl. Acad. Sci. U.S.A.* **71**:345.

Haye, B., Champion, S., and Jacquemin, C., 1973, Control by TSH of a phospholipase A_2 activity, a limiting factor in the biosynthesis of prostaglandins in the thyroid, *FEBS Lett.* **30**:253.

Henby, C. N., 1974, Reduction of prostaglandin E_2 to prostaglandin $F_{2\alpha}$ by an enzyme in sheep blood, *Biochim. Biophys. Acta* **348**:145.

Herbaczynska-Cedro, K., and Vane, J., 1973, Contribution of intrarenal generation of prostaglandin to autoregulation of renal blood flow in the dog, *Circ. Res.* **33**:428.

Horton, E. W., 1972, The prostaglandins, *Proc. R. Soc. London Ser. B* **182**:411.

Itskovitz, H. D., and Odya, C., 1971, Intrarenal formation of angiotensin I, *Science* **174**:58.

Jaffe, B. M., Smith, J. W., Newton, W. T., and Parker, C. W., 1971, Radioimmunoassay for prostaglandins, *Science* **171**:494.

Jones, R. L., 1970, A prostaglandin isomerase in cat plasma, *Biochem. J.* **119**:64P.

Jones, R. L., and Cammock, S., 1973, Purification, properties, and biological significance of prostaglandin A isomerase, *Adv. Biosci.* **9**:61.

Jones, R. L., Cammock, S., and Horton, E. W., 1972, Partial purification and properties of cat plasma prostaglandin A isomerase, *Biochim. Biophys. Acta* **280**:588.

Kaplan, L., Lee, S. C., and Levine, L., 1975, Partial purification and some properties of human erythrocyte prostaglandin 9-ketoreductase and 15-hydroxyprostaglandin dehydrogenase, *Arch. Biochem. Biophys.* **167**:287.

Katzen, D. R., Pong, S. S., and Levine, L., 1975, Distribution of prostaglandin E 9-ketoreductase and NAD^+-dependent and $NADP^+$-dependent 15-hydroxyprostaglandin dehydrogenase in the renal cortex and medulla of various species, *Res. Commun. Chem. Path. Pharmacol.* **12**:781.

Kirton, K. T., Cornette, J. C., and Barr, K. L., 1972, Characterization of antibody to prostaglandin $F_{2\alpha}$, in: *Prostaglandins in Fertility Control 2* (S. Bergstrom, K. Green, and B. Samuelsson, eds.), pp. 60–68, WHO, Karolinska Institutet, Stockholm.

Kunze, H., and Vogt, W., 1971, Significance of phospholipase A for prostaglandin formation, *Ann. N.Y. Acad. Sci.* **180**:123.

Lands, W., Lee, R., and Smith, W., 1971, Factors regulating the biosynthesis of various prostaglandins, *Ann. N.Y. Acad. Sci.* **180**:107.

Lands, W. E. M., LeTellier, P. R., Rome, L., and Vanderhoek, J. Y., 1974, Regulation of prostaglandin synthesis, in: *Prostaglandin Synthetase Inhibitors* (H. J. Robinson and J. R. Vane, eds.), pp. 1–7, Raven Press, New York.

Larsson, C., and Änggård, E., 1973, Regional differences in the formation and metabolism of prostaglandins in the rabbit kidney, *Eur. J. Pharmacol.* **21**:30.

Lee, S. C., and Levine, L., 1974*a*, Prostaglandin metabolism. I. Cytoplasmic reduced nicotinamide adenine dinucleotide phosphate-dependent and microsomal reduced nicotinamide adenine dinucleotide-dependent prostaglandin E 9-ketoreductase activities in monkey and pigeon tissues, *J. Biol. Chem.* **249**:1369.

Lee, S. C., and Levine, L., 1974*b*, Purification and properties of chicken heart prostaglandin Δ^{13}-reductase, *Biochem. Biophys. Res. Commun.* **61**:14.

Lee, S. C., and Levine, L., 1975*a*, Purification and regulatory properties of chicken heart prostaglandin E 9-ketoreductase, *J. Biol. Chem.* **250**:4549.

Lee, S. C., and Levine, L., 1975*b*, Prostaglandin metabolism. II. Identification of two 15-hydroxyprostaglandin dehydrogenase types, *J. Biol. Chem.* **250**:548.

Lee, S. C., Pong, S. S., Katzen, D., Wu, K. Y., and Levine, L., 1975, Distribution of prostaglandin E 9-ketoreductase and types I and II 15-hydroxyprostaglandin dehydrogenase in swine kidney medulla and cortex, *Biochemistry* **14**:142.

Leslie, C. A., and Levine, L., 1973, Evidence for the presence of a prostaglandin E_2-9-ketoreductase in rat organs, *Biochem. Biophys. Res. Commun.* **52**:717.

Levine, L., 1973, Antibodies to pharmacologically active molecules: Specificities and some applications of antiprostaglandins, *Pharmacol. Rev.* **25**:293.

Levine, L., and Van Vunakis, H., 1970, Antigenic activity of prostaglandins, *Biochem. Biophys. Res. Commun.* **41**:1171.

Levine, L., Gutierrez-Cernosek, R., and Van Vunakis, H., 1971, Specificities of prostaglandins B_1, $F_{1\alpha}$, and $F_{2\alpha}$ antigen–antibody reactions, *J. Biol. Chem.* **246**:6782.

Levine, L., Wu, K. Y., and Pong, S. S., 1975, Stereospecificity of enzymatic reduction of prostaglandin E_2 to $F_{2\alpha}$, *Prostaglandins* **9**:531.

Marrazzi, M. A., and Anderson, N. H., 1974, Prostaglandin dehydrogenase, in: *The Prostaglandin Dehydrogenase* (R. W. Ramwell, ed.), pp. 99–155, Plenum Press, New York.

Marrazzi, M. A., and Matschinsky, F. M., 1972, Properties of 15-hydroxyprostaglandin dehydrogenase: Structural requirements for substrate binding, *Prostaglandins* **1**:373.

McDonald-Gibson, W. J., McDonald-Gibson, R. G., and Greaves, M. W., 1972, Prostaglandin E_1 metabolism by human plasma, *Prostaglandins* **2**:251.

Miyamoto, T., Yamamoto, S., and Hayashi, O., 1974, Prostaglandin synthetase system resolution into oxygenase and isomerase, *Proc. Natl. Acad. Sci. U.S.A.* **71**:3645.

Nakano, J., Änggård, E., and Samuelsson, B., 1969, 15-Hydroxy-prostanoate dehydrogenase: Prostaglandins as substrates and inhibitors, *Eur. J. Biochem.* **11**:386.

Nugteren, D., and Hazelhof, E., 1973, Isolation and properties of intermediates in prostaglandin biosynthesis, *Biochim. Biophys. Acta* **326**:448.

Pace-Asciak, C., 1975*a*, Prostaglandin 9-hydroxydehydrogenase activity in the adult rat kidney, *J. Biol. Chem.* **25**:2789.

Pace-Asciak, C., 1975*b*, Activity profiles of prostaglandin 15- and 9-hydroxydehydrogenase and 13-reductase in the developing rat kidney, *J. Biol. Chem.* **250**:2795.

Pace-Asciak, C., and Miller, D., 1973, Prostaglandins during development. I. Age-dependent activity profiles of prostaglandin 15-hydroxydehydrogenase and 13,14-reductase in lung tissues from late prenatal, early postnatal, and adult rats, *Prostaglandins* **4**:351.

Pace-Asciak, C., and Miller, D., 1974, Prostaglandins during development. II. Identification of prostaglandin 9-hydroxydehydrogenase activity in adult rat kidney homogenates, *Experientia* **30**:590.

Polet, H., and Levine, L., 1971, Serum prostaglandin A_1 isomerase, *Biochem. Biophys. Res. Commun.* **45**:1169.

Polet, H. and Levine, L., 1975*a*, Metabolism of prostaglandins E, A, and C in serum, *J. Biol. Chem.* **250**:351.

Polet, H., and Levine, L., 1975*b*, Partial purification and characterization of prostaglandin A isomerase from rabbit serum, *Arch. Biochem. Biophys.* **168**:96.

Pong, S. S., and Levine, L., 1976*a*, Stimulation of the microsomal prostaglandin synthetase system of rabbit renal medulla by cytoplasmic components, *Prostaglandins* **11**:477.

Pong, S. S., and Levine, L., 1976*b*, Biosynthesis of prostaglandins in rabbit renal cortex, *Res. Commun. Chem. Path. Pharmacol.* **13**:122.

Pong, S. S., and Levine, L., 1976*c*, Prostaglandin synthetase systems of rabbit tissues and their inhibition by nonsteroidal anti-inflammatory drugs, *J. Pharmacol. Exp. Ther.* **196**:226.

Powell, W. S., Hammarstrom, S., and Samuelsson, B., 1975, Occurrence and properties of a prostaglandin $F_{2\alpha}$ receptor in bovine corpora lutea, *Eur. J. Biochem.* **56**:73.

Rao, C. V., 1974, Prostaglandin receptors in the bovine corpus luteum cell membranes, *Prostaglandins* **6**:533.

Rose, A. J., and Collins, A. J., 1974, The effect of pH on the production on prostaglandin E_2 and $F_{2\alpha}$, and a possible pH dependent inhibitor, *Prostaglandins* **8**:271.

Russell, P. T., Alam, N., and Clary, P., 1973, Impaired placental conversion of prostaglandin E_1 to A_1 in toxemia of pregnancy, *Fed. Proc.* **32**:804.

Samuelsson, B., 1965, On the incorporation of oxygen in the conversion of 8,11,14-eicosatrienoic acid to prostaglandin E_1, *J. Am. Chem. Soc.* **87**:3011.

Samuelsson, B., 1967, Biosynthesis and metabolism of prostaglandins, *Prog. Biochem. Pharmacol.* **3**:59.

Samuelsson, B., 1972, Biosynthesis of prostaglandins, *Fed. Proc.* **31**:1442.

Samuelsson, B., Granstrom, E., Green, K., and Hamberg, M., 1971, Metabolism of prostaglandins, *Ann. N.Y. Acad. Sci.* **180**:138.

Schlegel, W., and Greep, R. O., 1975, Prostaglandin 15-hydroxydehydrogenase from human placenta, *Eur. J. Biochem.* **56**:245.

Schlegel, W., Demers, L., Hildebrandt-Stark, H. E., Behrman, H. R., and Greep, R. O., 1974, Partial purification of human placental 15-hydroxyprostaglandin dehydrogenase: Kinetic properties, *Prostaglandins* **5**:417.

Shen, T. Y., Ham, E. A., Cirillo, V. J., and Zanetti, M., 1974, Structure-activity relationship of certain prostaglandin synthetase inhibitors, in: *Prostaglandin Synthetase Inhibitors* (H. J. Robinson and J. R. Vane, eds.), pp. 19–31, Raven Press, New York.

Sih, C. J., and Takeguchi, C., 1973, Biosynthesis, in: *The Prostaglandins* (P. W. Ramwell, ed.), Vol. 1, pp. 83–100, Plenum Press, New York.

Smith, W. L., and Lands, W., 1971, Stimulation and blockade of prostaglandin biosynthesis, *J. Biol. Chem.* **246**:6700.

Smith, W. L., and Lands, W., 1972, Oxygenation of polyunsaturated fatty acids during prostaglandin biosynthesis by sheep vesicular gland, *Biochemistry* **11**:3276.

Tai, H. H., Tai, C. L., and Hollander, C. S., 1974, Regulation of prostaglandin metabolism: Inhibition of 15-hydroxyprostaglandin dehydrogenase by thyroid hormones, *Biochem. Biophys. Res. Commun.* **57**:457.

Takeguchi, C., Kohno, E., and Sih, C. J., 1971, Mechanism of prostaglandin biosynthesis. I. Characterization and assay of bovine prostaglandin synthetase, *Biochemistry* **10**:2372.

van Dorp, D. A., 1967, Aspects of the biosynthesis of prostaglandins, *Prog. Biochem. Pharmacol.* **3**:71.

van Dorp, D. A., Beerthuis, R. K., Nugteren, D. H., and Vonkeman, H., 1964, Enzymatic conversion of all-cis-polyunsaturated fatty acids into prostaglandins, *Nature (London)* **203**: 839.

Vane, J. R., 1971, Inhibition of prostaglandin synthesis as a mechanism of action for aspirin-like drugs, *Nature (London) New Biol.* **231**:232.

Vogt, W., Bartels, J., Kunze, H., and Meyer, U., 1969, A possible physiological role of phospholipase A for the formation and release of prostaglandin, in: *Abstracts of the 4th International Congress of Pharmacology,* Basle, p. 378.

Westbrook, C., and Jarabak, J., 1975, Purification and partial characterization of an NADH-linked Δ^{13}-15-ketoprostaglandin reductase from human placenta, *Biochem. Biophys. Res. Commun.* **66**:541.

Yalow, R. S., 1973, Radioimmunoassay methodology: Application to problems of heterogeneity of peptide hormones, *Pharmacol. Rev.* **25**:161.

Yoshimoto, A., Ito, H., and Tomita, K., 1970, Factor requirements of the enzyme synthesizing prostaglandin in bovine seminal vesicles, *J. Biochem. (Tokyo)* **68**:487.

Yu, S. C., and Burke, G., 1972, Antigenic activity of prostaglandins: Specificities of prostaglandin E_1, A_1, and $F_{2\alpha}$ antigen-antibody reactions, *Prostaglandins* **2**:11.

Zusman, R. M., Caldwell, B. V., and Speroff, L., 1972, Radioimmunoassay of the prostaglandins, *Prostaglandins* **2**:41.

NOTE ADDED IN PROOF

Recently, prostaglandin-15-hydroxydehydrogenase and prostaglandin-9-ketoreductase have been purified from chicken kidney (A. Hassid and L. Levine, submitted to *Prostaglandins,* 1977). Both enzymes exist in multiple forms as determined by isoelectric focusing. The dehydrogenases catalyze the transformation of the functional group at C-15 but not the functional group at C-9. The preferred cofactors in these reactions are NAD^+ or NADH. The 9-ketoreductases catalyze the reversible transformation of the functional group at C-9 and also the oxidation or reduction of the C-15 functional group. The preferred cofactors are $NADP^+$ or NADPH. Bradykinin does not affect the activities of any of the three prostaglandin 9-ketoreductases. Flavin mononucleotide and the flavonoid, quercetin, as well as indomethacin, ethacrynic acid, and furosemide, inhibit all three 9-ketoreductases. An inhibitor of 9-ketoreductase isolated from chicken breast muscle also inhibits the three separable reductases, but the pattern of inhibition of the reductase that focuses at pH 5.7 differs from that of the reductases focusing at pH 7.8 and 8.2.

The Endocrine System: Interaction of Prostaglandins with Adenylyl Cyclase– Cyclic AMP Systems

James G. Kenimer, Vivian Goldberg, and Melvin Blecher

Departments of Biochemistry and Physiology
Georgetown University Medical Center
Washington, D.C. 20007

I. INTRODUCTION

Higgins and Braunwald (1972) in a review of the biochemical, physiological, and clinical considerations of the prostaglandins made the observation that "prostaglandins . . . are ubiquitous in mammalian tissues and have potent physiologic activities, probably through an as yet incompletely characterized interaction with adenylyl cyclase in the endocrine, reproductive, nervous, digestive, hemostatic, respiratory, cardiovascular and renal systems." It is the intent of this chapter to review those recent advances pertaining to this "incompletely characterized interaction" of prostaglandins and the adenylyl cyclase–cyclic AMP systems in various of the organs comprising the endocrine system. The reader should be forewarned that although progress has been made in anwering some of the questions relating to this interaction, much is still unknown and consequently no satisfactory hypothesis has been formulated to unify all of the divergent findings.

A number of earlier reviews have dealt in part with this topic and are highly recommended. They include Ramwell and Shaw (1970), Weeks (1971), Hittelman and Butcher (1973), Mashiter and Field (1974a), and Labrie et al. (1975).

II. THYROID

The prostaglandins have been shown to mimic most, if not all, of the effects of thyroid-stimulating hormone (TSH) on the thyroid gland (see Mashiter and Field, 1974b, for review). These effects coupled with the demonstration by Horton (1969) of the presence of prostaglandins in thyroid tissue strongly suggest that prostaglandins play a potentially important role in thyroid metabolism. Considerable evidence now exists which indicates that the effects of both TSH and prostaglandins are mediated primarily via the adenylyl cyclase–cyclic AMP system (Mashiter and Field, 1974a,b).

The initial demonstration of an elevation in cyclic AMP levels in response to stimulation by prostaglandin was by Kaneko et al. (1969), who reported that PGE_1 (85 μM) increased the cyclic AMP concentration of incubated dog thyroid slices almost tenfold. Zor et al. (1969a) extended these studies to show that PGE_2 (28 μM) also increased cyclic AMP accumulation in dog thyroid slices, whereas $PGF_{1\alpha}$ and PGB_1 (both 28 μM) had no effect. PGA_1 (28 μM) increased cyclic AMP levels, but to a lesser extent than PGE_1. Of interest was their observation that although $PGF_{1\alpha}$ did not increase cyclic AMP accumulation, it did stimulate the oxidation of [1-^{14}C]-D-glucose to an extent almost equal to that of PGE_1; PGB_1 was without effect on glucose oxidation. Other investigators have also reported the prostaglandin stimulation of cyclic AMP accumulation in thyroid slices (Dekker and Field, 1970; Van Sande and Dumont, 1973).

Results obtained using broken-cell thyroid preparations have been more variable. Ahn and Rosenberg (1970) reported fivefold stimulation of adenylyl cyclase activity by PGE_1 (30 μM) in canine thyroid homogenates, and also demonstrated PGE_1 stimulation of the accumulation of [^{14}C]adenine. Burke (1970) reported that PGE_1, PGE_2, $PGF_{1\alpha}$ and $PGF_{1\beta}$ (2.5–50 μM) all stimulated adenylyl cyclase activity in a mitochondrial fraction of ovine thyroid homogenate, but that only PGE_1 and PGE_2 (2.5 μM) consistently stimulated the oxidation of [1-^{14}C]-D-glucose in ovine thyroid slices. These results are seemingly in partial disagreement with those of Zor et al. (1969), mentioned previously. Field et al. (1971) and Mashiter and Field (1974b) discuss these discrepancies, and others, in the effects of prostaglandins on thyroid gland metabolism. Their evaluation urges caution in the interpretation of such experiments since wide variations of effects can be achieved depending on the type and concentration of prostaglandin used.

Burke (1970) suggested that prostaglandins might compete with TSH for a common adenylyl cyclase receptor site on the thyroid cell. This interpretation was based on data which showed that the effects of submaximal doses of prostaglandins and TSH were additive with respect to adenylyl cyclase activation, in ovine thyroid tissue slices, whereas prostaglandins showed no augmentation of maximal TSH stimulation. Wolff and Jones (1971), however, reported that PGE_1 did not stimulate adenylyl cyclase activity in some of their bovine thyroid membrane preparations, all of which were responsive to TSH. Additional indirect evidence has been provided by Wolff and Cook (1973) which support the concept of separate receptors for prostaglandins and TSH. They were able to demonstrate that (1) the TSH and PGE_1 responses reacted differently to the presence of ITP and GTP, (2) K^+ enhanced only the TSH effect, (3) additive effects could be produced with maximal concentrations of each compound, and (4) the responsiveness of PGE_1 appeared to be more labile. In addition, Moore and Wolff (1973) have showed that TSH displaces $[^3H]PGE_1$ from a beef thyroid membrane preparation only at high concentrations, and have presented evidence that this is a nonspecific effect shared by many other proteins. Combined with the obvious molecular differences in the two compounds, these results provide strong indications that the receptors for prostaglandin and TSH are separate and distinct.

Burke and his colleagues investigated the possibility that prostaglandins might be obligatory intermediates in the action of TSH on the thyroid. They demonstrated that the prostaglandin antagonists, 7-oxa-13-prostynoic acid, 7-oxa-15-hydroxyprostynoic acid (Sato *et al.* 1972*b*), and polyphloretin phosphate (Burke and Sato, 1971; Sato *et al.* 1972*a*), blocked both TSH and PGE_1 effects on thyroidal adenylyl cyclase activity and cyclic AMP formation. Utilizing a radioimmunoassay to measure cellular prostaglandin levels, they also demonstrated that TSH specifically induced an increase in tissue prostaglandin levels in isolated bovine thyroid cells (Yu *et al.* 1972). This effect could be abolished if the cells were preincubated with aspirin (100 μg/ml) or indomethacin (10 μg/ml), two inhibitors of prostaglandin synthesis (Burke, 1972). These results were considered supportive of the obligatory intermediate hypothesis.

Subsequent results have, however, supported an alternative explanation for these findings. Burke (1973*a*) demonstrated that both TSH and dibutyryl cyclic AMP elicited a dose-related increase in mouse thyroid prostaglandin levels and that the effects of both were abolished by indomethacin and aspirin. This observation was in agreement with the suggestion of Shio *et al.* (1971) that increasing cyclic AMP levels would effect an activation of prostaglandin synthesis and/or release. In addition, Mashiter *et al.* (1974) demonstrated that aspirin and indomethacin did not inhibit the stimulatory effects of TSH on cyclic AMP synthesis in dog thyroid slices. Wolff and Moore (1973) also found no effect of indomethacin on TSH stimulation of cyclic AMP levels in beef thyroid membranes. These results are obviously incompatible with the initial thesis of prostaglandins operating as an obligatory intermediate in TSH action. Thus while it had been previously established that prostaglandins would activate thyroidal adenylyl cyclase, evidence now also supported the concept that an increase in cyclic AMP levels would, in turn, increase prostaglandin levels.

The effect of purine nucleotides on the prostaglandin-responsive adenylyl cyclase in thyroid tissue has also received attention. Kowalski *et al.* (1972) reported that with [α-^{32}P]ATP as a substrate there was no PGE$_2$ effect on adenylyl cyclase activity in bovine thyroid plasma membranes which were highly responsive to TSH and NaF. In contrast, with AMP-PNP as a substrate (AMP-PNP is a substrate for adenylyl cyclase, but not for plasma membrane ATPase), a dose-related response to PGE$_2$ was seen which could be inhibited by 7-oxa-13-prostynoic acid. However, in the presence of 0.1 mM GTP, a stimulatory PGE$_2$ effect could be reproducibly demonstrated even using [α-^{32}P]ATP as a substrate. Thyroid membrane preparations from Burke's laboratory have consistently required the presence of a purine nucleotide for exhibition of prostaglandin-stimulated adenylyl cyclase with ATP as the substrate (Kowalski *et al.*, 1972; Burke, 1973*b*). Wolff and Jones (1971) reported prostaglandin stimulatable activity in bovine thyroid plasma membranes in the absence of added nucleotides; however, it was also observed that a significant increase in the magnitude of the stimulation accompanied addition of the nucleotides. This effect was shown, however, to be nonspecific in that the response could be elicited by all purine, and certain pyrimidine, nucleotide triphosphates. This observation was confirmed by Burke (1973*b*). It appears, therefore, that a specific nucleotide effect is not a factor in prostaglandin stimulation of thyroid adenylyl cyclase.

Although the obligatory intermediate theory has been substantially disproven, and separate receptors for TSH and prostaglandins appear to exist, a significant number of experimental results still indicate some type of interaction between TSH and prostaglandins in the thyroid (see Mashiter and Field, 1974*a*). For example, Sato *et al.* (1972*b*) using isolated thyroid cells demonstrated that, in the presence of ineffective or minimally effective concentrations of one stimulator (PGE$_1$ or TSH), adenylyl cyclase activation and cyclic AMP formation induced by a maximally effective concentration of the other stimulator were markedly reduced. Combinations of maximally effective doses of each were shown to have effects equal to that of the weaker agonist (Kowalski *et al.*, 1972; Burke, 1973*b*; Sato *et al.*, 1972*b*). In all reports except that of Sato *et al.* (1972*b*), maximal effects of TSH were greater than those of prostaglandins.

Kendall-Taylor (1972), however, reported that the presence of PGE$_1$ did not alter the doze–response curve of adenylyl cyclase activation of ^{131}I release in response to TSH in intact mouse thyroid glands *in vitro*. Mashiter and Field (1974*b*) using dog thyroid slices have shown that a combination of maximal doses of TSH or PGE$_1$, with submaximal doses of the other, did produce additive results. In addition, they reported that maximal doses of each agent in combination produced additive results. Wolff and Cook (1973) also reported additivity of TSH- and PGE$_1$-stimulated responses at maximal concentrations of each. Burke (1974) has discussed these apparent discrepancies and has presented additional data to support his observations that the potent TSH effect is consistently blunted in the presence of PGE$_1$. Methodological differences may be the explanation behind these discrepancies and resolution thus awaits investigations by other laboratories.

Although no scheme has been proposed which can unify all the divergent findings in this area, Burke *et al.* (1973) has proposed, as a working model, a theory of prostaglandin–TSH interaction in the thyroid. This postulate envisions an intracellular cyclic AMP–prostaglandin mediated negative feedback system modulating TSH effects on thyroid function. TSH interaction with its receptor on the thyroid membrane would result in an increased intracellular cyclic AMP level which would act to increase prostaglandin levels. The prostaglandins would then interact with some segment of the TSH–adenylyl cyclase receptor apparatus in a manner such that a diminished response to TSH resulted. Although prostaglandins themselves are stimulators of adenylyl cyclase, since they are less potent than TSH they can behave, in effect, as inhibitors of the TSH-induced stimulus.

III. PITUITARY

A. Introduction

A substantial literature exists in which it has been shown that *in vivo* administration of prostaglandins results in increased plasma titers of the anterior pituitary hormones, with the exception of TSH (Table I). Elucidation of the mechanism(s) through which this effect is achieved has proven to be an elusive goal. It is our intent in this section to review the literature with respect to prostaglandin stimulation of each of the pituitary hormones, with emphasis on the following questions: (1) What is the primary site of action of the prostaglandins? (2) Is cyclic AMP involved? (3) Do prostaglandins play a mediator role at the pituitary level in the hypothalamic neurohormone-mediated stimulation of pituitary hormone release?

Table I. *In Vivo* Prostaglandin Stimulation of Anterior Pituitary Hormone Release

Pituitary hormone released	Test animal	Prostaglandin				Reference
		Type	Dose	Injection route[a]	Response[b]	
LH	Rat, ovex	E_2	1 mg	1	+	Lau and Saksena (1974)
LH	Rat, ovex	E_2	5 µg/100 g	2,3,4	+	Harms et al. (1974)
LH	Sheep	$E_1, F_{1\alpha}, F_{2\alpha}$	50 µg/100 g	2,3,4	−	Carlson et al. (1973)
		$F_{2\alpha}$	0.75–100 µg/hr	5	+	
LH and FSH	Rat, male	E_2	5 µg	3	+	Ojeda et al. (1974a)
		$F_{1\alpha}, F_{2\alpha}$	5 µg	3	−	
FSH	Rat, male	E_2	500 µg	2	+	T. Sato et al. (1974)
Prolactin	Rat, ovex	$E_1, E_2, F_{2\alpha}$	670 µg	2	+	T. Sato et al. (1974)
Prolactin	Human, female	$F_{2\alpha}$	30 mg	6	+	Yue et al. (1974)
Prolactin	Rat, ovex	E_1, E_2	50 µg/100 g	2	−	Ojeda et al. (1974b)
		E_1	5 µg	3	+	
		$E_2, F_{1\alpha}, F_{2\alpha}$	5 µg	3	−	
GH	Pigeon	E_1	15 µg	2	+	McKeown et al. (1974)
GH	Sheep	E_1	20 µg/kg	2	+	Hertelendy et al. (1972)
GH	Rat, male	E_1, E_2	100 µg	2	+	Labrie et al. (1975)
GH	Human, male	E_1	50–140 ng/kg/min	5	+	Ito et al. (1971)
ACTH	Rat, female	$E_1, F_{1\alpha}, F_{2\alpha}$	1.0, 0.5, 0.5 µg, respectively	3	+	Hedge and Hanson (1972)
		$E_1, F_{1\alpha}, F_{2\alpha}$	1.0, 0.5, 0.5	2,4	−	
ACTH	Rat, male	E_1	10 µg	2	+	DeWied et al. (1969)
		$F_{1\alpha}, F_{2\alpha}$	10 µg	2	−	
TSH	Rat, female	$A, B, E, F_{1\alpha}$	20 µg/100 g	2,3,4	−	Brown and Hedge (1974)

[a] 1, Subcutaneous; 2, intravenous injection; 3, intrahypothalamic; 4, intrapituitary; 5, intravenous; 6, intraamniotic injection.
[b] +, Stimulation; −, no effect.

The involvement of cyclic AMP has been suggested by several lines of evidence. *In vitro* experiments with pituitary preparations have shown that the prostaglandins can stimulate cyclic AMP accumulation (Zor *et al.*, 1969*b*, 1970, 1974; Deery and Howell, 1973; Sato *et al.*, 1974; Borgeat *et al.*, 1975). Borgeat *et al.* (1975) reported the order of potency of prostaglandin stimulation of cyclic AMP accumulation in rat anterior hemipituitaries to be $PGE_1 \cong PGE_2 > PGA_1 \cong PGA_2 > PGF_{1\alpha} \cong PGE_{1\alpha}$. The stimulatory effect of 10^{-7} to 10^{-6} M PGE_2 was detectable at 2 min and reached maximal levels at approximately 30 min. Theophylline (10 mM) potentiated the prostaglandin effect.

In a recent report, Bergeron and Bardew (1975) showed that of several fatty acids tested only the direct precursors of PGE_1 (*cis*-8,11,14-eicosatrienoic acid) and PGE_2 (arachidonic acid) increased intracellular cyclic AMP levels in rat anterior pituitary. Both aspirin and indomethacin eliminated completely the prostaglandin-precursor-induced cyclic AMP increase, suggesting that prior conversion to one of the prostaglandins is essential for activity. The effects of these two fatty acids were seen with 1 min, and reached a maximum at 15 min. When compared with the data discussed previously on the time course of stimulation with PGE_2 (Borgeat *et al.*, 1975) these results suggest that fatty acid availability, and not conversion to the prostaglandin, is the rate-limiting factor in this event.

In addition to the evidence which shows *in vitro* stimulation of pituitary cyclic AMP levels by prostaglandin, it has also been shown that cyclic AMP or theophylline will stimulate the release of all the adenohypophyseal hormones (see Borgeat *et al.*, 1975). However, conflicting results have been obtained in nearly all cases when measuring *in vitro* effects of prostaglandins on release of the various pituitary hormones. These conflicts and possible explanations will be discussed in the following sections.

B. LH/FSH

The possible involvement of prostaglandins in gonadotropin secretion was suggested initially by studies in which inhibitors of prostaglandin synthesis were seen to block ovulation (see Labrie *et al.*, 1975; Sato *et al.*, 1975*a*). Several direct experiments subsequently showed that *in vivo* administration of prostaglandins stimulated the secretion of LH and/or FSH (see Table I; Castracane and Saksena, 1974; Tsafriri *et al.*, 1973).

In vitro studies have, however, produced conflicting data. For example, Zor *et al.* (1970) demonstrated that PGE_1 (0.1–40 µg/ml) increased cyclic AMP levels in whole, intact anterior rat pituitaries, but was without significant effect on LH release during a 20-min incubation period. Other prostaglandins (A, B, and $F_{1\alpha}$) exhibited an even smaller elevating effect on cyclic AMP levels, also without an effect on LH release. In contrast, a crude hypothalamic extract, at a concentration which increased cyclic AMP to a level comparable to that achieved with 2 μg/ml PGE_1, stimulated LH release by 400%. PGE_1 (10 μg/ml) was also shown to have no effect on the cyclic AMP levels of hypothalamic fragments.

Results conflicting with those of Zor *et al.* (1970), cited previously, have been reported. Makino (1973) found that PGE_1 (0.5 μg) enhanced the release of both LH and FSH in isolated rat hemipituitaries. Incubations in their studies were for 4 hr. Other workers (Dowd *et al.*, 1973; Ratner *et al.*, 1974) have also reported the prostaglandin-stimulated release of LH from hemipituitary preparations. In all of these studies, prostaglandin was also shown to increase cyclic AMP levels. Ratner *et al.* (1974) demonstrated a correlation between cyclic AMP stimulation and LH release which was PGE_1 dose dependent. Combined with earlier work (Ratner, 1970) in which LH release was shown to be stimulated by dibutyryl cyclic AMP and theophylline, they postulated that the PGE_1 effect on LH release in the pituitary was mediated via the pituitary adenylyyl cyclase–cyclic AMP system. The authors, however, were aware that proof of this concept could be achieved only by study of the effect of prostaglandin on the specific LH-producing cells in the pituitary.

Varavudhi and Chobdieng (1972), following stereotaxic implantation of $PGF_{2\alpha}$ in the anterior pituitary and the median eminence of pseudopregnant rats, reported consistent ovarian luteolytic effects in those rats with the anterior pituitary implant, but not those with the median eminence implant. This was interpreted as favoring the hypothesis of a direct stimulatory action of $PGF_{2\alpha}$ on release of a pituitary luteolytic agent, possibly LH. Harms *et al.*, (1974), however, obtained opposite results using rats in which permanent cannulas were implanted in the third ventricle or in each lobe of the anterior pituitary of ovariectomized rats. Intraventricular injections of PGE_2 (5–50 μg/ 100 g body weight) markedly increased plasma LH levels and slightly increased FSH levels. Other prostaglandins (E_1, $F_{2\alpha}$, and $F_{1\alpha}$) were ineffective at similar doses. Intrapituitary injection of the prostaglandins had an insignificant effect on LH or FSH titers. Similar results were obtained by the same group of investigators (Ojeda *et al.*, 1974*a*) following intraventricular injection of PGE_1 and PGE_2 into conscious, free-moving male rats. PGE_2 was reported to be a much more potent stimulator than PGE_1. Spies and Norman (1973) obtained similar results in which third ventricular injection of PGE_1 (5–20 μg) increased plasma LH levels, whereas intrapituitary injections had no effect. In their studies, however, PGE_1 was a more potent stimulator than PGE_2. Recently, Sato *et al.* (1975*a*) have reported that doses of 50–100 μg of PGE_1 and PGE_2 injected directly into the pituitary would produce a two- to threefold increase in plasma LH levels 10 to 45 min after injection; $PGF_{2\alpha}$ was ineffective. They also reported that *in vitro* incubation of rat hemipituitaries with PGE_1 (2–20 μg), PGE_2 (0.002–0.02 μg), or $PGF_{2\alpha}$ (200 μg) would produce a significant increase in LH release into the incubation medium.

The previous *in vivo* and *in vitro* results are conflicting with regard to establishing whether the major prostaglandin site of action is the hypothalamus or the pituitary. Chobsieng *et al.* (1975) recently published additional *in vivo* evidence which considerably strengthens the hypothalamic-site postulate. Using an antiserum prepared against, and specific for, synthetic luteinizing hormone releasing hormone (LHRH) they were able to show that intraperitoneal administration of the anti-LHRH 2 hr before intravenous administration of PGE_2 would completely prevent the PGE_2-induced rise in serum LH levels in both male and female rats. Additional evidence favoring a hypothalamic site is the demonstration by Sato *et al.* (1975*b*) that PGE_2 (700 μg) injected into the jugular vein of ovariectomized rats pretreated with estrogen and progesterone would cause a decrease in the total LHRH content of the hypothalamus and an increase in the plasma levels of LH. Electron microscopic examination showed that pituitary morphological changes induced following PGE_2 injection were very similar to those found in LHRH treated animals. Animals given hypothalamic lesions showed morphological changes of a far lesser magnitude than did those with an intact hypothalamus. Ojeda *et al.* (1975) using a radioimmunoassay for LHRH showed that PGE_2 (5 μg) injected into the third ventricle of conscious ovariectomized rats resulted in a significant rise in plasma LHRH levels. They also showed that the pharmacological blockage of adrenergic, cholinergic, dopaminergic, or serotoninergic receptors failed to prevent PGE_2-induced LH release. Intraventricular injection of indomethacin or 5,8,11,14-eicosatetraynoic acid, two inhibitors of prostaglandin synthesis, decreased plasma LH levels, supporting the view that prostaglandins play a physiological role in the control of gonadotropin secretion. Eskay *et al.* (1975) also reported that PGE_2 (5–20 μg), infused into the lateral ventricle of adult male rats, caused a two- to threefold increase in LHRH in hypophyseal portal plasma. The release of LH, FSH, and prolactin but not that of TSH was stimulated.

Carlson *et al.* (1974) suggested that prostaglandins might play an *obligatory* role in LH release. This suggestion was based on data which showed that indomethacin pretreatment prevented the estradiol-

17β stimulation of LH release in female sheep, suggesting that endogenous synthesis of prostaglandin might be involved in LH release. In contrast, Naor *et al.* (1975) suggested that the stimulatory action of LHRH on pituitary cyclic AMP production and LH release is *not* mediated by the prostaglandins. They reported that LHRH did not affect either the levels of prostaglandins or the activity of prostaglandin synthetase in rat hemipituitaries, and that aspirin and indomethacin, while reducing prostaglandin content and prostaglandin synthetic activity, had no effect on LHRH-induced LH release. These results are consistent with a role for prostaglandins in the hypothalamic secretion of LHRH, but not in the LHRH-induced secretion of pituitary LH.

In summary, prostaglandins E_1 and E_2 stimulate pituitary LH release predominantly via interaction with the hypothalamus, with resultant release of LHRH. Direct effects of prostaglandins on the pituitary may occur but are not the major site of action. Prostaglandins other than the E series are ineffective. The role, if any, of cyclic AMP in the mediation of the hypothalamic mechanism of action of the prostaglandins has not been investigated. The prostaglandins are not involved in the stimulatory action of LHRH on pituitary cyclic AMP production and resultant LH release.

C. ACTH

DeWied *et al.* (1969) measured ACTH release in response to prostaglandins in rats pretreated with pentobarbital plus chlorpromazine, pentobarbital, atropine plus chlorpromazine, pentobarbital plus morphine, chlorpromazine plus morphine, or dexamethasone plus pentobarbital. The same experiments were carried out in neurohypophysectomized rats pretreated with pentobarbital, in rats bearing lesions in the median eminence of the hypothalamus, and in isolated anterior pituitaries *in vitro*. The rate of corticosteroid production *in vitro* and corticosterone titers of plasma were used as measurements of ACTH release. PGE_1 (10 μg/rat) significantly stimulated ACTH release in intact drug-pretreated rats, and morphine reduced the response. Neurohypophysectomy also reduced the response to PGE_1, while rats with lesions in the median eminence were unresponsive. Pituitaries incubated *in vitro* also were unresponsive to PGE_1. Prostaglandins $F_{1\alpha}$ and $F_{2\alpha}$ were ineffective in all systems at similar doses. In contrast, a crude corticotropin releasing factor (CRF), isolated from calf brain, was active in all assay systems. Flack and Ramwell (1972) have argued that these results support the concept that prostaglandins do not have a direct effect on pituitary release of ACTH, and that the observed increase in adrenal activity following administration of prostaglandins is due to cardiovascular stress. An alternative explanation is that the effect of prostaglandins on the release of ACTH from the pituitary is secondary to an effect of the former on the hypothalamus, resulting in a release of CRF.

Peng *et al.* (1970) reported that PGE_1 (0.5–2.0 μg/rat) depleted adrenal ascorbic acid content at 1 hr following intravenous injection into pentobarbital-anesthetized rats. In contrast, PGE_1 or $PGF_{2\alpha}$ at 5 μg/rat was without effect. In rats hypophysectomized 24 hr prior to injection, PGE_1 at 5 μg/rat also was ineffective. Morphine was shown to have an inhibitory effect on the PGE_1-induced depletion of adrenal ascorbic acid. These results are consistent with an indirect effect of PGE_1 on pituitary ACTH secretion, possibly involving a site in the central nervous system. Hedge and Hanson (1972) obtained additional evidence to support this concept. The stereotaxic microinjection of prostaglandins (1.0 μg PGE_1, 0.5 μg $PGF_{1\alpha}$, and 0.5 μg $PGF_{2\alpha}$) directly into the hypothalamic median eminence of pentobarbital-anesthetized, dexamethasone-pretreated female rats resulted in increased ACTH secretion as estimated by plasma corticosterone levels. Similar doses were ineffective when injected into nearby regions of the basal hypo-thalamus, the anterior pituitary, or the tail vein. Morphine pretreat-ment abolished the positive responses. Incubation of PGE_1 and $PGF_{1\alpha}$ in rat plasma showed that the $PGF_{1\alpha}$ is much more rapidly degraded, a fact which explains the apparent discrepancy between the activities of PGE_1 and $PGF_{1\alpha}$ following intravenous injection as compared with their activities following intrahypothalamic injection. These results are consistent with a hypothalamic site of action of prostaglandin stimula-tion of pituitary ACTH secretion.

A stimulatory effect of PGE_1 on rat pituitaries *in vitro* has, however, been reported by Vale *et al.* (1971), and Hedge and Hanson (1972) reported a paradoxical response to 0.5 μg (but not to higher doses) of PGE_1 injected into the pituitary. The possibility of a direct pituitary role in prostaglandin ACTH stimulation cannot, therefore, be ruled out.

In order to investigate the possible relationship between prosta-glandin and cyclic AMP in the hypothalamic-pituitary-mediated release of ACTH, Hedge and Hanson (1972) conducted several experiments in rats primed with theophylline. It had been previously demonstrated (Hedge, 1971) that intrapituitary injection of cyclic AMP would evoke ACTH secretion only if preceded by theophylline. Prostaglandins were therefore tested under the same conditions, but still were without effect, further supporting the hypothesis that prostaglandins are not involved at the level of the pituitary in ACTH secretion. Intrahypothal-amic injection of prostaglandin following theophylline priming resulted in a response no different from that observed without theophylline.

These experiments were conducted at a submaximal prostaglandin dosage such that a potentiating effect should have been seen if it existed. These results were interpreted as suggesting that prostaglandin stimulation of the hypothalamus is not mediated by cyclic AMP. Zor *et al.* (1970) previously showed that prostaglandins are without effect on the adenylyl cyclase activity in hypothalamic tissue.

More recently, Hedge (personal communications) has expanded previous results by reporting that PGA_1 and PGB_1 (1 μg) also act to stimulate ACTH secretion following microinjection into the basomedial region of the hypothalamus. In contrast, neither prostaglandin increased ACTH when given via the tail vein or following microinjection into the anterior pituitary or the lateral region of the basal hypothalamus. It was also shown that PGA_1, and PGE_1, $PGF_{1\alpha}$, but not PGA_1, (2 μg) would partially inhibit the pituitary release of ACTH in response to a crude CRF preparation. These data are the first to suggest that prostaglandins might function in the pituitary as a modulator of the response to CRF.

D. Prolactin

The ability of prostaglandins to increase circulating prolactin levels has been demonstrated in rats and in humans (see Table I). Harms *et al.* (1974) and Ojeda *et al.* (1974*b*) presented evidence which suggests that this effect is mediated primarily by the hypothalamus. Ovariectomized rats were stereotaxically implanted with permanent cannulas either in the third ventricle or in each lobe of the pituitary. Intraventricular injections of PGE_1 (5 μg) significantly increased plasma prolactin levels at 15 and 30 min, whereas intrapituitary injections of the same dose showed only a slight increase in plasma prolactin at 15 min. Estrogen pretreatment of the rats was without influence on the effects of intraventricular injection of PGE_1, but it eliminated the small effect seen following intrapituitary injection. Other prostaglandins (PGE_2, $PGF_{1\alpha}$, and $PGF_{2\alpha}$) were ineffective in stimulating prolactin increases under all conditions tested. These results with PGE_2 and $PGF_{2\alpha}$ are in apparent discrepancy with those of Vermouth and Deis (1972), who reported a rise in circulating prolactin levels following two intraperitoneal injections of 300 μg $PGF_{2\alpha}$ in pregnant rats, and with those of Sato *et al.* (1974), who reported elevated prolactin titers following intravenous injections of $PGF_{2\alpha}$ and PGE_2 (670 μg) in spayed rats primed with estrogen and progesterone. The results with $PGF_{2\alpha}$ may be explained by the suggestion of Vermouth and Deis (1972) that the $PGF_{2\alpha}$-mediated release of prolactin might be a response secondary to the lowering of progesterone.

Control of pituitary prolactin secretion is mediated by two specific neurohormones which are produced in the hypothalamus and which reach the pituitary via the hypophyseal portal vessels. One of these neurohormones, prolactin inhibiting factor (PIF), inhibits pituitary prolactin release, and the other, prolactin releasing factor (PRF), exerts a stimulatory effect. The net effect of hypothalamic influence is that of chronic inhibition of pituitary prolactin release (Labrie *et al.*, 1975).

Ojeda *et al.* (1974*d*) have evidence which suggests that cyclic AMP is involved in the hypothalamic control of prolactin secretion by modifying the release of PIF and/or PRF. They demonstrated that third ventricular injections of cyclic AMP or dibutyryl cyclic AMP (0.1 M) decreased plasma prolactin levels measured at 30 and 60 min, whereas injections of dibutyryl cyclic AMP into the pituitary at the same dose had no effect. These results are consistent with cyclic AMP involvement in the hypothalamic mechanism controlling pituitary prolactin secretion. However, in the absence of concurrent measurements of PIF or PRF, the participation of either of these neurohormones in the cyclic nucleotide effect cannot be established. Dopamine, which has been shown by MacLeod and Lehmeyer (1974) to decrease *in vitro* pituitary prolactin release, or some other unknown factor could be responsible. These results are in disagreement with those of Pelletier *et al.* (1972), who reported a stimulation by N^6-monobutyryl cyclic AMP of prolactin as well as growth hormone and ACTH secretion by rat anterior pituitary *in vitro*.

Ojeda *et al.* (1974*c*) have evidence which suggests that the involvement of PGE_1 in the hypothalamic control of pituitary prolactin release may be that of a mediator of the dopamine stimulation of PIF release. According to their postulate, interaction of dopamine with its receptors on the PIF-secreting cells of the hypothalamus results in an increased cyclic AMP production and a subsequent rise in PIF release. The increased cyclic AMP levels would also result in increased PGE_1 synthesis; the latter would act on the PIF-secreting cells to decrease their sensitivity to dopamine activation, thus lowering PIF output and enhancing prolactin secretion. This hypothesis is similar to that proposed by Burke *et al.* (1973) to explain the action of prostaglandins in the thyroid.

E. Growth Hormone

MacLeod and Lehmeyer (1970) reported that PGE_1 and PGE_2 (10^{-6} M) increased the release of growth hormone (GH) into the incubation medium following incubation of intact whole anterior rat pituitaries for 7 hr. PGE_1 increased GH synthesis, but not its release, and $PGF_{2\alpha}$ had no effect. All the above prostaglandins at high concentrations (10^{-4} M) increased pituitary adenylyl cyclase activity, but had no effect on cyclic nucleotide phosphodiesterase activity. Enhanced GH release was also observed in response to dibutyryl cyclic AMP (1 mM), NaF, and theophylline, suggesting a participatory action of cyclic AMP.

Kudo *et al.* (1972) using dispersed pituitary cells demonstrated that PGE_1 (10^{-5} M) stimulated both GH and TSH release as well as adenylyl cyclase activity following a 15-min incubation. However, neither dibutyryl cyclic AMP nor phosphodiesterase inhibitors influenced GH release, although they did stimulate TSH release. Crude stalk–median eminence extracts exhibited variable effects on the pituitary cell release of GH and TSH. Some preparations had high GH-releasing ability, while others had high TSH-releasing ability. In all cases, cyclic-AMP-stimulating ability paralleled TSH release but not GH release. The authors thus suggested that pituitary GH release might not involve cyclic AMP.

Other results, however, support the hypothesis of the involvement of cyclic AMP in pituitary GH secretion. Cooper *et al.* (1972) noted a close correlation between increased pituitary cyclic AMP content and release of GH following 45-min incubations of ox pituitary slices with PGE_2 (10^{-6} M). Theophylline (0.5 mM) did not stimulate GH release; however, it did sensitize the tissue to PGE_2. Ratner *et al.* (1973) reported that the addition of 7-oxa-13-prostynoic acid (a prostaglandin antagonist) significantly inhibited rat hemipituitary GH release and cyclic AMP accumulation in response to PGE_1. Theophylline (6.7 mM) and dibutyryl cyclic AMP (10 mM) also increased GH release, in both the presence and the absence of the antagonist. These results suggest that the inhibition caused by the antagonist occurs at a point prior to cyclic AMP production, and are consistent with the involvement of the adenylyl cyclase–cyclic AMP system in the mechanism of action of prostaglandin-stimulated pituitary GH release.

Evidence in support of this hypothesis has been provided in other reports. Labrie *et al.* (1975) demonstrated a close correlation between the order of potency of various prostaglandins in stimulating GH release from rat anterior pituitary cells in culture and the potency observed with respect to cyclic AMP accumulation. They also observed *in vivo* stimulation of GH release following injection of PGE_1 or PGE_2 into unanesthetized rats. Similar positive *in vivo* results have been obtained in sheep (Hertelendy *et al.*, 1972), pigeon (McKeown *et al.*, 1974), and man (Henzl *et al.*, 1973; Ito *et al.*, 1971). It has also been demonstrated that a close parallelism exists between changes of cyclic AMP levels and GH release induced by both the hypophyseal neurohormone growth hormone releasing hormone (GHRH) and prostaglandins (Borgeat *et al.*, 1973). Somatostatin, a growth hormone release inhibiting hormone, was shown to inhibit PGE_2-induced accumulation of pituitary cyclic AMP (Borgeat *et al.*, 1974). These results and others have prompted these authors to suggest a mediator role of E prostaglandins in the pituitary secretion of GH (Borgeat *et al.*, 1975). Confirmation of this hypothesis awaits direct measurement of the influence of the hypothalamic factors controlling GH release on the availability of intrapituitary prostaglandins.

F. Thyroid Stimulating Hormone

Conflicting results have been obtained with regard to the effects of prostaglandins on the stimulation of pituitary thyroid stimulating hormone (TSH) secretion.

In vitro experiments using dispersed pituitary cells (Vale *et al.*, 1972; Kudo *et al.*, 1972) and rat hemipituitaries (Vale *et al.*, 1971; Dupont *et al.*, 1972) have shown a significant effect of prostaglandins on the release of TSH.

Using dispersed hog pituitary cells, Kudo *et al.* (1972) reported that a 30-min incubation with PGE_1 (10^{-5} M) increased TSH levels (McKenzie mouse bioassay) by about 50%; dibutyryl cyclic AMP (10^{-3} M) and puromycin (50 μg/ml) caused only small (20 and 34%, respectively) stimulations of TSH release. However, there is some doubt as to the physiological integrity of these cells, since they were unresponsive to the hypothalamic thyrotropin releasing factor (TRF).

Cultured rat pituitary cells (Vale *et al.*, 1972) released TSH (McKenzie mouse bioassay) in response to 3 days of incubation with TRF (10^{-8} M) or PGE_2 (10^{-5} M); however, the prostaglandin also caused profound morphological changes in these cells, adding a degree of doubt to these results.

Also using the McKenzie bioassay, but with rat hemipituitary preparations, Dupont *et al.* (1972) reported in abstract form that 50 μM PGE_1 stimulated TSH release two- to threefold in the first 45 min of incubation. Indomethacin (10–20 μM) had minimal effects on basal TSH release, but led to a 50–100% inhibition of TRH (10^{-7} M) induced TSH release. There was no effect of indomethacin or TSH release induced by PGE_1 or dibutyryl cyclic AMP. Also in an abstract, Vale *et al.* (1971) reported that PGE_1 (2.8 μM) stimulated the release of TSH from rat hemipituitaries. $PGF_{2\alpha}$ (up to 30 μM) had no effect. 7-Oxa-13-prostynoic acid (30 μM) decreased the amount of TSH secreted in response to TRF. These results prompted the authors of both abstracts to speculate that protaglandins might play an essential role in the stimulation of TSH secretion.

In contrast to the results and conclusions described above, other investigators, using both *in vitro* and *in vivo* techniques, have observed no effects of prostaglandins on the release of TSH.

Thus Tal *et al.* (1974) using highly specific radioimmunoassays for cyclic AMP and TSH noted a dissociation between levels of cyclic AMP and TSH release following addition of TRH or PGE$_1$ to rat anterior pituitaries. TRH (1 μg/ml) markedly increased TSH secretion without enhancement of cyclic AMP accumulation. PGE$_1$ (10^{-4}–10^{-6} M), on the other hand, although failing to stimulate TSH secretion, did elevate cyclic AMP levels. The putative prostaglandin inhibitor, 7-oxy-13-prostynoic acid, did not modify the effect of TRH or of the prostaglandins on cyclic AMP levels. These results are interpreted as being consistent with the lack of an intermediary role of prostaglandins or cyclic AMP in TRH-induced TSH secretion.

Recent *in vivo* results by Brown and Hedge (1974) support the aforementioned conclusions. A variety of prostaglandins (PGA$_1$, PGB$_1$, PGE$_1$, and PGE$_{1\alpha}$), when given alone intravenously (20 μg/100 g body weight) or by stereotaxic injection directly into the medial basal hypothalamus or anterior pituitary (2 μg/rat) were without effect on TSH secretion. However, a potentiation (85–100%) of subsequent stimulation by TRH followed intrapituitary administration of the prostaglandins. This potentiation effect did not occur if administration of the TRH preceded administration of the prostaglandin. Intravenous injection of prostaglandin prior to TRH administration suppressed TRH-induced TSH secretion, but other evidence indicated that this was a peripheral effect rather than a direct effect in the pituitary. Similar studies utilizing intrahypothalamic injections were not reported. These results are interpreted as indicating that prostaglandins may act to prime TRH receptors to TRH stimulation, thereby acting as a mediator of TRH-induced TSH secretion.

G. Summary

At this time, it is obvious that a concise statement as to the effect of the prostaglandins on the physiological release of the pituitary hormones cannot be formulated. Many areas of discrepancy still exist. Until it becomes technically feasible to investigate separately the various cell types in the pituitary, it is unlikely that the final answers regarding the involvement of the adenylyl cyclase–cyclic AMP system in the prostaglandin mechanism of action can be attained.

Current knowledge does support a number of conclusions:

1. Prostaglandin-induced release of LH, FSH, prolactin, and ACTH occurs via prostaglandin interaction at a hypothalamic site and subsequent release of hypothalamic neurohormones. Very little is known regarding the possible participation of cyclic AMP in this response.
2. LH and FSH are released only in response to the E series of prostaglandins. ACTH release is stimulated by all prostaglandins tested. Prolactin release in response to prostaglandins other than the E series has not been adequately investigated.
3. GH is the only pituitary hormone released via a direct action of prostaglandin on the pituitary. Cyclic AMP appears to be involved.
4. TSH release does not appear to be stimulated by prostaglandins.

IV. PANCREAS

Results from several laboratories have implicated cyclic AMP in the process of insulin secretion by the β cells of the pancreas (Sussman and Vaughan, 1967; Atkins and Matty, 1971). Howell and Montague (1973) have shown that cyclic AMP probably plays an important role in the mediation of effects of hormones and adrenergic agents on insulin release, but that short-term release effected by such compounds as glucose or amino acids does not involve cyclic AMP.

In vivo experiments measuring the increase in plasma insulin levels following injection of prostaglandins have been few and conflicting. Thus, Bressler *et al.* (1968) reported an increase in mouse plasma insulin and blood glucose levels following injection of 2.5 and 5.0 μg PGE$_1$ (366 and 808% increase, respectively); these responses were decreased by the β-adrenergic blocker, MJ 1999. In contrast, Spellacy *et al.* (1971) reported that in pregnant humans infusion of PGE$_2$ or PGF$_{2\alpha}$ in increasing graded doses (0.3 μg/min to 2.4 μg/min for PGE$_2$; 2.5 μg/min to 20 μg/min for PGF$_{2\alpha}$) over a 300-min period had no effect on plasma insulin or blood glucose levels.

Data from several laboratories, derived from *in vitro* experiments with various pancreatic preparations, on the involvement of prostaglandins and cyclic AMP on insulin production seem to be in conflict.

Vance *et al.* (1971) reported that PGE$_1$ (1 μg/ml) did not affect release of immunoreactive insulin or glucagon from isolated pancreatic islets from male rats, although the adenylyl cyclase activity of homogenates of these islets has been shown by Kuo *et al.* (1973) and by Howell and Montague (1973) to be substantally stimulated by various prostaglandins (PGE$_1$, PGE$_2$, PGA$_1$, and PGF$_{1\alpha}$) at 10^{-5} M. Such results suggest that cyclic AMP is not involved in insulin production in this islet preparations.

The aforementioned lack of a prostaglandin effect in isolated rat pancreatic islets may have been due to the absence of glucose from incubation mixtures. Johnson *et al.* (1973) using isolated pancreatic islets observed that PGE$_1$ (10^{-5} M) increased the release of insulin into the medium only if high (300 mg/dl) concentrations of glucose were present; at 30 mg/dl glucose, PGE$_1$ was without influence on insulin secretion. PGE$_{2\alpha}$ and the endoperoxide, PGG$_2$, were only a tenth as effective as PGE$_1$. In parallel experiments, PGE$_1$ caused a large (two- to threefold) increase in cyclic AMP accumulation, and this did not appear to be influenced significantly by the level of glucose present. The finding that PGE$_1$ could elevate cyclic AMP levels under conditions (30 mg/dl glucose) in which it did not stimulate the secretion of insulin suggested to these authors that insulin secretion is dependent on glucose concentration by a mechanism independent of cyclic AMP concentration.

Since the discovery by Rodbell *et al.* (1971) that GTP has pronounced effects on the activation by glucagon of liver plasma membrane adenylyl cyclase, hormonal effects on adenylyl cyclases of a wide variety of other tissues have also been observed to be influenced by GTP and its analogues. Within the present context, Johnson *et al.* (1974) studied the effects of PGE$_1$ and GTP on the activation of adenylyl cyclases of rat pancreatic islet homogenates and cell membranes derived therefrom. In homogenates, PGE$_1$ alone (10 μM) had only a slight (20%) stimulatory effect on adenylyl cyclase, while GTP (10 μM) alone was without effect. When they were present together at the same concentrations, cyclase activity doubled. Results with islet membranes were essentially the same. In the presence of GTP, PGE$_1$ and PGE$_2$ also stimulated cyclase, but the effect was only half that seen with PGE$_1$; PGF$_{2\alpha}$ had no influence on cyclase under any condition. Cyclic AMP phosphodiesterase was unaffected by any of these agents.

Since GTP had such a profound influence on the activation of islet cyclase by prostaglandins, it would seem of interest to determine if GTP could replace glucose as the requirement for a prostaglandin effect on insulin secretion. Such experiments have not yet been reported.

Secretin and isoproterenol are known to affect pancreatic secretions, and both are known to stimulate adenylyl cyclase in this tissue.

That these agents operate via receptors different than those responding to prostaglandins has been shown by experiments in which PGE_1, secretin, and isoproterenol exhibited additive effects on cyclase when tested at their maximal concentrations (D. G. Johnson, personal communication).

V. ADRENAL

PGE_1 and PGE_2 stimulate steroidogenesis in a variety of rat (Flack *et al.*, 1969; Flack and Ramwell, 1972; Saruta and Kaplan, 1972) and beef (Saruta and Kaplan, 1972) adrenal preparations. Saruta and Kaplan (1972) noted that stimulation by PGE_1 was similar to that by ACTH in (1) needing calcium, (2) being inhibited by puromycin but not by actinomycin D, (3) increasing cyclic AMP levels, and (4) not exhibiting an additive effect with exogenous cyclic AMP, PGE, $PGF_{1\alpha}$, and $PGF_{2\alpha}$ were ineffective.

Other reports concerning the effect of prostaglandins on adrenal cyclic AMP have presented conflicting results. Taunton *et al.* (1969) reported that PGE_1 (5–10 μg/ml) neither influenced the adenylyl cyclase activity of an adrenal tumor homogenate preparation nor affected the ACTH- or fluoride-induced increase in adenylyl cyclase activity in the same preparation. In addition, Hurko *et al.* (1974) reported that an active adenylyl cyclase preparation from bovine adrenal medulla was not affected by PGE_1 over a wide range of concentrations (10^{-4}–10^{-12} M). There may be species differences in connection with *in vitro* affects of prostaglandins on adenylyl cyclase. Thus in contrast to the abovementioned negative results with bovine adrenal preparations, Dazord *et al.* (1974) reported that PGE_1 and PGE_2 not only specifically bound to subcellular preparations of human and ovine adrenal glands and to purified membrane preparations from ovine glands but also stimulated adenylyl cyclase activity (115–159% in the human adrenal preparations and 17–25% in the sheep preparations) in the process. Several results indicated that ACTH and prostaglandin receptors are distinct entities: ACTH did not inhibit the binding of labeled prostaglandins to subcellular membrane preparations, and ACTH and prostaglandins had an additive effect on adenylyl cyclase activity at maximal concentrations of each. Based on these results and the observation that indomethacin (up to 1 μg/ml) had no effect on ACTH stimulation of adenylyl cyclase activity Dazord *et al.* (1974) concluded that prostaglandins do not mediate the steroidogenic effect of ACTH.

VI. OVARIES

The relationship of gonadotropins, prostaglandins, adenylyl cyclase, and steroidogenesis in the ovary is complex and does not conform to either the second-messenger or prostaglandin-mediated model of hormone action. The early finding of Marsh *et al.* (1966) that luteinizing hormone (LH) stimulated steroidogenesis as well as cyclic AMP accumulation in the bovine corpus luteum suggested that gonadotropin action in this tissue was mediated by cyclic AMP and conformed to the second-messenger model originally postulated by Sutherland and Rall. When prostaglandins were found to stimulate steroidogenesis (Pharriss *et al.*, 1968; Speroff and Ramwell, 1970), it no longer seemed that the second-messenger model accounted for the actions of LH. When Kuehl (1970) observed that PGE_1 and PGE_2, as well as LH, could stimulate adenylyl cyclase and that the effects of both the prostaglandins and LH could be inhibited by 7-oxa-13-prostynoic acid (OPA), it was suggested that prostaglandins mediate LH effects on cyclic AMP synthesis and steroidogenesis. To satisfy this model, the system must have the following characteristics: (1) The effects of LH on steroidogenesis must be mimicked by the prostaglandins and exogenous cyclic AMP (dibutyryl cyclic AMP). (2) Prostaglandins and LH must stimulate cyclic AMP accumulation in intact- and broken-cell preparations. (3) Prostaglandins and LH should not exhibit additive effects on cyclic AMP synthesis or steroidogenesis. (4) Inhibitors of prostaglandin synthesis should abolish the gonadotropin effects on cyclic AMP accumulation and steroidogenesis. (5) Gonadotropin-induced production of cyclic AMP and prostaglandins should have parallel time courses. Similarly, inhibitors of adenylyl cyclase should abolish the effects of prostaglandins on steroidogenesis. This model requires that the gonadotropins and prostaglandins activate a common adenylyl cyclase. It is assumed that adenylyl cyclase reaction is the only step controlled by both prostaglandins and gonadotropins and that prostaglandins do not have a regulatory action on subsequent reactions in steroidogenesis. A survey of the literature shows that the experimental findings are not adequately explained by this model.

A. Effects of Cyclic AMP and Prostaglandins on Steroidogenesis

The first requirement for prostaglandin mediation of gonadotropin effects is that both prostaglandins and cyclic AMP mimic the gonadotropin effects on steroidogenesis. Cyclic AMP (20 mM) stimulated steroidogenesis in bovine corpus luteum (Marsh and Savard, 1964), and dibutyryl cyclic AMP (2 mM) mimicked LH effects on steroidogenesis in cultured porcine granulosa cells (Channing and Seymour, 1970).

Prostaglandins stimulate ovarian progesterone synthesis in spontaneously ovulating species, as was first observed in minced ovaries from pseudopregnant rats (Pharris *et al.*, 1968). Later, PGE_1, PGE_2 (10 μg/ml) (Speroff and Ramwell, 1970), PGA_1, PGA_2, and $PGF_{2\alpha}$ (Hansel *et al.*, 1973; Sellner and Wickersham, 1970) were found to promote steroidogenesis in bovine corpus luteum. Channing (1972*a*) observed that LH (0.1 μg/ml) and PGE_2 (10 μg/ml) had similar effects on morphological luteinization and steroid synthesis in cultured monkey follicles. These observations have been extended to the mouse ovary in culture (Neal *et al.*, 1975) where $PGF_{2\alpha}$ (30 μg/ml), PGE_2 (30 μg/ml), and human chorionic gonadotropin (HCG) (0.4 IU/ml) had similar effects on progesterone synthesis. In corpora lutea from nonpregnant women, PGE_2 stimulated progesterone synthesis (Marsh and LeMaire, 1974).

Rabbit ovarian tissue apparently is not stimulated by prostaglandins. $PGF_{2\alpha}$ (10 μg/ml) inhibited steroidogenesis in rabbit corpora lutea slices (O'Grady *et al.*, 1973), and PGE_1 (1–100 μg/ml) inhibited LH effects in chopped ovaries (Bedwani and Horton, 1971). Similarly, rabbit granulosa cells in culture did not respond to PGE_2 (1 μg/ml) or $PGF_{2\alpha}$ (1 μg/ml) with increased progestin synthesis (Erickson and Ryan, 1975).

B. Effects of Prostaglandins on Adenylyl Cyclase

A second requirement for prostaglandin mediation of gonadotropin effects is that prostaglandins stimulate cyclic AMP formation. Marsh (1970) reported that PGE_2 and LH had similar effects on the adenylyl cyclase of homogenized bovine corpus luteum. Later, Kuehl *et al.* (1970) found stimulation of cyclic AMP accumulation in intact mouse ovaries with PGE_1 and PGE_2. The E prostaglandins had stimulatory effects on porcine granulosa cells (Kolena and Channing, 1972), immature rat ovaries (Mason *et al.*, 1973; Lamprecht *et al.*, 1973), rat follicles in culture (Tsafriri *et al.*, 1972), and human corpus luteum (Marsh and LeMaire, 1974). The E series also stimulated adenylyl cyclase in homogenized corpora lutea from pseudopregnant rabbits (Jonsson *et al.*, 1972).

C. Additivity of LH and Prostaglandins on Adenylyl Cyclase

A third requirement for prostaglandin mediation of gonadotropin action is that both agents activate the same adenylyl cyclase. Operationally effective doses of each agent should not produce additive effects on adenylyl cyclase activity. However, additivity has been observed in mouse ovaries (Kuehl *et al.,* 1970), rabbit corpus luteum homogenates (Jonsson *et al.,* 1972), porcine granulosa cells (Kolena and Channing, 1972), and immature rat ovaries (Lamprecht *et al.,* 1973). Additivity of LH and PGE$_2$ action was also observed when these agents were added sequentially. Rat follicles previously incubated with LH for 18 hr did not produce additional cyclic AMP during a subsequent incubation with LH. These follicles did produce additional cyclic AMP when PGE$_2$ was added following the preincubation with LH (Lamprecht *et al.,* 1973).

The observation that PGE and LH have different time courses suggests that these agents activate separate receptors. In whole ovaries from the immature rats, and in isolated preovulatory rat follicles, PGE$_2$ (10 μg/ml) induced cyclic AMP accumulation peaked within 15 min while LH (10 μg/ml) induced cyclic AMP accumulation became maximal at 60 min (Selstam *et al.,* 1974; Nilsson *et al.,* 1974). These effects may be due to metabolism of the prostaglandins or to the presence of separate adenylyl cyclase enzymes.

The explanation of the additivity and time course effects will be explained when the LH and PGE receptors are characterized in isolated cell types. Analogues of LH and PGE which bind to the receptor but which do not activate the adenylyl cyclase may be useful in these studies.

D. Effects of Inhibitors of Prostaglandin Synthesis on LH Action

If prostaglandins are the mediators of gonadotropin action, then prostaglandin antagonists and inhibitors of prostaglandin synthetase should inhibit LH-induced cyclic AMP accumulation and steroidogenesis. Early findings of Kuehl *et al.* (1970) indicated that 7-oxa-13 prostynoic acid (a prostaglandin antagonist) inhibited the effects of LH, PGE$_2$, and PGE$_1$ on cyclic AMP accumulation in mouse ovaries. These findings were not confirmed, however, in similar studies using immature rat ovaries (Lamprecht *et al.,* 1973).

In monkey granulosa cell cultures, OPA (50 μg/ml) antagonized the LH, HCG, and PGE$_2$ effects on morphological luteinization and steroidogenesis, but also caused necrotic changes in the cells (Channing, 1972*b*). Shorter exposure (2 hr) of rat follicles to the drug did not reduce the LH or PGE$_2$ effects on morphological luteinization (Ellsworth and Armstrong, 1974).

The reported effects of indomethacin, a prostaglandin synthetase inhibitor, are unambiguous and demonstrate the independence of prostaglandin effects from those of gonadotropins. Although indomethacin treatment abolished the synthesis of prostaglandins that normally follows the administration of HCG or LH or mating in rabbits (Yang *et al.*, 1973), the follicles in these animals were nevertheless luteinized (Grinwich *et al.*, 1972; O'Grady *et al.*, 1972). Similarly, the addition of indomethacin was without effect on LH- or PGE$_2$-stimulated synthesis of cyclic AMP in intact ovaries (Lamprecht *et al.*, 1973). A series of nonsteroidal anti-inflammatory agents, including flufenamate, meclofenamate, and mefanamate, antagonized PGE$_2$, but not LH, action on cyclic AMP accumulation (Zor *et al.*, 1973).

The failure of the prostaglandin antagonists and synthetase inhibitors to abolish LH effects is the strongest evidence that prostaglandins do not mediate gonadotropin action. Other explanations are needed to clarify these relationships.

E. Adenylyl Cyclase in Follicles from Immature and Adult Rats

In recent studies, investigators recognized that the ovarian follicle is an emergent tissue, and have designed experiments in several species to test the effects of gonadotropins during the differentiation and maturity of the follicle. A coherent relationship among gonadotropins, prostaglandin synthetase, adenylyl cyclase, and steroidogenesis is becoming evident from such studies.

The gonadotropins and prostaglandins have different actions on ovarian tissue in the prepubertal rat. In the neonate, PGE$_2$, but not LH, induces cyclic AMP accumulation. LH-induced activity does not become evident until day 10, and the ovaries of a 15-day-old rat already exhibit adult-type sensitivity to LH and PGE$_2$ (Lamprecht *et al.*, 1972).

Mason *et al.* (1973) confirmed the findings concerning the sensitivity of 21-day-old rat ovaries to LH and PGE$_2$. When these animals were treated with pregnant mare serum (PMS) and HCG in order to force superovulation and luteinization of the follicles, ovarian slices were sensitive to LH, but did not synthesize cyclic AMP in response to PGE$_2$ (7.5 μg/ml). Similarly, LH stimulated adenylyl cyclase in ovarian 600g pellets prepared from superovulated immature rat ovaries through day 15 of pseudopregnancy (Nugent *et al.*, 1975).

In the human, the corpora lutea during normal menses were found to be sensitive to LH- and PGE$_2$-induced cyclic AMP accumulation and steroidogenesis, but became refractory during pregnancy (Marsh and LeMaire, 1974).

F. Adenylyl Cyclase, Prostaglandin Synthesis, and Steroidogenesis in Preovulatory Follicles

The events which transpire between the LH surge and ovulation have been investigated in the rabbit and the rat. The time courses of syntheses of cyclic AMP, steroids, and prostaglandins were followed in the rabbit. At estrus, follicles were sensitive to LH-induced cyclic AMP accumulation (Marsh *et al.*, 1972, 1973). Follicles had high rates of LH-induced cyclic AMP synthesis 30 and 60 min after HCG injection. The cyclic AMP content of rabbit follicles *in vivo* rose within 15 min of HCG injection, peaked at 2 hr, and declined to the estrous level by 8 hr (Goff and Major, 1975). At the time of ovulation, follicles could not be induced to produce cyclic AMP by the addition of LH (Marsh *et al.*, 1973).

Similarly, steroid synthesis became refractory to LH stimulation as the follicle matured. Estrous follicles were sensitive to LH-induced steroidogenesis (predominantly 17-hydroxyprogesterone and testosterone) (Mills and Savard, 1973). Two hours after coitus, the follicular steroid synthesis (predominantly progesterone and 17-hydroxyprogesterone) increased above that seen in estrus, but no further increase was obtained when these follicles were incubated with LH. Follicles obtained at the time of ovulation did not synthesize steroids in response to LH.

Prostaglandin synthesis was followed in rabbit follicles *in vivo*. Prostaglandin concentrations in the preovulatory rabbit follicle remained at estrous levels until 5 hr after HCG injection in the case of PGE, and 9 hr in the case of PGF (Yang *et al.*, 1973; LeMaire *et al.*, 1973). Prostaglandin synthesis occurred hours later than the rise of cyclic AMP levels, suggesting an inverse relationship between cyclic AMP and prostaglandin synthesis. Both LH and cyclic AMP (20 mM) stimulated PGE and PGF synthesis during a 5-hr incubation of follicles obtained from estrous rabbits (Marsh *et al.*, 1974). Cyclic AMP had similar effects on mouse ovary (Kuehl *et al.*, 1972).

Progesterone secretion appears to be incompatible with prostaglandin synthesis in follicular tissues. The increase of PGF levels in cultured rabbit granulosa cells was accompanied by a decline in progesterone secretion, in both the absence and the presence of LH-FSH (Challis *et al.*, 1974). PGE_2 (1 μg/ml) or $PGF_{2\alpha}$ (1 μg/ml) failed to induce progesterone production in cultured rabbit granulosa cells (Erickson and Ryan, 1974). Interestingly, cells cultured in the presence of dibutyryl cyclic AMP had the highest progesterone synthesis rate, but no PGF accumulation (Challis *et al.*, 1974). Since the cells cultured with dibutyryl cyclic AMP were comprised solely of differentiated granulosa cells, while the other cultures had a mixture of cell types, it was suggested that PGF was synthesized by the undifferentiated cells.

Preovulatory prostaglandin synthesis follows cyclic AMP synthesis in the rabbit follicle and coincides with a preovulatory suspension of steroidogenesis. This finding correlates well with the observation that steroid synthesis in rabbit follicular tissue does not occur when prostaglandins are present. Prostaglandins appear to be required for expulsion of the ovum since indomethacin-treated rabbits had luteinized but unruptured follicles (O'Grady *et al.,* 1972; Grinwich *et al.,* 1972).

In the rat, the temporal relationships among cyclic AMP, steroid, and prostaglandin synthesis have not been completely delineated. As in the rabbit, the interval between LH surge and the occurrence of elevated ovarian prostaglandin content was 4–6 hr (Lindner *et al.,* 1974). Although the prostaglandins could mimic LH action in many preovulatory events, including cyclic AMP accumulation, steroidogenesis, and ovum maturation, the only event which was blocked by indomethacin was the release of the mature ovum from the follicle (Tsafriri *et al.,* 1972, 1973; Lindner *et al.,* 1974; Neal *et al.,* 1975). The finding that preovulatory events proceed normally despite the absence of endogenous prostaglandins suggests that the actions of exogenous prostaglandin may be pharmacological rather than mimetic of the *in vivo* events.

Ovarian prostaglandin synthesis appears to be a transient event preceding ovulation, suggesting that the enzymes which synthesize prostaglandins have a short lifetime. Alternatively, the prostaglandins may be produced by a cell type which disappears as the follicle matures into corpus luteum. The prostaglandin synthetic capacity appears to be diminished in pregnancy, as the conversion of radiolabeled arachidonate to prostaglandins was less than 0.1% in rat ovarian homogenates (Carminati *et al.,* 1975). However, data concerning ovarian prostaglandin content or synthesis during the estrus cycle are lacking, and speculation on the site of synthesis is premature.

Increased cellular levels of cyclic nucleotides coincide with decreased prostaglandin synthesis in the rabbit granulosa cell and rabbit follicle. In rat uterus, elevated cyclid GMP levels were associated with decreased PGF synthesis (Ham *et al.,* 1975). Correlations with cyclic GMP and prostaglandin levels have not been reported for the ovary, and these studies are needed to further characterize the regulation of synthesis of prostaglandins in this tissue.

To summarize, prostaglandins can mimic gonadotropin effects on adenylyl cyclase and on progesterone synthesis (except in rabbits). While the hypothesis that prostaglandins mediate gonadotropin effects is attractive, the finding that inhibitors of prostaglandin synthesis have no effect on gonadotropin action on adenylyl cyclase, steroidogenesis, and ovum maturation eliminates this model as an explanation. The time courses of cyclic AMP, steroid, and prostaglandin synthesis in the rabbit and rat suggest that cyclic AMP precedes prostaglandin synthesis. A complex relationship between cyclic nucleotide levels and prostaglandin synthesis is indicated and deserves investigation. Finally, prostaglandins do appear to be required for the expulsion of the ovum from the mature follicle.

VII. REFERENCES

Ahn, C. S., and Rosenberg, I. N., 1970, Iodine metabolism in thyroid slices: Effects of TSH, dibutytyl cyclic 3',5'-AMP, NaF, and prostaglandin E₁, *Endocrinology* **86**:396.

Atkins, T., and Matty, A. J., 1971, Adenyl cyclase and phosphodiesterase activity in the isolated islets of Langerhans of obese mice and their lean litter mates: The effect of glucose, adrenaline and drugs on adenyl cyclase activity, *J. Endocrinol.* **51**:67.

Bedwani, J. R., and Horton, E. W., 1971, Interaction between prostaglandins and gonadotrophins in the rabbit ovary, *Br. J. Pharmacol.* **43**:794.

Bergeron, L., Barden, N., 1975, Stimulation of cyclic AMP accumulation by arachidonic and 8,11,14-eicosatrienoic acids in rat anterior pituitary gland, *Mol. Cell. Endocrinol.* **2**:253.

Borgeat, P., Labrie, F., Poirier, G., Chavancy, G., and Shally, A. V., 1973, Stimulation of adenosine 3',5'-cyclic monophosphate accumulation in anterior pituitary gland by purified growth hormone-releasing hormone, *Trans. Assoc. Am. Physicians* **86**:284.

Borgeat, P., Labrie, F., Drouin, J., Belanger, A., Immer, H., Sestanj, K., Nelson, V., and Gotz, M., 1974, Inhibition of adenosine 3',5'-monophosphate accumulation in anterior pituitary gland *in vitro* by growth hormone-release inhibiting hormone, *Biochem. Biophys. Res. Commun.* **56**:1052.

Borgeat, P., Labrie, F., and Garneau, P., 1975, Characteristics of action of prostaglandins on cyclic AMP accumulation in rat anterior pituitary gland, *Can. J. Biochem.* **53**:455.

Bressler, R., Vargas-Cordon, M., and Lebovitz, H. E., 1968, Tranylcypromine: A potent insulin secretagogue and hypoglycemic agent, *Diabetes* **17**:617.

Brown, M. R., and Hedge, G. A., 1974, *In vivo* effects of prostaglandins on TRH-induced TSH secretion, *Endocrinology* **95**:1392.

Burke, G., 1970, Effects of prostaglandins on basal and stimulated thyroid function, *Am. J. Physiol.* **218**:1445.

Burke, G., 1972, Aspirin and indomethacin abolish thyrotropin-induced increase in thyroid cell prostaglandins, *Prostaglandins* **2**:413.

Burke, G., 1973a, Effects of thyrotropin and N⁶, O²′-dibutyryl cyclic 3',5'-adenosine monophosphate on prostaglandin levels in thyroid, *Prostaglandins* **3**:291.

Burke, G., 1973b, Comparative effects of purine nucleotides on thyrotropin- and prostaglandin E₁-responsive adenylate cyclase in thyroid plasma membranes, *Prostaglandins* **3**:537.

Burke, G., 1974, Effects of prostaglandin antagonists on the induction of cyclic 3',5'-adenosine monophosphate formation and thyroid hormone secretion *in vitro*, *Endocrinology* **94**:91.

Burke, G., and Sato, S., 1971, Effects of long-acting thyroid stimulator and prostaglandin antagonists on adenyl cyclase activity in isolated bovine thyroid cells, *Life Sci.* **10**:969 (Part II).

Burke, G., Chang, L.-L., and Szabo, M., 1973, Thyrotropin and cyclic nucleotide effects on prostaglandin levels in isolated thyroid cells, *Science* **180**:872.

Carlson, J. C., Barcikowski, B., and McCracken, J. A., 1973, Prostaglandin F₂α and the release of LH in sheep, *J. Reprod. Fertil.* **34**:357.

Carlson, J. C., Barcikowski, B., and McCracken, J. A., 1974, Evidence for an obligatory role of prostaglandins in luteinizing hormone release, *Fed. Proc.* **33**:A-54 (abst.).

Carminati, P., Luzzani, F., Soffientini, A., and Lerner, L. J., 1975, Influence of day of pregnancy on rat placental, uterine, and ovarian prostaglandin synthesis and metabolism, *Endocrinology* **97**:1071.

Castracane, V. D., and Saksena, S. K., 1974, Prostaglandins of E series and LH release in fertile male rats, *Prostaglandins* **7**:53.

Challis, J. R. G., Erickson, G. F., and Ryan, K. J., 1974, Prostaglandin F production *in vitro* by granulosa cells from rabbit pre-ovulatory follicles, *Prostaglandins* **10**:183.

Channing, C. P., 1972a, Stimulatory effects of prostaglandins upon luteinization of rhesus monkey granulosa cell cultures, *Prostaglandins* **2**:331.

Channing, C. P., 1972b, Effects of prostaglandin inhibitors, 7-oxa-13-prostynoic acid and eicosa-5,8,11,14-tetraynoic acid, upon luteinization of rhesus monkey granulosa cells in culture, *Prostaglandins* **2**:351.

Channing, C. P., and Seymour, J. F., 1970, Effects of dibutyryl cyclic-3',5'-AMP and other agents upon luteinization of porcine granulosa cells in culture, *Endocrinology* **70**:165.

Chobsieng, P., Naor, Z., Koch, Y., Zor, U., and Lindner, H. R., 1975, Simulatory effect of prostaglandin E$_2$ on LH release in the rat: Evidence for hypothalamic site of action, *Neuroendocrinology* **17**:12.

Cooper, R. H., McPherson, M., and Schofield, J. G., 1972, The effect of prostaglandins on ox pituitary content of adenosine 3',5'-cyclic monophosphate and the release of growth hormone, *Biochem. J.* **127**:143.

Dazord, A., Morera, A. M., Bertrand, J., and Saez, J. M., 1974, Prostaglandin receptors in human and ovine adrenal glands: binding and stimulation of adenyl cyclase in subcellular preparations, *Endocrinology* **95**:352.

Deery, D. J., and Howell, S. L., 1973, Rat anterior pituitary adenyl cyclase activity: GTP requirement of prostaglandin E$_1$ and E$_2$ and synthetic luteinizing hormone-releasing hormone activation, *Biochim. Biophys. Acta* **329**:17.

DeWied, D., Witter, A., Versteg, D. H. G., and Mulder, A. H., 1969, Release of ACTH by substances of central nervous system origin, *Endocrinology* **85**:561.

Dekker, A., and Field, J. B., 1970, Correlation of effects of thyrotropin, prostaglandins and ions on glucose oxidation, cyclic-AMP, and colloid droplet formation in dog thyroid slices, *Metabolism* **19**:453.

Dowd, A. J., Hoffman, D. C., and Speroff, L., 1973, Direct effects of prostaglandins and prostaglandin inhibitors on pituitary LH release demonstrated by *in vitro* perfusion, *Endocrinol. Soc. 55th Annu. Meeting,* p. 1973 A-135, Chicago (abst.).

Dupont, A., Chavancy, G., and Labrie, R., 1972, Prostaglandins as mediators of thyrotropin-releasing hormone action, *Adv. Biosci. Suppl.* **9**:34.

Ellsworth, L. R., and Armstrong, D. T., 1974, Effect of indomethacin and 7-oxa-13-prostynoic acid on luteinization of transplanted rat ovarian follicles induced by luteinizing hormone and prostaglandin E$_2$, *Prostaglandins* **7**:165.

Erickson, G. F., and Ryan, K. J., 1975, The effect of LH/FSH, dibutyryl cyclic AMP, and prostaglandins on the production of estrogens by rabbit granulosa cells *in vitro, Endocrinology* **97**:108.

Eskay, R. L., Warberg, J., Mical, R. S., and Porter, J. C., 1975, Prostaglandin E$_2$-induced release of LHRH into hypophysial portal blood, *Endocrinology* **97**:816.

Field, J., Dekker, A., Zor, U., and Kaneko, T., 1971, *In vitro* effects of prostaglandins on thyroid gland metabolism, *Ann. N.Y. Acad. Sci.* **180**:278.

Flack, J. D., and Ramwell, P. W., 1972, A comparison of the effects of ACTH, cyclic AMP, dibutyryl cyclic AMP, and PGE$_2$ on corticosteroidogenesis *in vitro, Endocrinology* **90**:371.

Flack, J. D., Jessup, R., and Ramwell, P. W., 1969, Prostaglandin stimulation of rat corticosteroidogenesis, *Science* **163**:691.

Goff, A. K., and Major, P. W., 1975, Concentrations of cyclic AMP in rabbit ovarian tissue during the preovulatory period and psuedopregnancy after the induction of ovulation by administration of human chorionic gonadotropin, *J. Endocrinol.* **65**:73.

Grinwich, D. L., Kennedy, T. G., and Armstrong, D. T., 1972, Dissociation of ovulatory and steroidogenic actions of luteinizing hormone in rabbits with indomethacin, and inhibitor of prostaglandin synthesis, *Prostaglandins* **1**:89.

Ham, E. A., Cirillo, V. J., Zanetti, M. E., and Kuehl, F. A., 1975, Estrogen-directed synthesis of specific prostaglandins in uterus, *Proc. Natl. Acad. Sci. U.S.A.* **72**:1420.

Hansel, W., Concannon, P. W., and Lukaszewska, J. H., 1973, Corpora lutea of large domestic animals, *Biol. Reprod.* **8**:222.

Harms, P. G., Ojeda, S. R., and McCann, S. M., 1974, Prostaglandin-induced release of pituitary gonadotropins: Central nervous system and pituitary sites of action, *Endocrinology* **94**:1459.

Hedge, G. A., 1971, ACTH secretion due to hypothalamo-pituitary effects of adenosine-3',5'-monophosphate and related substances, *Endocrinology* **89**:500.

Hedge, G. A., and Hanson, S. D., 1972, The effects of prostaglandins on ACTH secretion, *Endocrinology* **91**:925.

Henzl, M. R., Ortega, E., Cortes-Galleyos, V., Tomlinson, R. V., and Segre, E. J., 1973, Prostaglandin E$_2$ and the luteal phase of the menstrual cycle. Effects on blood progesterone, estradio, cortisol, and growth hormone, *J. Clin. Endocrinol. Metab.* **36**:784.

Hertelendy, F., Todd, H., Ehrhart, K., and Blute, R., 1972, Studies on growth hormone secretion. IV. *In vivo* effects of prostaglandin E$_1$, *Prostaglandins* **2**:79.

Higgins, C. B., and Braunwald, E., 1972, The prostaglandins: Biochemical, physiological, and clinical considerations, *Am. J. Med.* **53**:92.

Hittleman, K. J., and Butcher, R. W., 1973, Cyclic AMP and the mechanism of action of the prostaglandins, in: *The Prostaglandins: Pharmacological and Therapeutic Advances* (M. F. Cuthbert, ed.), p. 151, Heinemann, London.

Horton, E. W., 1969, Hypothesis on physiological roles of prostaglandins, *Physiol. Rev.* **49**:122.

Howell, S. L., and Montague, W., 1973, Adenylate cyclase activity in isolated rat islets of Langerhans. Effects of agents which alter rates of insulin secretion, *Biochim. Biophys. Acta* **320**:44.

Hurko, O., Elster, P., and Wurtman, R. J., 1974, Adenylyl cyclase activity in bovine adrenal medulla, *Endocrinology* **94**:591.

Ito, H., Momose, G., Katayama, T., Takagishi, H., Ito, L., Nakajima, H., and Takei, Y., 1971, Effect of prostaglandin on the secretion of human growth hormone, *J. Clin. Endocrinol. Metab.* **32**:857.

Johnson, D. G., Fujimoto, W. Y., and Williams, R. H., 1973, Enhanced release of insulin by prostaglandins in isolated pancreatic islets, *Diabetes* **22**:658.

Johnson, D. G., Thompson, W. J., and Williams, R. H., 1974, Regulation of adenylyl cyclase from isolated pancreatic islets by prostaglandins and guanosine 5′-triphosphate, *Biochemistry* **13**:1920.

Jonsson, H. T., Shelton, V. L., and Baggett, B., 1972, Stimulation of adenyl cyclase by prostaglandins in rabbit corpus luteum, *J. Reprod. Biol.* **7**:107.

Kaneko, T., Zor, U., and Field, J. B., 1969, Thyroid-stimulating hormone and prostaglandin E_1 stimulation of cyclic 3′,5′-adenosine monophosphate in thyroid slices, *Science* **163**:1062.

Kendall-Taylor, P., 1972, Comparison of the effects of various agents on thyroidal adenyl cyclase activity with their effects on thyroid hormone release, *J. Endocrinol.* **54**:137.

Kolena, J., and Channing, C. P., 1972, Stimulatory effects of LH, FSH, and prostaglandins upon cyclic 3′,5′-AMP levels in porcine granulosa cells, *Endocrinology* **90**:1543.

Kowalski, K., Sato, S., and Burke, G., 1972, Thyrotropin- and prostaglandin E_2- responsive adenyl cyclase in thyroid plasma membranes, *Prostaglandins* **2**:441.

Kudo, C. F., Rubinstein, D., McKenzie, J. M., and Beck, J. C., 1972, Hormonal release by dispersed pituitary cells, *Can. J. Physiol. Pharmacol.* **50**:860.

Kuehl, F. A., Humes, J. L., Tarnoff, J., Cirillo, V. J., and Ham, E. A., 1970, Prostaglandin receptor site: Evidence for an essential role in the action of luteinizing hormone, *Science* **169**:883.

Kuehl, F. A., Cirillo, C. V., Ham, E. A., and Humes, J. L., 1972, The regulatory role of prostaglandins on the cyclic 3′,5′-AMP system, *Adv. Biosci.* **9**:155.

Kuo, W-N, Hodgins, D. S., and Kuo, J. G., 1973, Adenylate cyclase in islets of Langerhans: Isolation of islets and regulation of adenylate cyclase activity by various hormones and agents, *J. Biol. Chem.* **248**:2705.

Labrie, F., Pelletier, G., Borgeat, P., Drovin, J., Ferland, L., and Belanger, A., 1975, Mode of action of hypothalamic regulatory hormones in the adenohypophysis, *Front. Neuroendocrinol.* (in press).

Lamprecht, S. A., Zor, U., Tsafriri, A., and Lindner, H. R., 1973, Action of prostaglandin E_2 and of luteinizing hormone on ovarian adenylate cyclase, protein kinase, and ornithine decarboxylase activity during postnatal development and maturity in the rat, *J. Endocrinol.* **57**:217.

Lau, I. F., and Salsena, S. K., 1974, Prostaglandin E_2 and LH release in ovariectomized rats, *Prostaglandins* **7**:49.

LeMaire, W., Yang, N. S. T., Behrman, H. H., and Marsh, J. M., 1973, Preovulatory changes in the concentration of prostaglandin in rabbit Graafian follicle, *Prostaglandins* **3**:367.

Lindner, H. R., Tsafriri, A., Lieberman, M. E., Zor, U., Koch, Y., Bauminger, S., and Barnea, A., 1974, Gonadotropin action on cultured Graafian follicles: Induction of maturation division of the mammalian oocyte and differentiation of the luteal cell, *Recent Prog. Horm. Res.* **30**:79.

MacLeod, R. M., and Lehmeyer, J. E., 1970, Release of pituitary growth hormone by prostaglandins and dibutyryl adenosine cyclic 3′,5′-monophosphate in the absence of protein synthesis, *Proc. Natl. Acad. Sci. U.S.A.* **67**:1172.

MacLeod, R. M., and Lehmeyer, J. E., 1974, Studies on the mechanism of the dopamine-mediated inhibition of prolactin secretion, *Endocrinology* **94**:1077.

Makino, T., 1973, Studies of the intracellular mechanism of LH release in the anterior pituitary, *Am. J. Obstet. Gynecol.* **115**:606.

Marsh, J. M., 1970, The stimulatory effect of protaglandin E_2 on adenyl cyclase in the bovine corpus luteum, *FEBS Lett.* **7**:283.

Marsh, J. M., and LeMaire, W. J., 1974, Cyclic AMP accumulation and steroidogenesis in the human corpus luteum: Effect of gonadotropins and prostaglandins, *J. Clin. Endocrinol. Metab.* **38**:99.

Marsh, J. M., and Savard, K., 1964, The effect of 3',5'-AMP on progesterone synthesis in the corpus luteum, *Fed. Proc.* **23**:462.

Marsh, J. M., Butcher, R. W., Savard, K., and Sutherland, E. W., 1966, The stimulatory effect of luteinizing hormone on adenosine 3',5'-monophosphate accumulation in corpus luteum slices, *J. Biol. Chem.* **241**:5436.

Marsh, J. M., Mills, T. M., and LeMaire, W. J., 1972, Cyclic AMP synthesis in rabbit Graafian follicles and the effect of luteinizing hormone, *Biochim. Biophys. Acta* **273**:389.

Marsh, J. M., Mills, T. M., and LeMaire, W. J., 1973, Preovulatory changes in the synthesis of cyclic AMP by rabbit Graafian follicles, *Biochim. Biophys. Acta* **304**:197.

Marsh, J. M., Yang, N. S. T., and LeMaire, W. J., 1974, Prostaglandin synthesis in rabbit Graafian follicles *in vitro:* Effect of luteinizing hormone and cyclic AMP, *Prostaglandins* **7**:269.

Mashiter, K., and Field, J. B., 1974*a,* The thyroid gland, in: *The Prostaglandins* (P. W. Ramwell, ed.), Vol. 2, pp. 49–73, Plenum Press, New York.

Mashiter, K., and Field, J. B., 1974*b,* Prostaglandins and the thyroid gland, *Fed. Proc.* **33**:78.

Mashiter, K., Mashiter, G. D., and Field, J. B., 1974, Effects of prostaglandin E_1, ethanol and TSH on the adenylate cyclase activity of beef thyroid plasma membranes and cyclic AMP content of dog thyroid slices, *Endocrinology* **94**:370.

Mason, N. R., Schaffer, R. J., and Toomey, R. E., 1973, Stimulation of cyclic AMP accumulation in rat ovaries in vitro, *Endocrinology* **93**:34.

McKeown, B. A., John, T. M., and George, J. C., 1974, The effect of prostaglandin E_1 on plasma growth hormone, free fatty acids, and glucose levels in the pigeon, *Prostaglandins* **8**:303.

Mills, T. M., and Savard, K., 1973, Steroidogenesis in ovarian follicles isolated from rabbits before and after mating, *Endocrinology* **92**:788.

Moore, W. V., and Wolff, J., 1973, Binding of prostaglandin E_1 to beef thyroid membranes, *J. Biol. Chem.* **248**:5705.

Naor, Z., Koch, Y., Bauminger, S., and Zor, U., 1975, Action of luteinizing hormone-releasing hormone and synthesis of prostaglandins in the pituitary gland, *Prostaglandins* **9**:211.

Neal, P., Baker, T. G., McNatty, K. P., and Scaramuzzi, R. J., 1975, Influence of prostaglandins and human chorionic gonadotropin on progesterone concentration and oocyte maturation in mouse ovarian follicles maintained in organ culture, *J. Endocrinol.* **65**:19.

Nilsson, L., Rosberg, S., and Ahren, K., 1974, Characteristics of the cyclic 3',5'-AMP formation in isolated ovarian follicles from PMSG-treated immature rats after stimulation *in vitro* with gonadotropins and prostaglandins, *Acta Endocrinol.* **77**:559.

Nugent, C. L., Lopata, A., and Gould, M. K., 1975, The effects of exogenous gonadotropins on ovarian adenylate cyclase activity, *Endocrinology* **97**:581.

O'Grady, J. P., Caldwell, B. V., Auletta, E. J., and Speroff, L., 1972, The effects of an inhibitor of prostaglandin synthesis (indomethacin) on ovulation, pregnancy, and pseudopregnancy in the rabbit, *Prostaglandins* **1**:97.

O'Grady, J. P., Kohorn, E. I., Glass, R. H., Caldwell, B. V., Brook, W. A., and Speroff, L., 1973, Inhibition of pregesterone synthesis *in vitro* by prostaglandin F_2 alpha. *J. Reprod. Fertil.* **30**:153.

Ojeda, S. R., Harms, P. G., and McCann, S. M., 1974*a,* Effect of third ventricular injections of prostaglandins (PG's) on gonadotropin release in conscious free moving male rats, *Prostaglandins* **8**:545.

Ojeda, S. R., Harms, P. G., and McCann, S. M., 1974*b,* Central effect of prostaglandin E_1 (PGE₁) on prolactin release, *Endocrinology* **95**:613.

Ojeda, S. R., Harms, P. G., and McCann, S. M., 1974*c*, Possible role of cyclic AMP and prostaglandin E₁ in the dopaminergic control of prolactin release, *Endocrinology* **95**:1694.

Ojeda, S. R., Krulich, L., and McCann, S. M., 1974*d*, Effect of an intraventricular injection of cyclic AMP on plasma prolactin and LH levels of ovariectomized, estrogen treated rats, *Neuroendocrinology* **16**:342.

Ojeda, S. R., Wheaton, J. E., and McCann, S. M., 1975, Prostaglandin E₂-induced release of luteinizing hormone-releasing factor (LRF), *Neuroendocrinology* **17**:283.

Pelletier, G., Lemay, A., Beraud, G., and Labrie, F., 1972, Ultrastructural changes accompanying the stimulatory effect of N^6-monobutyryl adenosine 3′,5′-monophosphate on the release of growth hormone (GH), prolactin (Prl) and adrenocorticotropin hormone (ACTH) in rat anterior pituitary gland *in vitro*, *Endocrinology* **91**:1355.

Peng, T.-C., Six, K. M., and Munson, P. L., 1970, Effects of prostaglandin E₁ on the hypothalamo-hypophyseal-adrenocortical axis in rats, *Endocrinology* **80**:202.

Pharriss, B. B., Wyngardon, L. J., and Gutknecht, G. D., 1968, Biological interactions between prostaglandins and luteotropins in the rat, in: *Gonadotropins* (E. Rosemberg, ed.), pp. 121–129, Geron-X, Los Altas.

Ramwell, P. W., and Shaw, J. E., 1970, Biological significance of the prostaglandins, *Recent Prog. Horm. Res.* **26**:139.

Ratner, A., 1970, Stimulation of luteinizing hormone release *in vitro* by dibutyryl-cyclic-AMP and theophylline, *Life Sci.* **9**:1221.

Ratner, A., Wilson, M. C., and Peake, G. T., 1973, Antagonism of prostaglandin-promoted pituitary cyclic AMP accumulation and growth hormone secretion *in vitro* by 7-oxa-13-prostynoic acid, *Prostaglandins* **3**:413.

Ratner, A., Wilson, M. C., Sprivastava, L., and Peake, G. T., 1974, Stimulatory effects of prostaglandin E₁ on rat anterior pituitary cyclic AMP and leuteinizing hormone release, *Prostaglandins* **5**:165.

Rodbell, M., Kraus, H. M. J., Pohl, S. L., and Birnbaumer, L., 1971, The glucagon-sensitive adenyl cyclase system in plasma membranes of rat liver. IV. Effects of guanyl nucleotides on binding of ¹²⁵I-glucagon, *J. Biol. Chem.* **246**:1872.

Saruta, T., and Kaplan, N. M., 1972, Adrenocortical steroidogenesis: The effects of prostaglandins, *J. Clin. Invest.* **51**:2246.

Sato, A., Ohaya, T., Kotani, M., Harada, A., and Yamada, T., 1974, Effects of biogenic amines on the formation of adenosine 3′,5′-monophosphate in porcine cerebral cortex, hypothalamus and anterior pituitary slices, *Endocrinology* **94**:1311.

Sato, S., Kowalski, K., and Burke, G., 1972*a*, Effects of a prostaglandin antagonist, polyphloretin phosphate, on basal and stimulated thyroid function, *Prostaglandins* **1**:345.

Sato, S., Szabo, M., Kowalski, K., and Burke, G., 1971*b*, Role of prostaglandin in thyrotropin action on thyroid, *Endocrinology* **90**:343.

Sato, T., Jyujo, T., Tesaka, T., Ishikawa, J., and Igarashi, M., 1974, Follicle stimulating hormone and prolactin release induced by prostaglandins in rat, *Prostaglandins* **5**:483.

Sato, T., Hirono, M., Jyujo, T., Iesaka, T., Taya, K., and Igarashi, M., 1975*a*, Direct action of prostaglandins on the rat pituitary *Endocrinology* **96**:45.

Sato, T., Jyujo, T., Kawari, Y., and Asai, T., 1975*b*, Changes in LH-releasing hormone content of the hypothalamus and electron microscopy of the anterior pituitary after prostaglandin E₂ injection in rats, *Am. J. Obstet. Gynecol.* **122**:637.

Sellner, R. G., and Wiekersham, E. W., 1972, Effects of prostaglandins on steroidogenesis by rabbit ovarian tissue *in vitro*, *Biol. Reprod.* **7**:107.

Selstam, G., Leljekvist, J., Rosberg, S., Gronquist, L., Perklev, T., and Ahren, K., 1974, Comparison between the effect of luteinizing hormone and prostaglandin E₁ on ovarian cyclic AMP, *Prostaglandins* **6**:303.

Shio, H., Shaw, J., and Ramwell, P., 1971, Relation of cyclic AMP to the release and actions of prostaglandins, *Ann. N.Y. Acad. Sci.* **185**:327.

Spellacy, W. N., Buhi, W. C., and Holsinger, K. K., 1971, The effect of prostaglandin F₂α and E₂ on blood glucose and plasma insulin levels during pregnancy, *Am. J. Obstet. Gynecol.* **111**:239.

Speroff, L., and Ramwell, P. W., 1970, Prostaglandin stimulation of *in vitro* progesterone synthesis, *J. Clin. Endocrinol.* **30**:345.

Spies, H. G., and Norman, R. L., 1973, Luteinizing hormone release and ovulation induced by the intraventricular infusion of prostaglandin E_1 into pentobarbital-blocked rats, *Prostaglandins* **4**:131.

Sussman, K. E., and Vaughn, C. D., 1967, Insulin release after ACTH, glucagon and adenosine-3′,5′-phosphate (cyclic AMP) in the perfused isolated rat pancreas, *Diabetes* **16**:449.

Tal, E., Szabo, M., and Burke, G., 1974, TRH and prostaglandin action on rat anterior pituitary: dissociation between cyclic AMP levels and TSH release, *Prostaglandins* **5**:175.

Taunton, O. D., Roth, J., and Pastan, I., 1969, Studies on the adrenocorticotropic hormone-activated adenyl cyclase of a functional adrenal tumor, *J. Biol. Chem.* **244**:247.

Tsafriri, A., Lindner, H. R., Zor, U., and Lamprecht, S. A., 1972, *In vitro* induction of meiotic division in follicle-enclosed rat oocytes by LH, cyclic AMP, and prostaglandin E_2, *J. Reprod. Fertil.* **31**:39.

Tsafriri, A., Koch, Y., and Lindner, H. R., 1973, Ovulation rate and serum LH levels in rats treated with indomethacin or prostaglandin E_2, *Prostaglandins* **3**:461.

Vale, W., Rivier, C., and Guillemin, R., 1971, A "prostaglandin receptor" in the mechanisms involved in the secretion of anterior pituitary hormones, *Fed. Proc.* **30**:363 (abst.).

Vale, W., Grant, G., Amoss, M., Blackwell, R., and Guillemin, R., 1972, Culture of enzymatically dispersed anterior pituitary cells: Functional validity of a method, *Endocrinology* **91**:562.

Van Sande, J., and Dumont, J. E., 1973, Effects of thyrotropin prostaglandin E_1 and iodide on cyclic 3′,5′-AMP concentration in dog thyroid slices, *Biochim. Biophys. Acta* **313**:320.

Vance, J. E., Buchanan, K. D., and Williams, R. H., 1971, Glucagon and insulin release. Influence of drugs affecting the autonomic nervous system, *Diabetes* **20**:78.

Varavudhi, P., and Chobdieng, P., 1972, Biological evidence for the direct stimulating effect of $PGF_{2\alpha}$ on the release of pituitary luteolytic agent(s) of pseudopregnant rats, *Prostaglandins* **2**:199.

Vermouth, N. T., and Deis, R. P., 1972, Prolactin release induced by prostaglandin $F_{2\alpha}$ in pregnant rats, *Nature (London)* **238**:248.

Weeks, J. R., 1971, Biological significance of prostaglandins with special reference to their effects on metabolism, *Naunyn-Schmiedebergs Arch. Pharmakol.* **269**:347.

Wolff, J., and Cook, G. H., 1973, Activation of thyroid membrane adenylate cyclase by purine nucleotides, *J. Biol. Chem.* **248**:350.

Wolff, J., and Jones, A. B., 1971, The purification of bovine thyroid plasma membranes and the properties of membrane-bound adenyl cyclase, *J. Biol. Chem.* **246**:3939.

Wolff, J., and Moore, W. V., 1973, The effect of indomethacin on the response of thyroid tissue to thyrotropin, *Biochem. Biophys. Res. Commun.* **51**:34.

Yang, N. S. T., Marsh, J. M., and LeMaire, W. J., 1973, Prostaglandin changes induced by ovulatory stimuli in rabbit Graafian follicles, The effect of indomethacin, *Prostaglandins* **4**:395.

Yu, S.-C., Chang, L., and Burke, G., 1972, Thyrotropin increases prostaglandin levels in isolated thyroid cells, *J. Clin. Invest.* **51**:1038.

Yue, D. K., Smith, I. D., Turtle, J. R., and Spearman, R. P., 1974, Effect of prostaglandin $F_{2\alpha}$ on the secretion of human prolactin, *Prostaglandins* **8**:387.

Zor, U., Kaneko, T., Lowe, I. P., Bloom, G., and Field, J. B., 1969a, Effect of thyroid-stimulating hormone and prostaglandins on thyroid adenyl cyclase activation and cyclic adenosine 3′,5′-monophosphate, *J. Biol. Chem.* **244**:5189.

Zor, U., Kaneko, T., Schneider, H. P. G., McCann, S. M., Lowe, I. P., Bloom, G., Borland, B., and Field, J. B., 1969b, Stimulation of anterior pituitary adenyl cyclase activity and adenosine 3′, 5′-cyclic phosphate by hypothalamic extract and prostaglandin E_1, *Proc. Natl. Acad. Sci. U.S.A.* **63**:918.

Zor, U., Kaneko, T., Schneider, H. P. G., McCann, S. M., and Field, J. B., 1970, Further studies of stimulation of anterior pituitary cyclic adenosine 3′,5′-monophosphate formation by hypothalamic extract and prostaglandins, *J. Biol. Chem.* **245**:2883.

Zor, U., Bauminger, S., Lamprecht, S. A., Koch, Y., Chobsieng, P., and Lindner, H. R., 1973, Stimulation of cyclic AMP production in the rat ovary by luteinizing hormone: independence of prostaglandin mediation, *Prostaglandins* **4**:499.

Zor, U., Chayoth, R., Kaneko, T., Schneider, H. P. G., McCann, S. M., and Field, J. B., 1974, Effect of hormone treatment and age on the stimulation of rat anterior pituitary adenylate cyclase and cyclic adenosine 3′,5′-monophosphate by hypothalamic extract, *Metabolism* **23**:549.

4

Induction of Labor with Prostaglandins

Tom P. Barden

Department of Obstetrics and Gynecology
University of Cincinnati Medical Center
Cincinnati, Ohio 45267

I. HISTORICAL BACKGROUND

The earliest reported observations of the biological activity of prostaglandins were those of Kurzrok and Lieb in 1930. They found that the effect of human semen on the motility of isolated nonpregnant uterine strips was influenced by the reproductive history of the patients. Semen relaxed strips from patients with previous successful pregnancy, and stimulated those from women with a history of sterility. Many subsequent studies revealed that the normal response of nonpregnant uterine strips to human seminal fluid is relaxation. Bergström and Sjövall (1957, 1960) were the first to isolate pure crystalline prostaglandins E_1 and E_2 from sheep vesicular glands. These and other workers (Bergström and Samuelsson, 1962; Bygdeman *et al.*, 1966; Bygdeman and Samuelsson, 1966; Hamberg and Samuelsson, 1965; Samuelsson, 1963*a,b*) soon identified 13 different prostaglandins in human seminal fluid. In 1964, Bygdeman reported that isolated strips of nonpregnant uterus responded to PGE compounds by relaxation. In contrast, most strips from pregnant uteri were stimulated by PGEs and both nonpregnant and pregnant strips were stimulated by PGFs. Sandberg *et al.* (1964) reported that the inhibitory effect of PGEs on nonpregnant myometrium *in vitro* was most pronounced on strips from the isthmus area, in contrast to those from the corpus. Embrey and Morrison (1968) working with pregnant myometrium *in vitro*, found that PGE_2 and $PGF_{2\alpha}$ are stimulatory to upper segment myometrium, but relatively inactive with lower segment strips.

As *in vitro* studies of prostaglandins progressed, there also was growing evidence of the physiological role of prostaglandins in parturition. In 1966, Karim identified prostaglandins E_1, E_2, $F_{1\alpha}$, and $F_{2\alpha}$ in human amniotic fluid obtained during labor. Karim and Devlin (1967) subsequently studied a large number of amniotic fluid samples throughout pregnancy. Prior to labor the fluid contained only PGE_1, but after the onset of labor $PGF_{1\alpha}$ and $PGF_{2\alpha}$ were also present. Karim (1968) reported the appearance of $PGF_{2\alpha}$ in blood of patients during labor, increasing levels during labor, and the highest levels shortly before and during uterine contractions. No prostaglandins were detected in blood prior to the onset of labor. Kloeck and Jung (1973) studied the production of prostaglandins by human myometrium *in vitro*. After myometrial stretching, the perfusion fluid contained increased levels of PGE and lower levels of PGF. These observations prompted exciting new hypotheses on the physiology of labor. Csapo (1973) and Csapo and Kivikoski (1974) suggested that prostaglandins are an intrinsic myometrial stimulant controlled by the degree of myometrial stretch. Their synthesis results in changes of uterine microcirculation which lead to reduction of fetoplacental endocrine function, provoking a regulatory imbalance which in turn releases inherent uterine contractility.

The results of *in vitro* studies led to early *in vivo* trials. In 1966, Embrey, as quoted by Pickles *et al.* (1966), instilled $PGF_{2\alpha}$ into the uterine cavity in two nonpregnant patients and subsequently observed "definite increase in uterine activity. . . ." This and more elaborate studies (Karim *et al.,* 1971; Roth-Brandel *et al.,* 1970) soon established that both PGE and PGF are stimulatory to the nonpregnant uterus *in vivo,* and that the response is not influenced by the phase of the menstrual cycle. These studies revealed that PGE_2 is approximately 10 times more potent *in vivo* then $PGF_{2\alpha}$. This observation applies when prostaglandins were administered by intravenous, intramuscular, subcutaneous, intravaginal, or oral routes.

The first report of the effect of prostaglandins on the pregnant uterus was by Bygdeman and associates in 1967. They observed that PGE_1 invariably stimulates uterine contractility in midpregnancy and at term. Most of the early *in vivo* studies were performed in patients undergoing midtrimester abortion. In such circumstances, Bygdeman *et al.* (1968) studied dose–response data to conclude that PGE_1 and PGE_2 were similar in potency. They observed a tendency for PGE to produce increasing uterine tone, as well as to increase the frequency and amplitude of contractions. They concluded that PGE would probably lack efficacy for termination of pregnancy at term because of the likelihood of hypertonus. Karim and Filshie (1970a) also observed that the initial response to intravenous infusion of PGE_2, 5 $\mu g/min$ in midpregnancy was an increase in tone to between 20 and 50 mm Hg, but within 20–30 min the tone decreased to 10–20 mm Hg and the frequency and amplitude of contractions increased. They concluded that the uterine activity induced by prostaglandins was similar to that of spontaneous abortion. In another report (Karim and Filshie, 1970*b*), the same workers reported that similar uterine activity was produced by an intravenous infusion of $PGF_{2\alpha}$, 50 $\mu g/min$, a dose 10 times that of PGE_2. Karim (1970) subsequently reported on successful use of PGE_2, intravenously, in the management of missed abortion, hydatidiform mole, and fetal death. More recent observations in these clinical applications will be considered later in this chapter.

Despite the early indications that prostaglandins might possess efficacy for termination of early pregnancy, further experience revealed that they do not compare favorably with vacuum curettage for interruption of first trimester pregnancy. During this period of pregnancy, intraamniotic injection of prostaglandin is prohibited by the small volume of fluid, while adequate dosage by intravenous infusion produces unacceptable side effects, especially nausea, vomiting, diarrhea, and local tissue reaction. Alternate routes of administration, such as oral, vaginal, and extraamniotic, were also found to produce excessive side effects (Embrey and Hillier, 1971; Karim and Sharma, 1971*b,c;* Wiqvist and Bygdeman, 1970).

In contrast to the lack of efficacy of prostaglandins in first trimester abortion, the technique of intraamniotic injection of $PGF_{2\alpha}$ or PGE_2 (Karim and Sharma, 1971*a*) rapidly became established as the method of choice for termination of second trimester pregnancy. Subsequent evidence indicated that coagulation disorders produced by intraamniotic injection of hypertonic saline (Stander *et al.,* 1971; Glueck *et al.,* 1973) were not triggered by prostaglandins (Brenner *et al.,* 1973; Glueck and Barden, 1975). In addition, the intraamniotic injection of prostaglandins produced few other side effects. To date, there is no evidence that other naturally occurring prostaglandins, or prostaglandin analogues, have a more favorable ratio of oxytocic action to side effects than $PGF_{2\alpha}$ for termination of second trimester pregnancy.

II. INDUCTION OF LABOR WITH INTRAVENOUS PROSTAGLANDINS

The use of prostaglandins for induction of labor in late pregnancy has remained somewhat controversial throughout its history, which started in 1967 with a series of reports from the Karolinska Institutet. Bygdeman *et al.* (1967, 1968) reported that the potencies of PGE_1 and

PGE_2 were similar, that the midpregnant uterus was somewhat more sensitive to both compounds than the uterus at term, and that intravenous infusion of PGE in mid or late pregnancy often produced an unphysiological elevation of uterine tone. Their conclusions were based on observations of uterine response to intravenous infusions of PGE at 4–8 μg/min. They concluded that these compounds would likely be more suitable for the treatment of uterine atony after delivery than for induction of labor. In a subsequent study, Bygdeman and associates (1970) reported that PGE_1 and PGE_2 were approximately 8 times more potent that $PGF_{2\alpha}$, but that the qualitative response of the uterus to PGEs and PGFs was similar. There was a consensus among investigators who reported at later dates (Karim *et al.,* 1970; Roth-Brandel, 1971; Roth-Brandel and Adams, 1970) that the intravenous dosage of PGE employed by Bygdeman and associates in these early studies exceeded the optimal rate required to produce laborlike uterine activity.

Karim and associates (1968) confirmed that $PGF_{2\alpha}$ was 5–10 times less active that PGE_2. They suggested that the threshold dose of $PGF_{2\alpha}$ was 5 μg/min, intravenously, while that of PGE_2 was 0.5 μg/min. In 1969, Karim and associates reported the successful induction of labor at or near term in 33 of 35 patients using intravenous infusion of $PGF_{2\alpha}$. Induction of labor was indicated for a variety of complications which included postmaturity, premature rupture of membranes, and fetal death. There were 5 nulliparas and 30 multiparas in the series. Uterine activity was monitored by an external tocographic method. $PGF_{2\alpha}$ was administered by intravenous infusion at the rate of 0.05 μg/kg/min. The average infusion to delivery time in the 33 successful cases was 9 hr and 35 min. There were no apparent side effects.

Embrey (1969) reported from the University of Oxford on a small number of term labor inductions using intravenous infusions of PGE_1, PGE_2, or $PGF_{2\alpha}$. He concluded that infusions of $PGF_{2\alpha}$, 2–8 μg/min, or PGE_2 2–6 μg/min, produced qualitatively similar uterine activity. The oxytocic effect of prostaglandins persisted for 30–60 min after the infusion was stopped. In subsequent reports, Embrey (1970*a,b,* 1971) reported almost uniform success of inductions at or near term using infusions of PGE_2, 0.5–2 μg/min. In these studies, uterine activity was recorded using an abdominal wall transducer or an extraamniotic intrauterine balloon. There were no instances of uterine hyperstimulation and only an occasional episode of vomiting.

Studies at the Karolinska Institutet in Stockholm proceeded from the initial reports of Bygdeman *et al.* (1967, 1968) to more elaborate observations by Roth-Brandel and Adams (1970). The latter, dose-finding study employed intravenous infusions of PGE_1, 0.4–1.2 μg/

min, PGE_2 0.7 μg/min, and $PGF_{2\alpha}$, 3.0–22.5 μg/min. They concluded that prostaglandin-induced uterine activity was characterized by relatively low-intensity, high-frequency contractions, often with incoordinated contraction complexes. Individual patient uterine response and side effects were noted to vary within a wide range. Although the study did not compare prostaglandins to oxytocin, the authors concluded that there was no evidence that prostaglandins are superior for induction of labor at term pregnancy. In a subsequent report, Roth-Brandel (1971) compared the effects of higher doses of PGE_1, ranging from 2.5 to 10 μg/min, in midpregnancy vs. term. Uterine activity was recorded from an intrauterine catheter. He concluded that the individual uterine response varied considerably from one case to another. The maximum dose which could be administered without major subjective side effects varied within a wide range, but generally the intensity of contractions was higher at term than in midpregnancy. He was unable to detect a correlation between the maximum tolerable dose of prostaglandin and the degree of induced uterine activity.

Karim *et al.* (1970) reported from Uganda on induction of labor with PGE_2 in 50 patients at or near term with various obstetrical and medical indications for induction of labor. An extraamniotic catheter was employed to record uterine activity. PGE_2 was administered intravenously at a fixed rate of 0.5 μg/min. In all cases, there was a resulting increase of uterine activity, generally starting in less than 15 min. The uterine response often was uncoordinated at first, but became more regular as the infusion continued. There were occasional episodes of transient hypertonus, but a similar pattern was seen in 4 out of 12 patients studied during spontaneous labor. Forty-six of the 50 patients in this series progressed to vaginal delivery with an average infusion to delivery time of approximately 10 hr. Side effects of vomiting and headache were rare and mild. The majority of patients in this study were African Bantu, who in contrast to most other ethnic groups have relatively late descent of the fetal presenting part during labor. They typically had a high, free fetal head and a long, closed cervix at the onset of the induction of labor. In this report, Karim was critical of Bygdeman's earlier conclusions that PGE_2 was not suitable for induction of labor because of the narrow therapeutic dosage range.

Beazley *et al.* (1970) reported from London on successful induction of labor at term with intravenous infusion of PGE_2 in 38 of 40 patients. As in Karim's study, patients were entered into the study for indicated induction of labor. The majority had toxemia or were post term, but there were also three with fetal death and five with abnormal glucose tolerance. Uterine activity was recorded from an extraamniotic catheter, inserted through the cervical os. Using fixed and variable infusion schedules, they gave PGE_2 in a dosage range from 0.5 to 2.0 $\mu g/min$, intravenously. Optimal uterine activity was easily achieved, but there was a fourfold difference in the required infusion rates. There was one episode of transient hypertonus, but no other side effects.

Roberts (1970) of Cardiff reported uniform success of induction of labor at term by intravenous PGE_2 or $PGF_{2\alpha}$ in a small series of patients. In contrast to previous studies, membranes were artificially ruptured before starting the infusion. One patient accidentally received PGE_2, 18 $\mu g/min$. The resulting uterine hypertonus resolved within 8 min of cessation of the infusion. In the same report, Roberts described the lack of antidiuretic effect of $PGF_{2\alpha}$, in contrast to oxytocin. The experiment was performed after water loading six patients who had been hospitalized for termination of pregnancy. An infusion of oxytocin, 128 mU/min, produced a pronounced antidiuresis which persisted for 4 hr after the infusion had been stopped. No antidiuresis was observed during infusion of $PGF_{2\alpha}$, 0.05 $\mu g/kg/min$. In a continuation of this study, Roberts and Turnbull (1971) reported on a series of 35 cases of induction of labor at term with infusions of PGE_2 or $PGF_{2\alpha}$. As before, membranes were ruptured initially and uterine activity was recorded from an intrauterine catheter. With experience, the investigators found the effective dose range of PGE_2 to be 0.75–1.5 $\mu g/min$. During the study it was noted that the infusion could be stopped after labor was well established. Most of the patients in this series developed local inflammation at the infusion site. In addition to the hypertonus produced by accidental overdosage of PGE_2, described in the earlier report, there were three more instances of uterine hypertonus during infusions of PGE_2. In each case of hypertonus, there was no alteration of fetal heart rate, and uterine activity decreased to normal within 5 min of stopping the infusion. In one patient, the hypertonus resolved while the infusion continued. The authors concluded that the margin of safety between an effective oxytocic dose and a dangerous overdose might be narrower with PGE that with oxytocin. Gillespie *et al.* (1971) commented that the frequency of uterine hyperactivity encountered by Roberts and Turnbull was probably due to the initial artificial rupture of membranes, which they suggested would increase uterine sensitivity to oxytocic drugs. In their earlier double-blind trial of PGE_2 and oxytocin (Beazley and Gillespie, 1971), they had found it necessary to reduce the infusion rate after spontaneous rupture of membranes to avoid uterine overstimulation.

Craft *et al.* (1971) compared PGE_2 with oxytocin for induction of labor at term in 30 patients. All patients had intact membranes. In contrast to other studies, each patient received epidural injections of bupivacaine for analgesia during the study. The patient groups were not matched for age, parity, or cervical status, and the study was not double-blind. The infusions of oxytocin were started at 1 mU/min and doubled every 10 min until contractions were detected, and then doubled every 20 min to achieve optimal activity. PGE_2 was started at 0.5 μg/min for 30 min, then increased every 15 min to 0.75, 1.0, 1.5, 2.0, 3.0, and 4.0 μg/min, stopping at the dose producing adequate uterine activity. When successful, the average induction to delivery time of each drug was approximately 9 hr. About half the patients receiving PGE_2 vomited during labor, and many had erythema and soreness of the infusion site. After 12 hr of infusion, all 15 patients given PGE_2 had progressively dilated, but in only 9 of 15 given oxytocin had there been cervical dilatation progress.

Beazley and Gillespie (1971) reported on a double-blind study of PGE_2 and oxytocin for induction of term labor in 300 patients. As in earlier studies, the patients were entered into the study on a basis of various obstetrical complications which necessitated induction of labor. Uterine activity was recorded from an extraamniotic catheter. The investigator was not aware of the choice of drug for each patient. The dosage schedule started at the intravenous infusion of 2.1 mU/minute of oxytocin or 0.21 μg/min of PGE_2, and the dose was doubled every hour until satisfactory uterine activity was achieved, or to maximal levels of 67 mU/min or 6.7 μg/min, respectively. Membranes were ruptured artificially when uterine activity became laborlike. By defining success as achieving cervical dilatation of 6 cm within 12 hr, there was 73% success in each of the groups. The induction to delivery times varied widely with more brief labors but also with more long labors in the oxytocin-induced group. One intrapartum fetal death occurred after induction of labor with PGE_2 for prolonged pregnancy. There was no indication that it resulted from unphysiological uterine activity. No uterine hypertonus was reported in this study. Gillespie (1972) studied the clinical variables that might have influenced the effective dose of oxytocin or PGE_2 in the 300-patient trial. He found no correlation of success with maternal weight, parity, or gestation.

From these early studies of prostaglandin for induction of labor in late pregnancy, it was possible to conclude that its oxytocic properties might have clinical efficacy, but it was yet to be established if these compounds had any advantage over synthetic oxytocin. It was becoming evident that the pharmacological properties of prostaglandins were distinctly different than those of oxytocin. The uterine response to a fixed dose of prostaglandin developed over a longer period than that resulting from oxytocin. There was growing evidence that the dosage range from initial response to uterine hyperstimulation might be narrower for prostaglandins than for oxytocin. But there was no evidence that the uterine response to an arbitrary dose of either drug was more predictable. Experimental design had been quite variable among investigators. The patient population varied considerably in size and characteristics. Most of the patients studied had been chosen for induction due to obstetrical complications, many of which were known to influence uterine activity. The results of studies were almost certainly influenced by clinical variables such as the preinduction clinical status of patients, timing of amniotomy, position of patients during labor, frequency of examinations, and use of concomitant drugs. Monitoring techniques had varied and in some instances may have influenced uterine response to drugs. Even when the population and techniques were similar, the criteria of success or failure were not uniform among studies. Although there had been occasional instances of uterine hyperstimulation by prostaglandins, the response was easily reversed by slowing or stopping the infusion. Other known side effects of prostaglandins were infrequent at the dosage levels necessary for induction of labor in late pregnancy. Thus the early studies had served to define the effective dose range of prostaglandins for induction of labor, but further studies were needed to establish efficacy.

In 1971, Karim (1971a) reported on a double-blind clinical trial of PGE_2, $PGF_{2\alpha}$, and oxytocin for induction of labor in 300 patients. The dosage range of the respective drugs was 0.3–1.2 μg/min for PGE_2, 2.5–10 μg/min for $PGF_{2\alpha}$, and 2–8 mU/min for oxytocin. The report failed to describe dosage schedules, side effects, or criteria of success. Induction was successful in 96% of patients given PGE_2, 67% of those given $PGF_{2\alpha}$, and only 56% of the oxytocin cases.

At the same conference, presented by the New York Academy of Sciences, Anderson *et al.* (1971) of Yale presented a preliminary report of their experience in a double-blind study comparing oxytocin with PGE_2 and $PGF_{2\alpha}$ for induction of labor. They were the first to suggest that studies of these drugs should utilize an inducibility scoring system. They referred to the system described by Bishop (1964) in which a score from 0 to 13 is based on cervical effacement, dilatation, position, and consistency, and the station of the fetal presenting part. The higher the score, the more factors favor ease of induction. In the Yale study, approximately half of the study patients were admitted for elective rather than indicated induction of labor. The study was confined to multiparous patients. Membranes were uniformly intact at the onset of study, and not ruptured until labor was well established. The dosage schedule of the drugs was twice revised during the study. The final schedule was 2.5, 5.0, 10.0, 20.0, and 40.0 μg/min of $PGF_{2\alpha}$, 0.3,

0.6, 1.2, 2.4, and 4.8 μg/min of PGE_2, and 1, 2, 4, 8, and 16 mU/min of oxytocin, with the dosage advanced at the respective levels after 30 min, 30 min, 1 hr, 4 hr, and 4 hr, until there was a satisfactory uterine response, or a complication necessitating a reduction or cessation of the infusion. Early in the study, PGE_2 was withdrawn because of instability of the preparation, and therefore the results of the study as subsequently reported (Anderson *et al.*, 1972) concerned $PGF_{2\alpha}$ and oxytocin. The labor characteristics were indistinguishable between the two groups. There were five instances of transient hypertonus in the $PGF_{2\alpha}$ group and none in the oxytocin group. There were no other side effects attributable to the drugs. Labor induction was successful in 85% of the oxytocin group and in 80% of the $PGF_{2\alpha}$ series. They concluded that $PGF_{2\alpha}$ was as effective as synthetic oxytocin when contrasted among patients with similar inducibility characteristics. There was some evidence that the uterus was more sensitive to alteration of dosage rates of $PGF_{2\alpha}$ than to oxytocin.

In contrast to these results, a similar study of term inductions with oxytocin vs. $PGF_{2\alpha}$ was reported by Spellacy and Gall (1972). The oxytocin dosage schedule was reduced to 0.5, 1, 2, 4, and 8 mU/min, but $PGF_{2\alpha}$ dosage was identical to that in the Yale study, as were all other apparent features of the study protocol. These investigators studied 222 patients admitted for elective induction of labor at term pregnancy. During the drug infusions, there was hypertonus in 23.6% of $PGF_{2\alpha}$ cases and 8.4% of oxytocin cases. There were nine (7.8%) caesarean sections for fetal distress in the prostaglandin group and none in the oxytocin group. But the incidence of fetal bradycardia was identical (14.9%) in both groups. There was no difference in the incidence of side effects, length of labor, or neonatal outcome between the groups. The overall success of induction was 74.5% of the $PGF_{2\alpha}$ group and 66.4% of the oxytocin group.

Rangarajan *et al.* (1971) reported on induction of term labor with intravenous $PGF_{2\alpha}$ vs. oxytocin in 40 patients who were distributed into two groups on a basis of cervical score prior to the induction. Uterine activity was recorded from an abdominal wall transducer. The dose of oxytocin was from 1.25 to 10.0 mU/min, with the rate doubled if necessary to achieve laborlike uterine activity, at 15–30 min intervals. $PGF_{2\alpha}$ was infused at 2.5–20.0 μg/min, with the dose doubled, as necessary, every 15–30 min. Prolonged uterine contractions were observed in eight (40%) of the patients who were given $PGF_{2\alpha}$. In two of these, there was associated fetal bradycardia which subsided after decreasing the rate of infusion. Successful induction of labor occurred in 19 of 20 patients who received oxytocin, but only 16 of 20 in the

$PGF_{2\alpha}$ group. However, the mean induction to delivery interval was shorter in patients with an unfavorable cervix who received $PGF_{2\alpha}$ than in those given oxytocin.

Vakhariya and Sherman (1972) reported on a double-blind study, comparing oxytocin and $PGF_{2\alpha}$, for elective induction of labor in 100 multiparas at term. Oxytocin was infused in the dosage range of 0.8–16 mU/min, with the dose doubled if necessary every 30 min. $PGF_{2\alpha}$ was administered in a similar fashion in the dosage range of 2.0–40.0 µg/min. Induction to delivery times were comparable in the two groups based on cervical condition at the onset of induction, with no significant differences between the drugs. Unexpected uterine hypertonus was observed in seven patients receiving $PGF_{2\alpha}$, contrasted to none in the oxytocin group. The condition of the newborns was comparable between the two groups. It was concluded that $PGF_{2\alpha}$ was as effective and safe as oxytocin for the induction of labor in multiparas at term.

Brown and associates (1973) compared oxytocin, $PGF_{2\alpha}$, and PGE_2 for induction of labor in 170 patients at term pregnancy. Groups were matched for age, parity, and gestation, but not for inducibility score. The dosage schedule was 0.25, 0.5, and 0.75 µg/min of PGE_2 and 2.5, 5.0, and 7.5 µg/min of $PGF_{2\alpha}$, with the dosage advanced after 30 min and 4½ hr respectively. Oxytocin was started at 2.8 mU/min and increased by 1.4 mU/min every 15 min during the first hour, 3.5 mU/min during the second hour, and 7 mU/min during the third hour, to a maximum of 49 mU/min. As in some of the earlier studies (Roberts, 1970; Roberts and Turnbull, 1971), amniotomy was performed before starting the infusion of drug. There was mild phlebitis at the infusion site of ten patients who received prostaglandin, but no other side effects. Hypertonus was not observed, but labor was monitored only by intermittent clinical observations. The average length of labor was approximately 8 hr, with no significant differences among groups. Induction was successful in 83% of the PGE_2 cases, 69% of the $PGF_{2\alpha}$ cases, and 88% of the oxytocin group. It was suggested that early amniotomy may serve to lower the effective dosages, and thus reduce the incidence of hypertonus and other side effects.

Naismith *et al*. (1973) reported on a series of 40 patients induced at term by oxytocin, PGE_2, or $PGF_{2\alpha}$. All were primigravidas who were at least 7 days past the estimated date of delivery. The study was not performed in double-blind fashion. The dosage schedule was oxytocin starting at 2.66 mU/min, with the dose doubled every 15 min to a maximum of 340 mU/min. PGE_2 and $PGF_{2\alpha}$ were started at 0.25 and 2.5 µg/min, respectively, with the dose doubled every 30 min to maximum levels of 8.0 and 80 µg/min. The groups were matched for age, gestation, and inducibility scores. Amniotomy was performed after labor was well established. Among the 20 patients who received prostaglandin, five failed to progress in 12 hr. Nineteen of the 20 patients given oxytocin made satisfactory progress in labor. There were no significant side effects in the study. It was concluded that oxytocin may have greater efficacy that PGE_2 or $PGF_{2\alpha}$ for induction of labor in primigravidae.

Johnson *et al.* (1974) compared induction of labor in 91 patients at term with amniotomy followed by intravenous $PGF_{2\alpha}$ to a retrospective group of patients matched for age, parity, and pelvic score, who were induced by amniotomy and intravenous infusion of oxytocin. They concluded that $PGF_{2\alpha}$ was as effective as oxytocin for the induction of labor. In contrast to most other studies comparing the drugs, they found less vomiting in patients receiving $PGF_{2\alpha}$ than in those given oxytocin.

Clegg *et al.* (1974) evaluated PGE_2 in 40 consecutive patients requiring augmentation of labor. The group was matched for age, parity, gestation, cervical dilatation, and analgesia with 40 patients given oxytocin for augmentation of labor. All patients in the series had ruptured membranes for at least 2 hr. PGE_2 was started at 0.285 μg/min for a minimum of 30 min, and then the dose was doubled every hour until adequate contractions were achieved. Oxytocin was started at 2 mU/min and the dose was doubled every hour as needed. In all cases, there was additional progress after the drug infusion was started, but those receiving prostaglandin required significantly less time and had less uterine activity to progress to delivery. With a continuous monitoring of all cases, there was twice the incidence of hypercontractility in the oxytocin group than in the PGE_2 group, 13 vs. 6. However, these responses may have been influenced by concomitant use of epidural analgesia in some of the patients. The results of the study suggest somewhat greater efficacy for PGE_2 than oxytocin for stimulation of labor.

Sharma *et al.* (1975) studied two dosage schedules of intravenous $PGF_{2\alpha}$ for induction of labor in 91 patients at term. After all patients received a dose of 2.5 μg/min for 1 hr, a low-dose group was given 4 hr increments of 5.0 and 10.0 μg/min, while a high-dose group was advanced through the same schedule at 1 hr intervals. In both groups, the maximum rate of infusion was 20 μg/min. Amniotomy was performed only after labor was well established with the cervix dilated to 3 cm or more. After amniotomy, uterine activity was monitored by an intrauterine catheter. Uterine hypertonus occurred in two patients, both in the high-dose group, and one case was complicated by fetal anencephaly. There were few gastrointestinal side effects encountered in either group. Labor was successfully induced in approximately 90% of the patients in each dosage group.

Calder and Embrey (1975) reported a double-blind trial of intravenous oxytocin or PGE₂ for induction of labor at term in 100 primigravidas with unfavorable pelvic scores. Thirty minutes after initial amniotomy, oxytocin or PGE_2 was infused intravenously in a dosage range of 8–64 mU/min or 0.1–4.0 μg/min, respectively, with the dose doubled as necessary every 30 min to achieve laborlike uterine activity. Uterine activity was monitored by an intrauterine catheter. Hypertonus, defined as intrauterine pressure over 30 mm Hg for over 2 min, was observed in two patients of each group. Induction to delivery time correlated with the initial pelvic scoring but there was no difference in the effectiveness of the drugs between patients of comparable groups. Although not statistically significant, there was a suggestion that PGE_2 was more effective in patients with unfavorable pelvic scores. Maternal temperature was over 37.5°C in 22 of 50 patients receiving PGE_2, in contrast to 7 of 50 in the oxytocin group. These investigators concluded that 1 U of oxytocin is equivalent to 85 μg of PGE_2. Anderson and Schooley (1975) studied the characteristics of uterine activity, recorded from an intrauterine catheter, in spontaneous and oxytocin- or $PGF_{2\alpha}$-induced labors. There were no significant differences in frequency, intensity, and duration of contractions during labor among 20 patients in each group.

Among recent reports concerning induction of labor with oxytocin or prostaglandins, there have been several related to side effects and potential complications. Gowenlock and associates (1975) studied the biochemical and hematological responses of 75 patients induced near term by intravenous oxytocin, PGE_2, or $PGF_{2\alpha}$. In the oxytocin group, there was a significant rise of plasma cortisol, not seen with prostaglandin infusions. Also, in the oxytocin group, there was evidence of water retention, not found in the prostaglandin group. There were no significant differences among the groups for platelet aggregation, hemoglobin, or blood cell count; serum creatinine, potassium bilirubin, GOT, or GPT. Calder et al. (1974b) studied the possible relationship of induction of labor by oxytocin or PGE_2 to bilirubin levels in the neonates. Despite comparable birth weight and gestational age of neonates in the induced vs. noninduced labor groups, there was a significant increase in hyperbilirubinemia in infants after either oxytocin or PGE_2 induction of labor. In a prospective study, Beazley and Alderman (1975) found no evidence that the use of oxytocin in labor increased the frequency of neonatal hyperbilirubinemia, but they did find that the incidence of hyperbilirubinemia increased sharply when the total dose of oxytocin exceeded 20 U. Blackburn *et al.* (1973) compared groups of neonates following induction of labor with $PGF_{2\alpha}$ or oxytocin. They found no significant differences of heart rate, blood pressure, electroencephalograms, blood gas acid–base status, blood counts, serum bilirubin, serum glucose, or sleep–wake patterns between the two groups. Roberts *et al.* (1970) compared the effect of intravenous infusion of oxytocin and $PGF_{2\alpha}$ on induced diuresis in a group of patients prior to midtrimester abortion. There was no effect of $PGF_{2\alpha}$ on renal function but oxytocin produced a marked antidiuresis.

Brudenell and Chakravarti (1975) reported on five cases of uterine rupture in labor, of which, three patients were being induced with oxytocin, one was being induced with PGE_2, and one was in spontaneous labor. Three of the patients had experienced a previous caesarean section. Lyneham and associates (1973) suggested a causal relationship between $PGF_{2\alpha}$ administered intraamniotically for termination of pregnancy, and the occurrence of convulsions and pathological electroencephalographic patterns. In two subsequent reports (Thiery *et al.*, 1974a; Van der Plaetsen *et al.*, 1974) from Gent, Belgium, a series of patients were studied by electroencephalographic techniques and found to have no changes during the administration of PGE_2.

Another potential complication of prostaglandin administration is related to changes in airway resistance. Kreisman *et al.* (1975) demonstrated significant airway obstruction in patients given intraamniotic $PGF_{2\alpha}$ or 15(S)-15-methyl PGE_2 for termination of midtrimester pregnancy. It is not apparent from the clinical experience to date that prostaglandins are capable of provoking acute asthma in asthmatics, when used in the dosage range appropriate for induction of labor. Anderson *et al.* (1972) noted that nine patients in their study had asthma, but none developed symptoms of respiratory distress during the infusion of prostaglandins.

Spellacy *et al.* (1971) investigated the influence of oxytocin or prostaglandin infusion on blood glucose and insulin levels. In the dosage levels appropriate for induction of labor at term, there were no significant changes of blood glucose and insulin during the infusions. Karim *et al.* (1971) reported that intravenous infusions of 40 μg/min PGE_2 and 300 μg/min $PGF_{2\alpha}$ produced no effect on heart rate or blood pressure of pregnant women. Fishburne *et al.* (1972) detected no changes of cardiac output, central venous pressure, arterial pressure, or heart rate of women during infusions of 50–200 μg/min of $PGF_{2\alpha}$. Novy *et al.* (1975) studied the effects of oxytocin and PGE_2 in pregnant rhesus monkeys. In the doses studied, which were sufficient to produce laborlike uterine activity, oxytocin and PGE_2 had virtually no effects on cardiac output or hemodynamic parameters.

We may conclude that the evidence now suggests comparable efficacy and safety of PGE_2 or $PGF_{2\alpha}$ and synthetic oxytocin, administered by intravenous infusion, for the induction of labor in late human pregnancy. The facility of induction of labor with either agent correlates closely with the initial pelvic inducibility status. Several studies have suggested, but not proven, that prostaglandins are more effective than oxytocin for induction of labor in patients with unfavorable inducibility status. Although evidence suggests that amniotomy performed before or early in the drug infusion will tend to shorten the induction to delivery time, it may also contribute to the development of excessive uterine activity and fetal distress. The earlier concern over prostaglandin-induced uterine hypertonus has subsided as further experience has revealed its infrequent occurrence with revised dosage schedules. The best results of prostaglandins for induction of labor have been observed in the studies using a narrow dose range with longer dose increments. Evidence now suggests that the dose–response characteristics of prostaglandins and oxytocin are distinctly different. With prostaglandins the optimal dose range is narrower, and the response time is longer. With the exception of the antidiuretic effect of oxytocin, which is not produced by administration of prostaglandins, maternal and neonatal cardiovascular, hematological, and biochemical responses to the two agents are apparently quite similar.

III. OTHER ROUTES OF ADMINISTRATION

In an effort to reduce the side effects associated with the use of intravenous prostaglandin, others have evaluated alternate routes of administration. In an extension of their earlier studies (Embrey, 1969, 1970a,b, 1971; Embrey and Hillier, 1971), the group at Oxford subsequently reported on the extraamniotic infusion of PGE_2 for the induction of labor at term in 40 patients (Calder *et al.*, 1974a). The infusions were started at 20 μg/hr, and increased every 15 min by 10 μg/hr, until labor was established, with a maximum rate of 150 μg/hr. Amniotomy was performed only after labor was well established. Other than an occasional instance of vomiting, there were no side effects. In the series of 40 patients, induction of labor was uniformly successful, with an average induction to delivery time of 9.3 hr. The same group of investigators (Embrey *et al.*, 1974) reported that extraamniotic infusion of PGE_2 or $PGF_{2\alpha}$ successfully terminated 9 of 10 cases of missed abortion, 12 fetal deaths, 2 anencephaly, and 1 hydatidiform mole. They used a 14–26 French-gauge Foley catheter, with a 25–50 ml balloon, inserted into the extraamniotic space through the cervical os.

PGE$_2$ was infused at 30 μg/hr initially, and increased by 15 μg/hr every 15 min until effective uterine action was established. The maximum dose ranged from 60 to 240 μg/hr.

Miller and Mack (1974) evaluated the use of extraamniotic infusion of PGE$_2$ in 69 patients, ranging from 26 to 41 weeks gestation. Intravenous oxytocin was used to complement the effect of PGE$_2$ in about one-third of the cases. Labor was successfully induced in all patients. The series included four cases of anencephaly. The mean induction to delivery time in those with normal infants was 14 hr, 48 min in primigravidas, and 9 hr, 35 min in multiparas. Gastrointestinal side effects were infrequent.

Despite the degree of success and lack of complications reported for the use of extraamniotic route for administration of prostaglandins, many clinicians are reluctant to invade the uterine cavity through the cervical os because of potential for infection. Thus there have been many subsequent studies to evaluate other routes of administration. As early as 1968, Wiqvist *et al.* of Stockholm reported no uterine effects of intravaginal administration of PGE$_2$ in a dosage range of 200–1000 μg. Subsequently, Karim (1971*a*) and Karim and Sharma (1971*c*) reported on the successful use of intravaginal prostaglandins for termination of 45 first and second trimester pregnancies, and for induction of labor at term in ten patients. They used lactose tablets impregnated with either 50 mg PGF$_{2\alpha}$ or 20 mg PGE$_2$. The vaginal tablets were inserted every 2½ hr. In the abortion group, 40 of 45 were complete, with the average induction to abortion interval approximately 13 hr. There were few side effects. In ten patients at term, they used either 2 mg PGE$_2$ or 5 mg PGF$_{2\alpha}$, intravaginally every 2 hr. All were successfully induced and there were no complications.

At present, intraamniotic injection of prostaglandins is the most efficacious and safe method for termination of second trimester pregnancy. However, the technique of intraamniotic injection is often impractical in cases of missed abortion or fetal death, due to a reduced volume of amniotic fluid. It is well recognized that if spontaneous labor does not develop within the 4 weeks following fetal death, there is a subsequent risk of hypofibrinogenemia (O'Driscoll and Lavelle, 1955). In addition, the mental duress of a patient with a dead fetus often prompts earlier intervention. Karim (1970) reported successful treatment of missed abortion and fetal death with intravenous PGE$_2$. In six cases of missed abortion in which PGE$_2$, 5 μg/min, was given, the pregnancies terminated in 2½–14 hr. Some of the patients experienced vomiting. Patients with fetal death were given PGE$_2$, 0.5–2.0 μg/min, which produced no side effects, and 14 of 15 successfully delivered in an average of 12 hr. A similar degree of success was reported by

Filshie (1971), who terminated seven fetal deaths, five missed abortions, and one hydatidiform mole using intravenous prostaglandin E_2, 0.5–2.5 μg/min. There were few side effects and delivery occurred in from 1½ to 19 hr. Pedersen *et al.* (1972) reported an induction to termination interval of 8–12 hr and only mild side effects with intravenous $PGF_{2\alpha}$ to induce seven missed abortions, six intrauterine fetal deaths, and four anencephalic fetuses. Gordon and Pipe (1975) compared oxytocin and PGE_2 for induction of labor after fetal death in 30 patients. The treatment was considered successful in 12 of 15 patients in each group. However, the average induction to delivery interval was 21.4 hr for the oxytocin group, and only 11.1 hr for the prostaglandin group. In a recent report from Yale University (Bailey *et al.*, 1975), intravaginal PGE_2 was employed to successfully terminate 20 patients with either missed abortion or fetal death. The suppositories, containing 20 mg PGE_2, were placed intravaginally, every 30 min to 8 hr on a basis of uterine response and patient tolerance. The average treatment to delivery time was 8.7 hr, with a range of 2.5–20.6 hr. Side effects were infrequent and mild, consisting of nausea, vomiting, diarrhea, headache, and pyrexia. In contrast, when the same dosage schedule of intravaginal PGE_2 was used in a series of 40 patients, with pregnancies being terminated between 16 and 20 weeks gestation, Bolognese and Corson (1974) reported that 75% had vomiting and/or diarrhea, and 55% had elevation of temperature over 100°F. A multicenter evaluation of intravaginal PGE_2 for termination of pregnancy complicated by missed abortion, hydatid mole, or fetal death is currently in progress in the United States.

Karim (1971*b*) was the first to report the effects of oral administration of PGE_2 and $PGF_{2\alpha}$ on the human uterus. In four patients in late pregnancy with fetal death, oral administration of 0.5 mg PGE_2 or 5 mg $PGF_{2\alpha}$ resulted in uterine contractions indistinguishable from normal spontaneous labor. The initial experience in earlier pregnancy revealed that effective uterine contractions developed following oral administration of 2.5 mg. PGE_2 or 25 mg $PGF_{2\alpha}$. Higher doses of either prostaglandin resulted in watery diarrhea and vomiting. In subsequent reports, Karim and Sharma (1971*b*, 1972) evaluated oral PGE_2 in 1000 patients in late pregnancy. The series consisted of 764 patients for induction of labor and 236 for stimulation during labor. PGE_2 was given orally, 0.5–2.0 mg/2 hr, for up to 48 hr. The patients were almost all African Bantus, who typically have a long, closed cervix and nonengagement of the fetal presenting part at onset of labor. Eighty-five percent of the patients had an inducibility score of 5 or less. Of the entire series, 65 patients developed nausea and 34 had diarrhea. In the labor induction group, 560 of 764 or 73.4% delivered vaginally within 24 hr, and 90% delivered within 48 hr.

The efficacy of oral PGE_2 for induction of labor was soon confirmed by other studies. Craft (1972) reported on 50 patients, 46 of whom were successfully induced by oral PGE_2 within 24 hr. The patients were given an oral solution containing 0.5 mg PGE_2 initially, and if necessary at 60 min, and then every 2 hr until delivery. Gastrointestinal side effects occurred in 18 of 50 patients, but they were generally mild. Barr and Naismith (1972) reported on induction of labor in 24 patients at term. In the ten given 0.5 mg PGE_2 initially, 1.0 mg at 30 min, and then 1 or 1.5 mg every 2 hr, there were no significant side effects except for one instance of hypertonus associated with fetal heart-rate bradycardia. Of 14 patients given oral $PGF_{2\alpha}$, nine had significant gastrointestinal side effects.

Corson and Bolognese (1974) reported uniform success in induction of labor in 20 patients given 0.5 mg of PGE_2 orally every hour, or with hourly increments of 0.5 mg until labor was satisfactory. Mean time to delivery was 5 hr, 47 min, with a range of 1 hr, 35 min to 12 hr, 13 min. Thiery *et al.* (1974*b*) reported successful induction of labor at term in 46 of 47 patients. They used 0.5 mg of PGE_2 initially, and then increased variably up to 2 mg, repeating the dose every 1-2 hr. Average induction to delivery time was 6 hr, 48 min for primigravidas and 4 hr, 21 min for multiparas. Murnagham *et al.* (1974) reported uniform success of induction of labor in 50 patients at term with oral PGE_2 after amniotomy. Monitoring intrauterine pressure, they were impressed by the low-intensity contractions which produced good progress. Basu and Rajan (1975) reported 96% success of induction in 100 patients given oral PGE_2 after amniotomy. There were 46 primigravidas in this series of patients. Yip *et al.* (1973) reported 80% success of induction of labor at term in 57 patients. In their series, 9 of 36 (25%) primigravidas were failures. Wilkin *et al.* (1974) reported successful term induction of labor in 18 of 20 patients with PGE_2 with a dosage of one 0.5-mg tablet orally every hour.

Kelly *et al.* (1973) compared oral PGE_2 and intravenous oxytocin for induction of labor in 98 patients. When possible, they ruptured membranes at least 30 min prior to starting the drug treatment. Oral PGE_2, 0.5 mg, was given initially and followed in 30 min by 0.5–2.0 mg, repeated every 2 hr until delivery. Oxytocin was administered by intravenous infusion, doubling the dose as needed every 30 min. The average dose of oxytocin was 16.2 mU/min, and the range was 1.4–130.6. "Incoordinate uterine action" was defined as resting tone over 15 mm Hg, contractions longer than the relaxation phase, and irregular and frequent contractions. All components of "incoordinate uterine action" were statistically more prevalent in the oxytocin group. Successful induction of labor was accomplished within 24 hr in 47 of 49 in the PGE_2 group, and in 46 of 49 in the oxytocin group. The frequency of fetal distress was comparable in the two treatment groups. Marzouk (1975) reported comparable success of induction, but a significant difference in side effects when he compared oral and intravenous PGE_2. The intravenous infusion was started at 0.25 μg/min, and gradually increased as high as 2.0 μg/min. The oral dosage schedule started at 0.5 mg and gradually increased to 2 mg every 2 hr, if necessary. Fifteen percent of 72 patients in the oral medication group had gastrointestinal side effects, but in contrast there were virtually no side effects in the intravenous group. Read and Martin (1974) also compared oral PGE_2 and intravenous oxytocin in a series of 187 parous patients at term. The mean induction to delivery time was 4 hr, 35 min and there was no clinically observable difference between the two groups. In contrast, Miller *et al.* (1975) in a similar study of 151 patients concluded that oral PGE_2 was comparable to intravenous oxytocin in patients with a high inducibility score, but that intravenous oxytocin was more efficacious when the score was low.

Obel and Larsen (1975) compared PGE_2 in liquid vs. oral tablet form. There was no significant difference in the results, but the tablet form was more convenient to administer. Elder and Stone (1974) subsequently confirmed the comparability of the liquid and tablet formulations of PGE_2.

Friedman and Sachtleben (1974) studied three dosage regimens of oral PGE_2 for induction of labor at term to evaluate the patterns of resulting cervical change. Those given 0.5 mg PGE_2 every hour had labor patterns closely resembling spontaneous labor at term; whereas those given 1 mg every hour, on an incremental increase of 0.5 every 30 min to a maximum of 3.0 mg, had dilatation and descent patterns characterized by shorter phases and more steeply inclined slopes. Induction was successful in 39 of 45 (86.7%) of patients. There was detectable hypertonus in six patients, and gastrointestinal side effects occurred in 40% but were clearly dose related. Fraser (1974) also reported a case of uterine hypertonus associated with oral PGE_2. The episode clinically resembled abruptio placentae, and occurred 80 min after the multiparous patient ingested 0.5 mg PGE_2. Lauersen and Wilson (1974) also studied three dose regimens of PGE_2 for induction of labor at term. The most effective schedule was 0.5 mg initially with an increase of 0.5 mg every 30 min to a maximum of 3.0, and then 0.5–3.0 mg every 30–60 min until delivery. They observed one instance of hypertonus in a patient on the 1.0 mg/hr regimen. Johnson *et al.* (1974) reported successful induction of labor in 42 of 50 patients using $PGF_{2\alpha}$ administered orally at 5 mg initially, which was subsequently repeated or doubled every hour to achieve satisfactory uterine activity. The series included 23 primigravidas and 27 multiparas. Of the eight failures, four had inadequate uterine activity with $PGF_{2\alpha}$ and four had excessive side effects of vomiting and diarrhea which necessitated ending the trial. Those who failed to respond were then successfully induced with oxytocin. In another study, Myatt and Elder (1975) describe a reduction in ADP-induced platelet aggregation in patients given oral PGE_2 for induction of labor. From the available reports at present, there is no evidence that such an antithrombotic effect adversely influences postpartum uterine hemostasis.

There is a well-established correlation between the "ripeness" or high inducibility status of the cervix and the subsequent ease of induction of labor. Thus, preinduction priming would likely facilitate subsequent induction of labor. This premise was investigated by two groups using oral PGE_2 for preinduction priming. Friedman and Sachtleban (1975) studied 30 patients with a Bishop score of less than 6 who were given either a placebo or 1.0 mg PGE_2, every 3 hr for three doses prior to induction of labor with intravenous oxytocin. Following the priming period, there was slight improvement of inducibility score in both groups. There was no detectable influence of PGE_2 priming on subsequent induction of labor with oxytocin. Weiss *et al.* (1975) reported that PGE_2 priming of 32 normal, unripe, multiparas at term produced an average of 3 points increase of inducibility score. They used either 1.0 mg prostaglandin orally every 3 hr for three doses, or 0.5 or 1.0 mg every 2 hr, on a basis of uterine activity. Nine of 32 (28%) of the PGE_2-treated patients delivered during the priming period. In contrast to 24 control series patients, PGE_2 priming had no apparent effect on subsequent induction of labor with intravenous oxytocin.

Thus present evidence supports the efficacy of PGE_2, administered orally, for induction of labor at term, but it does not establish any apparent benefit of its use prior to labor. The induction of labor with oral PGE_2 in low dosage schedules is associated with few side effects, and is effective except in patients with unfavorable status for induction. As with intravenous prostaglandins, the use of higher dosage schedules is associated with frequent side effects. These are predominantly gastrointestinal effects, but also include excessive uterine activity and fetal distress. Although serious side effects have been encountered infrequently, the actions of prostaglandins administered orally are less readily controlled than when given by intravenous infusion. In the latter instance, the effects of prostaglandin are easily controlled by stopping the infusion. As with transbuccal administration of oxytocin, oral prostaglandin is a convenient and generally effective method of inducing labor, but it lacks the relative safety of the intravenous infusion method. There is also no evidence that extraamniotic administration of prostaglandins has an advantage over the intravenous route for induction of labor at term. However, the intravaginal route of administration of prostaglandins has considerable promise in termination of pregnancies complicated by missed abortion, hydatidiform mole, or fetal death. In many of these cases, there is inadequate volume of amniotic fluid to permit intraamniotic injection, and relative lack of uterine responsiveness to prostaglandins of earlier pregnancy does not permit adequate intravenous administration without excessive side effects.

IV. CONCLUSIONS

PGF$_{2\alpha}$ or PGE$_2$ and synthetic oxytocin have equal efficacy and safety for induction of labor in late pregnancy when administered by intravenous infusion, following optimal dosage schedules for each drug. To avoid excessive uterine response, the longer dose–response time of prostaglandins necessitates less frequent dosage increments. In contrast to oxytocin, administration of prostaglandins does not inhibit diuresis. In other respects, the maternal and neonatal cardiovascular, hematological, and biochemical responses of the two agents are similar. The ease of induction of labor with either agent correlates closely with the initial status of the cervix. Intravenous infusions of prostaglandin or oxytocin are equally effective in patients with conditions favorable for induction, but some evidence suggests that prostaglandin may be more efficient in difficult inductions. In contrast, induction of labor by oral administration of prostaglandins is often unsatisfactory in patients with unfavorable cervical conditions. When conditions are favorable, the induction of labor with orally administered prostaglandins is generally as effective as by use of intravenous prostaglandin or oxytocin. Although serious side effects are infrequent with orally administered prostaglandins, this route prohibits rapid withdrawal of the drug, as is possible by stopping an intravenous infusion. There is growing evidence that intravaginal prostaglandins are an effective and safe means of inducing labor in earlier pregnancy complicated by missed abortion, hydatidiform mole, or fetal death.

The induction of labor by any method must be considered in relation to any risks imposed by the procedure. In considering the use of oxytocic agents, the most serious consequence of their injudicious use is producing uterine activity which is injurious to the fetus. During uterine contractions, intramyometrial vessels passing to and from the intervillous space of the placenta are compressed. Uterine blood flow and the intensity and duration of a contraction determine the magnitude of the temporary interruption in the fetus's life support system. In addition, the integrity of the umbilical cord circulation influences the fetal response to labor. Ultimately, the status of the fetus determines its ability to tolerate the events imposed by labor. With current methods of evaluating fetal response to labor, it appears that most fetuses are able to tolerate brief episodes of uterine hypertonus without serious consequences. On the basis of most definitions of hypertonus, careful monitoring of uterine activity reveals that it occurs quite often during spontaneous labor; as well as with prostaglandin- or oxytocin-induced labor. However, some fetuses are unable to tolerate even "normal" labor.

Modern obstetrics is rapidly becoming aware of the potential perinatal morbidity of labor. From this growing awareness, we are obliged to weigh the potential benefits and risks of induction of labor. When medical or obstetrical conditions indicate the induction of labor, the obstetrician must consider relative risks of the complication vs. the risks of the available methods. At present, it is not clear in what circumstances the administration of prostaglandins offers the best solution. Unless elective induction of labor offers clear advantages over the obvious alternative, it is not reasonable to accept any permanent morbidity from the method of induction. It appears unlikely that frequent use of elective intervention in normal pregnancy will provide a means to quality survival of the neonate.

V. REFERENCES

Anderson, G. G., and Schooley, G. L., 1975, Comparison of uterine contractions in spontaneous and oxytocin or $PGF_{2\alpha}$ induced labors, *Obstet. Gynecol.* **45**:284.

Anderson, G. G., Cordero, L., Hobbins, J., and Speroff, L., 1971, Clinical use of prostaglandins as oxytocin substances, *Ann. N.Y. Acad. Sci.* **180**:499.

Anderson, G. G., Hobbins, J. C., and Speroff, L., 1972, Intravenous prostaglandins E_2 and $F_{2\alpha}$ for the induction of term labor, *Am. J. Obstet. Gynecol.* **112**:382.

Bailey, C. D. H., Newman, C., Ellinas, S. P., and Anderson, G. G., 1975, Use of prostaglandin E_2 vaginal suppositories in intrauterine fetal death and missed abortion, *Obstet. Gynecol.* **45**:110.

Barr, W., and Naismith, W. C. M. K., 1972, Oral prostaglandins in the induction of labor, *Br. Med. J.* **2**:188.

Basu, H. K. and Rajan, K. T. J., 1975, Induction of labour with prostaglandin E_2 tablets, *J. Int. Med. Res.* **3**:73.

Beazley, J. M., and Alderman, B., 1975, Neonatal Hyperbilirubinemia following the use of oxytocin in labour, *Br. J. Obstet. Gynaecol.* **82**:265.

Beazley, J. M., and Gillespie, A. 1971, Double-blind trial of prostaglandin E_2 and oxytocin in induction of labor, *Lancet* **1**:152.

Beazley, J. M., Dewhurst, C. J., and Gillespie, A., 1970, The induction of labour with prostaglandin E_2, *J. Obstet. Gynaecol. Br. Commonw.* **77**:193.

Bergström, S., and Samuelsson, B., 1962, Isolation of prostaglandin E_1 from human seminal plasma, *J. Biol. Chem.* **237**:PC3005–PC3006.

Bergström, S., and Sjövall, J., 1957, The isolation of prostaglandin, *Acta Chem. Scand.* **11**:1086.

Bergström, S., and Sjövall, J., 1960, The isolation of prostaglandin E from sheep prostate glands, *Acta Chem. Scand.* **14**:1701.

Bishop, E. H., 1964, Pelvic scoring for elective induction, *Obstet. Gynecol.* **24**:266.

Blackburn, M. G., Mancusi-Ungaro, H. R., Jr., Orzalesi, M. M., Hobbins, J. C., and Anderson, G. G., 1973, Effect on the neonate of the induction of labor with prostaglandin $F_{2\alpha}$ and oxytocin, *Am. J. Obstet. Gynecol.* **116**:847.

Bolognese, R. J., and Corson, S. L., 1974, Prostaglandin E_2 vaginal suppository as a mid trimester abortifacient, *Am. J. Obstet. Gynecol.* **120**:281.

Brenner, W. E., Fishburne, J. I., McMillan, C. W., Johnson, A. M., and Hendricks, C. H., 1973, Coagulation changes during abortion induced by prostaglandin $F_{2\alpha}$, *Am. J. Obstet. Gynecol.* **117**:1080.

Brown, A. A., Hamlett, J. D., Hibbard, B. M., and Howe, P. D., 1973, Induction of labour by amniotomy and intravenous infusion of oxytocic drugs—A comparison between prostaglandin and oxytocin, *J. Obstet. Gynecol. Br. Commonw.* **80**:111.

Brudenell, M., and Chakravarti, S., 1975, Uterine rupture in labor, *Br. Med. J.* **2**:122.

Bygdeman, M., 1964, The effect of different prostaglandins on the human myometrium in vitro, *Acta Physiol. Scand.* **63(242)**:1.

Bygdeman, M., and Samuelsson, B., 1966, Analyses of prostaglandins in human semen. *Clin. Chim. Acta* **13**:465.

Bygdeman, M., Hamberg, M., and Samuelsson, B., 1966, The content of different prostaglandins in human seminal fluid and their threshold doses on the human myometrium, *Mem. Soc. Endocrinol.* **4**:49.

Bygdeman, M., Kwon, S. U., and Wiqvist, N., 1967, The effect of prostaglandin E_1 on human pregnant myometrium *in vivo.,* in: *Nobel Symposium 2, Prostaglandins* (S. Bergström and B. Samuelsson, eds.), pp. 93–96, Almqvist and Wiksell, Stockholm.

Bygdeman, M., Kwon, S. U., Mukherjee, T., and Wiqvist, N., 1968, Effect of intravenous infusion of prostaglandin E_1 and E_2 on the motility of the pregnant human uterus, *Am. J. Obstet. Gynecol.* **102**:317.

Bygdeman, M., Kwon, S. U., Mukherjee, T., Roth-Brandel, U., and Wiqvist, N., 1970, The effect of the prostaglandin F compounds on the contractility of the pregnant human uterus, *Am. J. Obstet. Gynecol.* **106**:567.

Calder, A. A., and Embrey, M. P., 1975, Comparison of intravenous oxytocin and prostaglandin E_2 for induciton of labour using automatic and nonautomatic infusion techniques, *Br. J. Obstet. Gynaecol.* **82**:728.

Calder, A. A., Embrey, M. P., and Hillier, K., 1974a, Extra-amniotic prostaglandin E_2 for the induction of labor at term, *J. Obstet. Gynaecol. Br. Commonw.* **81**:39.

Calder, A. A., Moar, V. A., Ounsted, M. K., and Turnbull, A. C., 1974b, Increased bilirubin levels in neonates after induction of labour by intravenous prostaglandin E_2 or oxytocin, *Lancet* **2**:1339.

Clegg, D. R., Flynn, A. M., and Kelly, J., 1974, A comparison of intravenous prostaglandin E_2 and intravenous oxytocin for the augmentation of labour complicated by delay, *J. Obstet. Gynaecol. Br. Commonw.* **81**:995.

Corson, S. L., and Bolognese, R. J., 1974, Oral prostaglandin E_2 for induction of labor, *J. Reprod. Med.* **12**:167.

Craft, I., 1972, Amniotomy and oral prostaglandin E_2 titration for induction of labour, *Br. Med. J.* **2**:191.

Craft, I. L., Cullum, A. R., May, D. T. L., Noble, A. D., and Thomas, D. J., 1971, Prostaglandin E_2 compared with oxytocin for the induction of labour, *Br. Med. J.* **3**:276.

Csapo, A. I., 1973, The prospects of PGs in postconceptional therapy, *Prostaglandins* **3**:245.

Csapo, A. I., and Kivikoski, A., 1974, The effect of uterine stretch on first trimester abortions induced by extraovular prostaglandin impact, *Prostaglandins* **6**:427.

Elder, M. G., and Stone, M., 1974. Induction of labour by low amniotomy and oral administration of a solution compared to a tablet of prostaglandin E_2, *Prostaglandins* **6**:427.

Embrey, M. P., 1969, The effect of prostaglandins on the human pregnant uterus, *J. Obstet. Gynaecol. Br. Commonw.* **76**:783.

Embrey, M. P., 1970*a,* Effect of prostaglandins on human uterus in pregnancy, *J. Reprod. Fertil.* **23**:372.

Embrey, M. P., 1970*b,* Induction of labour with prostaglandins E_1 and E_2, *Br. Med. J.* **2**:256.

Embrey, M., 1971, PGE compounds for induction of labour and abortion, *Ann. N.Y. Acad. Sci.* **180**:518.

Embrey, M. P., and Hillier, K., 1971, Therapeutic abortion by intrauterine installation of prostaglandins, *Br. Med. J.* **1**:588.

Embrey, M. P., and Morrison, D. L., 1968, The effect of prostaglandins on pregnant human myometrium *in vitro, J. Obstet. Gynaecol. Br. Commonw.* **75**:829.

Embrey, M. P., Calder, A. A., and Hillier, K., 1974, Extra-amniotic prostaglandins in the management of intrauterine fetal death, anencephaly and hydatidiform mole, *J. Obstet. Gynaecol. Br. Commonw.* **81**:47.

Filshie, G. M., 1971, The use of prostaglnadin E_2 in the management of intrauterine death, missed abortion and hydatidiform mole, *J. Obstet. Gynaecol. Br. Commonw.* **78**:87.

Fishburne, J. I., Brenner, W. E., Braaksma, J. T., Staurovsky, L. G., Mueller, R. A., Hoffer, J. L., and Hendricks, C. H., 1972, Cardiovascular and respiratory responses to intravenous infusion of prostaglandin $F_{2\alpha}$ in the pregnant woman, *Am. J. Obstet. Gynecol.* **114**:765.

Fraser, I. S., 1974, Uterine hypertonus after oral prostaglandin E_2, *Lancet* **2**:162.

Friedman, E. A., and Sachtleben, M. R., 1974, Oral prostaglandin E_2 for induction of labor at term, *Obstet. Gynecol.* **43**:178.

Friedman, E. A., and Sachtleben, M. R., 1975, Preinduction priming with oral prostaglandin E_2, *Am. J. Obstet. Gynecol.* **121**:521.

Gillespie, A., 1972, Factors affecting the dose of prostaglandin E_2 and sytocinon required to induce labor, *J. Obstet. Gynaecol. Br. Commonw.* **79**:135.

Gillespie, A., Dewhurst, C. J., and Beazley, J. M., 1971, Prostaglandin-induced labour, *Br. Med. J.* **2**:222.

Glueck, H. I., and Barden, T. P., 1975, The coagulation mechanism in labor at term induced with prostaglandin $F_{2\alpha}$, *Am. J. Obstet. Gynecol.* **121**:213.

Glueck, H. I., Flessa, H. C., Kisker, C. T., and Stander, R. W., 1973, Hypertonic saline-induced abortion, correlation of fetal death with disseminated intravascular coagulation, *J. Am. Med. Assoc.* **225**:28.

Gordon, H., and Pipe, N. G. J., 1975, Induction of labor after intrautering fetal death: a comparison between prostaglandin E_2 and oxytocin, *Obstet. Gynecol.* **45**:44.

Gowenlock, A. H., Taylor, D. S., and Sanderson, J. H., 1975, Biochemical haematological changes during the induction of labour at term with oxytocin, prostaglandin E_2 and prostaglandin $F_{2\alpha}$, *Br. J. Obstet. Gynecol.* **82**:215.

Hamberg, M., and Samuelsson, B., 1965, Isolation and structure of a new prostaglandin from human seminal plasma, *Biochim. Biophys. Acta* **106**:215.

Johnson, A., Hyatt, D., Newton, J., and Phillips, L., 1974, Experience with prostaglandin $F_{2\alpha}$ (free acid) for the induction of labour, *Prostaglandins* **7**:487.

Karim, S. M. M., 1966, Identification of prostaglandins in human amniotic fluid, *J. Obstet. Gynaecol. Br. Commonw.* **73**:903.

Karim, S. M. M., 1968, Appearance of prostaglandin $F_{2\alpha}$ in human blood during labour, *Br. Med. J.* **4**:618.

Karim, S. M. M., 1970, Use of prostaglandin E_2 in the management of missed abortion, missed labour, and hydatidiform mole, *Br. Med. J.* **3**:196.

Karim, S., 1971*a,* Action of prostaglandin in the pregnant woman, *Ann. N.Y. Acad. Sci.* **180**:483.

Karim, S. M. M., 1971*b,* Effects of oral administration of prostaglandins E_2 and $F_{2\alpha}$ on the human uterus, *J. Obstet. Gynaecol. Brit. Commonw.* **78**:289.

Karim, S. M. M., and Devlin, J., 1967, Prostaglandin content of amniotic fluid, during pregnancy and labour, *J. Obstet. Gynaecol. Br. Commonw.* **74**:230.

Karim, S. M. M., and Filshie, G. M., 1970*a,* Use of prostaglandin E_2 for therapeutic abortion, *Br. Med. J.* **3**:198.

Karim, S. M. M., and Filshie, G. M., 1970*b,* Therapeutic abortion using prostaglandin $F_{2\alpha}$, *Lancet* **1**:157.

Karim, S. M. M., and Sharma, S. D., 1971a, Second trimester abortion with single intra-amniotic injection of prostaglandins E_2 or $F_{2\alpha}$, *Lancet* **2**:47.

Karim, S. M. M., and Sharma, S. D., 1971b, Oral administration of prostaglandins for the induction of labour, *Br. Med. J.* **1**:260.

Karim, S. M. M., and Sharma, S. D., 1971c, Therapeutic abortion and induction of labour by the intravaginal administration of prostaglandin E_2 and $F_{2\alpha}$, *J. Obstet. Gynaecol. Br. Commonw.* **78**:294.

Karim, S. M. M., and Sharma, S. D., 1972, Oral administration of prostaglandin E_2 for the induction and acceleration of labor, *J. Reprod. Med.* **9**:346.

Karim, S. M. M., Trussell, R. R., Patel, R. C., and Hillier, K., 1968, Response of pregnant human uterus to prostaglandin $F_{2\alpha}$-induction of labour, *Br. Med. J.* **4**:621.

Karim, S. M. M., Trussel, R. R., Hillier, K., and Paster, R. C., 1969, Induction of labour with prostaglandin $F_{2\alpha}$, *J. Obstet. Gynaecol. Br. Commonw.* **76**:769.

Karim, S. M. M., Hillier, K., Trussell, R. R., Patel, R. C., and Tamusauge, S., 1970, Induction of labour with prostaglandin E_2, *J. Obstet. Gynaecol. Br. Commonw.* **77**:200.

Karim, S. M. M., Hillier, K., Somers, K., and Trussel, R. R., 1971, The effects of prostaglandins E_2 and $F_{2\alpha}$ administered by different routes on uterine activity and the cardiovascular system in pregnant and non-pregnant women, *J. Obstet. Gynaecol. Br. Commonw.* **78**:172.

Kelly, J., Flynn, A. M., and Bertrand, P. V., 1973, A comparison of oral prostaglandin E_2 and intravenous syntocinon in the induction of labour, *J. Obstet. Gynaecol. Br. Commonw.* **80**:923.

Kloeck, F. K., and Jung, H., 1973, *In vitro* release of prostaglandins from the human myometrium under the influence of stretching, *Am. J. Obstet. Gynecol.* **115**:1066.

Kreisman, H., Van de Wiel, W., and Mitchell, C. A., 1975, Respiratory function during prostaglandin-induced labor, *Am. Rev. Respir. Dis.* **11**:564.

Kurzrok, R., and Lieb, C. C., 1930, Biochemical studies of human semen II. The action of semen on the human uterus, *Proc. Soc. Exp. Biol. Med.* **28**:268.

Lauersen, N. H., and Wilson, K. H., 1974, Induction of labor with oral prostaglandin E_2, *Obstet. Gynecol.* **44**:793.

Lyneham, R. C., Low, P. A., McLeod, J. C., Shearman, R. P., Smith, I. D., and Korda, A. R., 1973, Convulsions and electroencephalogram abnormalities after intra-amniotic prostaglandin $F_{2\alpha}$, *Lancet* **2**:1003.

Marzouk, A. F., 1975, Oral and intravenous prostaglandin E_2 in induction of labour, *Br. J. Clin. Prac.* **29**:68.

Miller, A. W. F., and Mack, D. S., 1974, Induction of labor by extra-amniotic prostaglandins, *J. Obstet. Gynaecol. Br. Commonw.* **81**:706.

Miller, J. F., Welply, G. A., and Elstein, M., 1975, Prostaglandin E_2 tablets compared with intravenous oxytocin in induction of labour, *Br. Med. J.* **1**:14.

Murnaghan, G. A., Lamki, H., and Rashid, S., 1974, Induction of labour with oral prostaglandin E_2, *J. Obstet. Gynaecol. Br. Commonw.* **8**:141.

Myatt, L., and Elder, M. G., 1975, The effects on platelet aggregation of oral prostaglandin E_2 used for the induction of labour, *Br. J. Obstet. Gynaecol.* **82**:449.

Naismith, W. C. M. K., Barr, W., and MacVicar, J., 1973, Comparison of intravenous prostaglandin $F_{2\alpha}$ and E_2 with intravenous oxytocin in the induction of labour, *J. Obstet. Gynaecol. Br. Commonw.* **80**:531.

Novy, M. J., Thomas, C. L., and Lees, M. H., 1975, Uterine contractility and regional blood flow responses to oxytocin and prostaglandin E_2 in pregnant rhesus monkeys, *Am. J. Obstet. Gynecol.* **122**:419.

Obel, E. B., and Larsen, J. F., 1975, A study of comparative efficacy of oral prostaglandin E_2 as liquid formulation and tablet for induction of labour, *Acta. Obstet. Gynecol. Scand. Suppl.* **37**:35.

O'Driscoll, D. T., and Lavelle, S. M., 1955, Blood coagulation defects associated with missed abortion, *Lancet* **2**:1169.

Pedersen, P. H., Larsen, J. F., and Sorensen, B., 1972, Induction of labour with prostaglandin $F_{2\alpha}$ in missed abortion, fetus mortuus, and anencephalia, *Prostaglandins* **2**:135.

Pickles, V. R., Hall, W. J., Clegg, P. C., and Sullivan, T. J., 1966, Some experiments on the mechanism of action of prostaglandin on the guinea-pig and rat myometrium, *Mem. Soc. Endocrinol.* **14**:89.

Rangarajan, N. S., LaCroix, G. E., and Moghissi, K. S., 1971, Induction of labor with prostaglandin, *Obstet. Gynecol.* **38**:546.

Read, M. D., and Martin, R. H., 1974, A comparison between intravenous oxytocin and oral prostaglandin E_2 for the induction of labour in parous patients, *Current Med. Res. Opin.* **2**:236.

Roberts, G., 1970, Induction of labor using prostaglandins, *J. Reprod. Fertil.* **23**:370.

Roberts, G., and Turnbull, A. C., 1971, Uterine hypertonus during labour induced by prostaglandins, *Br. Med. J.* **1**:702.

Roberts, G., Anderson, A., McGarry, J., and Turnbull, A. C., 1970, Absence of antidiuresis during administration of prostaglandin $F_{2\alpha}$, *Br. Med. J.* **2**:152.

Roth-Brandel, U., 1971, Response of the pregnant human uterus to low and high doses of prostaglandin E_1 and E_2, *Acta Obstet. Gynecol. Scand.* **50**:159.

Roth-Brandel, U., and Adams, M., 1970, An evaluation of the possible use of prostaglandin E_1, E_2 and $F_{2\alpha}$ for induction of labour, *Acta Obstet. Gynecol. Scand. Suppl.* **49(5)**:19.

Roth-Brandel, U., Bygdeman, M., and Wiqvist, N., 1970, Effect of intravenous administration of prostaglandin E_1 and $F_{2\alpha}$ on the contractility of the non-pregnant human uterus in vivo, *Acta Obstet. Gynecol. Scand. Suppl.* **49(5)**:19.

Samuelsson, B., 1963a, Prostaglandins of human seminal plasma, *Biochem. J.* **89**:34P.

Samuelsson, B., 1963b, Isolation and identificaiton of prostaglandin from human seminal plasma, *J. Biol. Chem.* **238**:3229.

Sandberg, F., Ingelman-Sundberg, A., and Rydén, G., 1964, The effect of prostaglandin E_2 and E_3 on the human uterus and fallopian tubes *in vitro*, *Acta Obstet. Gynecol. Scand.* **43**:95.

Sharma, S. D., Hale, R. W., and Muller, J. P., 1975, Induction of term labor with intravenous prostaglandin $F_{2\alpha}$ (A comparison of two dosage schedules), *Prostaglandins* **10**:1019.

Spellacy, W. N., and Gall, S. A., 1972, Prostaglandin $F_2\alpha$ and oxytocin for term labor induction, *J. Reprod. Med.* **9**:300.

Spellacy, W. N., Buhi, W. C., and Holsinger, K. K., 1971, The effect of prostaglandin $F_{2\alpha}$ and E_2 on blood glucose and plasma insulin levels during pregnancy, *Am. J. Obstet. Gynecol.* **111**:239.

Stander, R. W., Flessa, H. C., Glueck, H. I., and Kisker, C. T., 1971, Changes in maternal coagulation factors after intra-amniotic injection of hypertonic saline, *Obstet. Gynecol.* **37**:660.

Thiery, M., Amy, J. J., and de Hemptinne, D., 1974a, Prostaglandins and convulsions, *Lancet* **1**:218.

Thiery, M., Yo Le Sian, A., de Hemptinne, D., Derom, R., Martens, G., Van Kets, H., and Amy, J. J., 1974b, Induction of labour with prostaglandin E_2 tablets, *J. Obstet. Gynaecol. Br. Commonw.* **81**:303.

Vakhariya, V. R., and Sherman, A. I., 1972, Prostaglandin $F_{2\alpha}$ for induction of labor, *Am. J. Obstet. Gynecol.* **113**:212.

Van der Plaetsen, L., Thiery, M., Amy, J. J., and de Hemptinne, D., 1974, Effect of prostaglandin E_2 therapy on the cerebral cortex, *Lancet* **1**:1226.

Weiss, R. R., Tejani, N., Israeli, I., Evans, M. I., Bhakthavathsalon, A., and Mann, L., 1975, Priming of the uterine cervix with oral prostaglandin E_2 in the term multigravida, *Obstet. Gynecol.* **46**:181.

Wilkin, D., Graham, F., Shields, M., and Craft, I., 1974, Selective induction of labour following administration of an oral prostaglandin E_2 0.5 mg tablet hourly, *Prostaglandins* **6**:405.

Wiqvist, N., and Bygdeman, M., 1970, Therapeutic abortion by local administration of prostaglandin, *Lancet* **2**:716.

Wiqvist, N., Bygdeman, M., Kwon, S. U., Mukherjee, T., and Roth-Brandel, U., 1968, Effect of prostaglandin E_1 on the midpregnant human uterus, *Am. J. Obstet. Gynecol.* **102**:327.

Yip, S. K., Ma, H. K., and Ng, K. H., 1973, Induction of labour with oral prostaglandin E_2, *J. Obstet. Gynaecol. Br. Commonw.* **80**:442.

5

Cardiovascular Responses to the Prostaglandin Precursors

John C. Rose and Peter A. Kot

Department of Physiology and Biophysics
Georgetown University Medical Center
Washington, D.C. 20007

I. INTRODUCTION

Dramatic reports have drawn attention to arachidonic acid and its possible role in the pathogenesis of human thromboembolic and cardiovascular disease (Silver *et al.*, 1974; Pirkle and Carstens, 1974; Furlow and Bass, 1974). At the same time, intermediates in the prostaglandin (PG) biosynthetic pathway have been found to possess significant biological activity (Hamberg and Samuelson, 1973; Hamberg *et al.*, 1974), and biologically active nonprostaglandin end products of arachidonic acid metabolism have been described (Turner *et al.*, 1975; Hamberg *et al.*, 1976).

A great deal of study has been devoted to the effects of PGs on the circulatory system and its components in a variety of mammalian species and in man. A summary of current information on the physiological and pharmacological cardiovascular properties of the PG precursors is now timely. The precursors of the monoenoic PGs (dihomo-γ-linolenic acid, DGLA), the bisenoic PGs (arachidonic acid, AA), and the trienoic PGs (5,8,11,14,17-eicosapentaenoic acid, EPA) are all available in nearly pure form for use in experimental studies, the last in very limited quantities. In addition, stable analogues of PG endoperoxides have become available in limited quantities for biological testing (Corey *et al.*, 1975; Bundy, 1975).

II. BLOOD PRESSURE

A. Arachidonic Acid

Since the studies of Jaques (1959) which demonstrated that AA actively affects mammalian smooth muscle, there have been studies of its ability to reduce systemic arterial pressure in several species. Jaques showed that 100 μg/kg produced a brief fall in the blood pressure of the anesthetized cat. Ichikawa and Yamada (1962) found the hypotensive effects of free and albumin-bound AA to be similar in the rabbit, while the dog was refractory to the depressor effect of AA.

Cohen *et al.* (1973) showed that AA had an acute antihypertensive effect in spontaneously hypertensive rats, and Larsson and Änggård (1973) demonstrated the blood pressure lowering effect of AA in rabbits. Both groups of investigators found that anti-inflammatory agents (indomethacin in both studies, also phenylbutazone in rats) inhibited or abolished the hypotensive effect of AA. The obvious interpretation was that AA exerts its antihypertensive or blood pressure lowering effect through the formation of prostaglandins. Subsequently, Deby *et al.* (1974) noted, in both rabbits and dogs, that AA-induced hypotension is potentiated by heparin and several amino acids, including tryptophan, histidine, cysteine, lysine, and arginine (Deby *et al.*, 1974; Deby and Damas, 1974). These investigators suggested that PG synthesis is affected by the circulating levels of certain amino acids.

A contrary report on the blood pressure effect of AA is that of Laborit and Valette (1975), who fed AA to rats made hypertensive by unilateral nephrectomy and salt and DOCA excess. These authors found that the hypertension was aggravated. No adequate explanation was offered.

In our studies in dogs anesthetized with sodium pentobarbital, and receiving no heparin, AA in a single 300 μg/kg dose intravenously produced a marked vasodepressor effect lasting 5 min, with no tachyphylaxis (Rose *et al.*, 1974). This response differed from the vasodepressor effect of PGE_2 in that the onset of effect was delayed (15 sec for AA, 4.5 sec for PGE_2). This delay in effect suggested, of course, that a biosynthetic conversion was taking place before the vasculature reacted. The hypotensive effect of AA in dogs persisted after ganglionic and β-adrenergic blockade (Kot *et al.*, 1975) and was totally blocked by aspirin in large doses (Rose *et al.*, 1974).

B. Dihomo-γ-linolenic Acid

The literature contains no early references to the blood pressure lowering effects of DGLA. In our studies in dogs, DGLA was one-sixth to one-eighth as potent as AA in its hypotensive effect (Rose *et al.*, 1975). Single intravenous doses of 2.0 mg/kg produced a uniquely biphasic depressor alteration in systemic arterial pressure with a predominant and marked depressor effect. This dose was roughly equidepressor with 5 μg/kg PGE_1 and 300 μg/kg AA.

The hypotensive effect of DGLA in dogs was not blocked by ganglion blockade or β-adrenergic blockade. Aspirin blocked the sustained depressor response to DGLA but not the initial drop in systemic arterial pressure of a few seconds duration.

C. 5,8,11,14,17-Eicosapentaenoic Acid

Studies of the circulatory effects of EPA have not been reported. Our own limited experiments in three dogs suggest that this material has no specific pharmacological activity in doses up to 6 mg/kg. A transient drop in systemic arterial pressure immediately following intravenous injection is the same as that following other fatty acids, such as linoleic and palmitic, when injected in similar large doses.

D. PG Endoperoxides

Three stable endoperoxides have been studied in our laboratory for their cardiovascular effects in dogs. They are the *azo* endoperoxide analogue (Corey *et al.*, 1975) and two *cyclic ether* endoperoxides (Bundy, 1975). Each of these is predominantly pressor (Rose *et al.*, 1976). Slight and transient drops in systemic arterial pressure following intravenous injection of the three analogues we have interpreted as due to the marked pulmonary vasoconstriction which momentarily reduces left ventricular output. The cyclic ether endoperoxides of Bundy (1975) are approximately twice as potent in their pressor effects as the azo compound of Corey *et al.* (1975) and all are more potent than $PGF_{2\alpha}$. Indomethacin pretreatment does not affect these responses.

III. CARDIAC EFFECTS

While the bisenoic prostaglandins (5 μg/kg) consistently produce positive inotropic effects on the dog heart, the cardiac effects of AA (300 μg/kg) are variable (Rose *et al.*, 1974) and of reflex origin (Kot *et al.*, 1975). In open-chest dogs anesthetized with sodium pentobarbital, myocardial contractile force, as measured with the Walton-Brodie strain gauge arch, was increased in the majority, but remained unchanged or decreased in others. The changes were coincident with the depressor response. When these animals were given hexamethonium to produce autonomic ganglion blockade, alterations of myocardial contractile force in response to intravenously administered AA were abolished.

The positive inotropic effect of PGE_2 and $PGF_{2\alpha}$ was not affected by ganglionic or β-adrenergic blockade. These results, therefore, suggest that exogenous AA (300 μg/kg) does not result in the biosynthesis of sufficient PGE_2 and/or $PGF_{2\alpha}$ to stimulate the myocardium, and that the myocardial effects of AA are mediated through activation of the baroreceptor reflex.

After ganglionic blockade, much larger doses of AA (900 μg/kg) were capable of producing a weak, direct positive inotropic effect on the dog heart. This dose must be near the LD_{50} for AA in the dog, and apparently results in the rapid biosynthesis of a myocardial-active product or products in small amount. This effect persisted in the face of effective β-adrenergic blockade (practolol), indicating lack of interaction with myocardial β-adrenergic receptors. Aspirin and indomethacin completely block the myocardial effect of large doses of AA.

DGLA, on the other hand, has a positive inotropic effect on the dog heart which is more pronounced than that produced by an equidepressor dose of PGE_1 (Rose *et al.*, 1975). Neither ganglionic nor β-adrenergic blockade affected this myocardial response.

A short and slight increase in myocardial contractile force persisted in the dog heart, in response to intravenous DGLA (2 mg/kg), even following pretreatment with aspirin. Thus DGLA probably exerts a weak direct effect on the myocardium, but the major myocardial response is due to substances formed in the pathways of monoenoic prostaglandin synthesis.

As in the case of blood pressure responses, cardiac responses to EPA appear to be nonspecific and common to other fatty acids given in large doses.

Unlike AA, but similar to the E and F PGs, the three endoperoxide analogues have direct positive inotropic effects on the dog heart (Rose *et al.*, 1976).

IV. RENAL CIRCULATION

The infusion of nonhypotensive doses of AA into the rabbit aorta by Larsson *et al.* (1974) resulted in elevated plasma renin activity in inferior vena caval blood. Conversely, indomethacin reduced plasma renin activity. These authors have also shown that AA increases the ratio of juxtamedullary to superficial cortical blood flow in the rabbit kidney, and causes an increase in the output of urinary prostaglandins, chiefly PGE₂ (Larsson and Änggård, 1974). Indomethacin given before the AA infusion caused opposite effects.

Tannenbaum *et al.* (1975) and Chang *et al.* (1975) infused AA into one renal artery of the dog. An increase in renal blood flow occurred which was less than that produced by PGE_2. These authors confirmed that the blood flow increase was primarily to the juxtamedullary or inner cortical zone. PGE_2 increased flow to all cortical zones. They concluded that the PG formed endogenously in the kidney affects the vascular resistance of only the inner cortical nephrons.

Bolger *et al.* (1976) have infused AA into one renal artery of the dog and demonstrated an increased renal blood flow with no change in glomerular filtration fraction, suggesting predominantly efferent arteriolar dilation. This occurs in the absence of any change in systemic arterial pressure, and is accompanied by marked diuresis and natriuresis, and increased renal vein plasma renin activity, renin secretion, and renal vein PGF. (PGE was not measured.) All of these responses are inhibited by pretreatment with indomethacin. The delay in onset of these responses following the AA infusion suggested to these authors that there is a renal medullary site of conversion of AA to an active intermediate or prostaglandin(s) which in turn causes renin secretion from the juxtaglomerular apparatus.

V. PULMONARY CIRCULATION

In the isolated canine lung lobe, perfused with either autologous blood or an artificial dextran-based perfusate, AA produced a pressor effect (Wicks *et al.*, 1975). $PGF_{2\alpha}$ had a similar pulmonary pressor effect in this preparation which was, however, diminished when an artificial perfusate was used.

Aspirin completely blocked the AA pulmonary pressor response but did not affect the $PGF_{2\alpha}$ response. The pulmonary pressor effect of AA was not blocked by pretreatment with phentolamine, propranolol, cyproheptadine, or atropine. These results suggest that vasoactive substance(s) derived from AA act directly on pulmonary vascular smooth muscle and that neither platelet nor plasma factors are involved, nor are adrenergic or cholinergic mechanisms.

In the same perfused canine lung lobe preparation, DGLA at 200 μg/kg evoked a pulmonary pressor response equivalent to that produced by AA at 100 μg/kg. This response was abolished by indomethacin. Thus DGLA seems uniquely potent in the pulmonary circulation, approximately 50% of the potency of AA (Wicks *et al.*, 1976).

EPA produced weak pulmonary vasoconstrictor responses only in large doses in limited experiments. These could be duplicated with comparable doses of other fatty acids. This suggested a nonspecific fatty acid effect on the pulmonary circulation (Wicks *et al.*, 1976).

The three stable PG endoperoxide analogues studied in our laboratory are powerful pulmonary vasoconstrictors, perhaps the most potent we have studied (Rose *et al.*, 1976). In this small group of experiments, indomethacin pretreatment did not modify the responses.

VI. OTHER REGIONAL CIRCULATIONS

Ryan and Zimmerman (1974) studied the effects of both AA and DGLA on the vasoconstrictor responses to norepinephrine in the dog paw, a largely cutaneous vascular bed. In a steady-flow perfusion system, both AA and DGLA depressed the response to norepinephrine. These effects were antagonized by PG synthetase inhibitors, again suggesting that they were mediated through the generation of prostaglandins.

When studied in an isolated oxygenated whole canine hind-limb preparation, AA produced a dose-related vasoconstrictor response that was blocked by either aspirin or indomethacin (Fitzpatrick *et al.*, 1975, 1977). This vasoconstrictor response was of the same magnitude whether the perfusate was autologous whole blood or a dextran-based artificial perfusate, again demonstrating that the AA effect was independent of platelet enzymes. This suggested that the vasoactive substance(s) generated from AA was not these primary prostaglandins but rather intermediates in PG synthesis.

In heparinized rats, Furlow and Bass (1974) have reported that carotid artery injection of AA, in doses exceeding 0.45 mg/kg, caused complete obstruction of the ipsilateral hemispheric microcirculation by platelet aggregates. These authors used ^{14}C-tagged inulin as an indicator to assess cerebral blood flow (CBF) and found a 50% reduction in CBF 5 min after the AA injections. Intravascular or cerebrospinal fluid pressures were not measured.

The systemic hypotensive response to AA in dogs (Rose *et al.*, 1974) seems at variance with the observed increased perfusion pressure caused by AA in the canine hind limb preparation. Observations by Änggård and Larsson (1974) provide a plausible explanation. Using the radioactive microsphere technique in rabbits, these investigators found that infusions of AA caused a marked increase in blood flow to the abdominal viscera, notably stomach, liver, and spleen, as well as diaphragm and kidney, and relative decreases in blood flow to skin and muscle. Thus an increased vascular resistance in the hind limb in the presence of a systemic hypotensive response could be explained by an overwhelming decrease in vascular resistance elsewhere, particularly in the splanchnic circulation.

VII. PLATELETS

AA is known to stimulate irreversible platelet aggregation of rabbit (Silver *et al.,* 1974; Vargaftig and Zirinis, 1973), dog (Rose *et al.,* 1974), rat (Furlow and Bass, 1974), and human (Silver *et al.,* 1973) platelets *in vitro.* Intravascular injection of AA has demonstrated aggregation or increased aggregability *in vivo* as well (Silver *et al.,* 1974; Furlow and Bass, 1974; Rose *et al.,* 1974). The interaction of AA with platelets has been implicated in sudden pulmonary thromboembolic death in humans (Bolger *et al.,* 1976).

The circulatory studies described in this review suggest that the cardiovascular responses to AA are not fully consistent with the effects of PGE_2 and $PGF_{2\alpha}$, although they are antagonized by known PG synthetase inhibitors. The circulating platelet has been proposed as a prime source of the cyclo-oxygenase of the PG synthetase system which converts AA into a cascade of cardiovascular and platelet active PGs and PG-like compounds (Hamberg *et al.,* 1973; Vargaftig *et al.,* 1974).

Johnson *et al.* (1977) have tested the concept that the platelets are obligatory for the vascular responses to AA in dog experiments. First, repeated injections of AA caused hypotension of reproducible magnitude despite lower circulating platelet counts. These experiments were also performed in splenectomized animals to avoid support of the falling platelet count by the spleen's platelet storage capacity. From these experiments, it was concluded that 70% of circulating platelet function is not essential to the vascular responses to AA.

When platelet-poor blood was exchanged in dogs, reducing circulating platelet values to 90% of control, the circulatory responses to AA were still unchanged.

Finally, when an artificial dextran-based blood-free perfusate was used in the organ and regional perfusion experiments described above (Wicks *et al.,* 1975; Fitzpatrick *et. al.,* 1975) the vasoactive response to AA was the same as when these systems were perfused with whole blood.

It appears that the vasoactive contents of platelets do not contribute to the vascular responses to AA. While occlusion of the microcirculation by platelet aggregates has been clearly demonstrated in both rabbits (Silver *et al.,* 1973) and rats (Furlow and Bass, 1974) following the injection of AA, a vasoactive effect independent of platelets may play a role in the lethality of these injections.

DGLA, in contrast to AA, has been shown by Willis *et al.* (1974) to prevent the irreversible aggregation of human platelets induced by collagen or epinephrine. These authors even suggest that oral ingestion of the monoenoic PG precursor may help prevent arterial thrombosis. At least, it can be concluded that the vasodepressor effects of DGLA described (Rose *et al.*, 1975) are not related to platelet aggregation.

VIII. SUMMARY

The bisenoic PG precursor, AA, is a potent vasoactive agent. Hemodynamic responses to this agent are not dependent on the presence of platelets. Vascular effects are blocked by PG synthetase inhibitors. Cardiac responses are indirect and reflex in origin. Effects are of different magnitudes in several regional and organ vascular beds. Data suggest that the AA circulatory effects are not due to the primary prostaglandins PGE_2 or $PGF_{2\alpha}$, but rather to potent intermediates in PG synthesis or other end products of AA metabolism.

The monoenoic PG precursor, DGLA, is less potent than AA, but exerts a powerful vasodepressor effect in experimental animals. It shows a positive inotropic effect on the dog myocardium, but cardiovascular effects are abolished by PG synthetase inhibitors.

EPA, the trienoic PG precursor, has not been studied sufficiently well, but indications are that circulatory responses to large amounts injected in dogs are minimal, and that this precursor is relatively pharmacologically inert.

Detailed pursuit of these phenomena is now required in view of reports suggesting that at least AA among these substances is a factor in mechanisms of human disease.

Finally, preliminary experiments show that the newly synthesized stable analogues of PG endoperoxides are systemic pressor agents and powerful pulmonary vasoconstrictors. Also, they directly increase myocardial contractile force in the dog. To the extent that comparisons can be made at this time, they are far more potent than the fatty acid precursors in their cardiovascular effects, and also more potent than the prostaglandins derived from them.

IX. REFERENCES

Änggård, E., and Larsson, C., 1974, Stimulation and inhibition of prostaglandin biosynthesis: Opposite effects on blood pressure and intrarenal blood flow distribution, in: *Prostaglandin Synthetase Inhibitors* (H. J. Robinson and J. R. Vane, eds.), Raven Press, New York.

Bolger, P. M., Eisner, G. M., Ramwell, P. W., and Slotkoff, L. M., 1976, The effect of prostaglandin synthesis on renal function and renin in the dog. *Nature (London)* **259**:244.

Bundy, G. L., 1975, The synthesis of prostaglandin endoperoxide analogs, *Tetrahedron Lett.* **24**:1957.

Chang, L. C. T., Splawinski, J. A., Oates, J. A., and Nies, A. S., 1975, Enhanced renal prostaglandin production in the dog. II. Effects on intrarenal hemodynamics, *Circ. Res.* **36**:204.

Cohen, M., Sztokalo, J., and Hinsch, E., 1973, The antihypertensive action of arachidonic acid in the spontaneous hypertensive rat and its antagonism by anti-inflammatory agents, *Life Sci.* **13**:317.

Corey, E. J., Nicolaou, K. C., Machida, Y., Malmsten, C. L., and Samuelsson, B., 1975, Synthesis and biological properties of a 9,11-azo-prostanoid, highly active biochemical mimic of prostaglandin endoperoxide, *Proc. Natl. Acad. Sci., U.S.A.* **72**:3355.

Deby, C., and Damas, J., 1974, Influence du tryptophane sur l'effet hypotenseur de l'acide arachidonique, *Arch. Int. Physiol. Biochim.* **82**:742.

Deby, C. Barac, G., and Bacq, C. M., 1974, Action de l'acide arachidonique sur la pression arterielle du lapin avant et après heparine, *Arch. Int. Pharmacodyn. Ther.* **208**:363.

Fitzpatrick, T. M., Rose, J. C., Kot, P. A., Johnson, M., and Ramwell, P. W., 1975, Effects of arachidonic acid on the canine hind limb preparation, *Fed. Proc.* **34**:353.

Fitzpatrick, T. M., Johnson, M., Kot, P. A., Ramwell, P. W., and Rose, J. C., 1977, Vasoconstrictor response to arachidonic acid in the isolated hind limb of the dog, *Brit. J. Pharmacol.* (in press).

Furlow, T. M., and Bass, N. H., 1974, Stroke in rats produced by carotid injection of sodium arachidonate, *Science* **187**:658.

Hamberg, M., and Samuelsson, B., 1973, Detection and isolation of an endoperoxide intermediate in prostaglandin synthesis, *Proc. Natl. Acad. Sci. U.S.A.* **70**:899.

Hamberg, M., Svensson, J., and Samuelsson, B., 1974, Prostaglandin endoperoxides: A new concept concerning the mode of action and release of prostaglandins, *Proc. Natl. Acad. Sci. U.S.A.* **71**:3824.

Hamberg, M., Svensson, J., and Samuelsson, B., 1976, Novel transformations of prostaglandin endoperoxides: formation of thromboxanes, *in: Advances in Prostaglandin and Thromboxane Research* (B. Samuelsson and R. Paoletti, eds.), Raven Press, New York.

Ichikawa, S., and Yamada, J., 1962, Biological actions of free and albumin-bound arachidonic acid, *Am. J. Physiol.* **203**:681.

Jaques, R., 1959, Arachidonic acid, an unsaturated fatty acid which produces slow contractions of smooth muscle and causes pain: Pharmacological and biochemical characterisation of its mode of action, *Helv. Physiol. Pharmacol. Acta* **17**:255.

Johnson, M., Wicks, T. C., Fitzpatrick, T. M., Kot, P. A., Ramwell, P. W., and Rose, J. C., 1977, Role of platelets in vascular responses to arachidonic acid in dogs, *Cardiovasc. Res.* (in press).

Kot, P. A., Johnson, M., Ramwell, P. W., and Rose, J. C., 1975, Effects of ganglionic and β-adrenergic blockade on cardiovascular responses to the bisenoic prostaglandins and their precursor arachidonic acid, *Proc. Soc. Exp. Biol. Med.* **149**:953.

Laborit, H., and Valette, N., 1975, The action of arachidonic acid on experimental hypertension in the rat, *Chem.-Biol. Interact.* **10**:239.

Larsson, C., and Änggård, E., 1973, Arachidonic acid lowers and indomethacin increases the blood pressure of the rabbit, *J. Pharm. Pharmacol.* **25**:653.

Larsson, C., and Änggård, E., 1974, Increased juxtamedullary blood flow on stimulation of intrarenal prostaglandin biosynthesis, *Eur. J. Pharmacol.* **25**:326.

Larsson, C., Weber, P., and Änggård, E., 1974, Arachidonic acid increases and indomethacin decreases plasma renin activity in the rabbit, *Eur. J. Pharmacol.* **28**:391.

Pirkle, H., and Carstens, P., 1974, Pulmonary platelet aggregates associated with sudden death in man, *Science* **185**:1062.

Rose, J. C., Johnson, M., Ramwell, P. W., and Kot, P. A., 1974, Effects of arachidonic acid on systemic arterial pressure, myocardial contractility and platelets in the dog, *Proc. Soc. Exp. Biol. Med.* **147**:652.

Rose, J. C., Johnson, M., Ramwell, P. W., and Kot, P. A., 1975, Cardiovascular and platelet responses in the dog to the monoenoic prostaglandin precursor dihomo-γ-linolenic acid, *Proc. Soc. Exp. Biol. Med.* **148**:1252.

Rose, J. C., Kot, P. A., Ramwell, P. W., Doykos, M., and O'Neill, W. P., 1976, Cardiovascular responses to three prostaglandin endoperoxide analogues in the dog, *Proc. Soc. Exp. Biol. Med.* **153**:209.

Ryan, M. J., and Zimmerman, B. G., 1974, Effect of prostaglandin precursors, dihomo-γ-linolenic acid and arachidonic acid on the vasoconstrictor response to norepinephrine in the dog paw, *Prostaglandins* **6**:179.

Silver, M. J., Smith, J. B., Ingerman, C. M., and Kocsis, J. J., 1973, Arachidonic acid induced human platelet aggregation and prostaglandin formation, *Prostaglandins* **4**:863.

Silver, M. J., Hoch, W., Kocsis, J. J., Ingerman, C. M., and Smith, J. B., 1974, Arachidonic acid causes sudden death in rabbits, *Science* **183**:1085.

Tannenbaum, J., Splawinski, J. A., Oates, J. A., and Nies, A. S., 1975, Enhanced renal prostaglandin production in the dog. I. Effects on renal function, *Circ. Res.* **36**:197.

Turner, S. R., Tainer, J. A., and Lynn, W. S., 1975, Biogenesis of chemotactic molecules and function for the arachidonate lipoxygenase system of platelets, *Nature (London)* **257**:680.

Vargaftig, B. B., and Zirinis, P., 1973, Arachidonic acid induced platelet aggregation accompanied by release of potential inflammatory mediators distinct from PGE$_2$ and PGF$_{2\alpha}$, *Nature New Biol.* **244**:114.

Vargaftig, B. B., Trainer, Y., and Chignard, M., 1974, Inhibition by sulfhydryl agents of arachidonic acid induced platelet aggregation and release of potential inflammatory substances, *Prostaglandins* **8**:133.

Wicks, T. C., Rose, J. C., Johnson, M., Ramwell, P. W., and Kot, P. A., 1975, Vascular responses to arachidonic acid in the perfused canine lung, *Circ. Res.* **38**:167.

Wicks, T. C., Ramwell, P. W., Rose, J. C., and Kot, P. A., 1976, Vascular responses to the monoenoic prostaglandin (PG) precursor dihomo-γ-linolenic acid (DGLA) in the perfused canine lung, *Physiologist* **19**:410.

Willis, A. L., Comai, K., Kuhn, D. C., and Paulsrud, J., 1974, Dihomo-γ-linolenate suppresses platelet aggregation when administered in vitro or *in vivo*, *Prostaglandins* **8**:509.

Role of Prostaglandins in Fever and Temperature Regulation

W. L. Veale, K. E. Cooper, and Q. J. Pittman

Division of Medical Physiology
Faculty of Medicine
The University of Calgary
Calgary, Alberta, Canada

I. INTRODUCTION

The remarkable ability of a warm-blooded animal to regulate its body temperature makes it relatively independent of the thermal conditions of the external environment. Every living organism produces heat which may be stored or given off to the environment. This condition allows an animal to reach a "steady state" of thermal exchange with its surroundings. In the homeothermic animal, heat production and heat loss are regulated to maintain internal body temperature within narrow limits over a wide range of environmental conditions. This thermohomeostasis is achieved in animals by two principal systems, namely behavioral and autonomic. In man these include regulation of heat loss, and of heat production. The heat loss regulation involves the choice of clothing, artificial environments, alterations in the skin blood flow, and variation in the level of evaporative heat loss by sweating. Heat production can be modified by altering the level of basal heat production by shivering or nonshivering thermogenesis, and by voluntary muscular activity. Deviations of about 2°C in central body temperature from the normal level, in general, do not seriously impair body function in man. Increases above this range can lead to convulsions, particularly in infants, and a further increase may cause death. If body temperature falls below normal limits (hypothermia), nervous system function is depressed, leading to a loss of consciousness and impairment of thermoregulation itself. As temperature falls even further, the coordinated contraction of heart muscle is disrupted.

Deep-body temperature of man and other mammals remains relatively constant throughout their life span, yet we can only speculate about the reference signal which establishes body temperature at or about 37°C in most animals. The constancy of this reference value is well documented, and an example of this is given by Tanner (1951), who found that the mean rectal and oral temperatures of 46 young adult humans were 37.11°C and 36.72°C, respectively, and that no measurement differed from the mean value by more than 0.5°C, with a standard deviation of 0.21°C. In discussing body temperature, we are interested, therefore, in what is frequently referred to as "normal temperature" and "fever." Recently we have suggested that the regulation of body temperature may be considered as a "normal" process, closely related to but distinctly different from the "pathological" state we know as fever (Cooper and Veale, 1974). At least two lines of experiments lend support to this concept. In the first place, the regulation of body temperature involves the monoaminergic synapses within the anterior hypothalamic preoptic area (AH/POA), and the exogenous application of these substances to that area produces temperature changes which are very much dependent on the ambient temperature in which the animal is maintained. On the other hand, when fever is produced either by a pyrogen or a prostaglandin, the hyperthermic response seems to be little influenced by the environmental temperature (Cooper and Veale, 1974; Veale and Whishaw, 1976). Second, if the tissue of the AH/POA of a mammal is destroyed, this animal essentially is rendered poikilothermic. Nevertheless, fever can still develop to pyrogens even though normal thermoregulatory responses do not appear to be available (Cooper *et al.,* 1976).

A. Temperature Regulation

1. Neurohumoral Factors

The brain stem, particularly the hypothalamus, is known to contain high concentrations of the monoamines norepinephrine and 5-hydroxytryptamine (Vogt, 1954). In 1963, Feldberg and Myers postulated a new theory to explain how body temperature is regulated by the hypothalamus. They suggested that the temperature is governed by a delicate balance in the release of endogenous monoamines in the anterior hypothalamus. That is, one biogenic amine is selectively released within the anterior hypothalamic area to activate heat production when heat gain is required and a second amine is released at the same site to inhibit heat production and to initiate heat loss when a decrease in temperature is needed. This theory was built primarily on evidence from experiments in which unanesthetized cats were injected centrally with 5-hydroxytryptamine. This produced a body temperature increase, and similar injections of norepinephrine produced a temperature decrease (Feldberg and Myers, 1964). Since these observations were made, a great deal of additional support, involving that many species, for the monoamine theory of thermoregulation has been obtained in various laboratories. For a complete review of this evidence, the reader is referred to recent articles by Veale and Cooper (1973), Feldberg (1975), and Myers (1974). That other putative neurotransmitter substances such as GABA, polypeptides, etc., do not figure more in this story may derive from the fact that their effects on body temperature have, in the brain, not been studied. There is evidence (Myers and Waller, 1973; Lomax, 1970; Cooper *et al.*, 1976) that cholinergic synapses may be involved in heat production.

2. Thermosensitive Cells

There are temperature sensors distributed on the body surface, within the viscera, and within the central nervous system. Concentrations of specifically thermosensitive neurons are located within certain areas of the spinal cord (Thauer, 1970) and midbrain (Nakayama and Hardy, 1969; Cabanac, 1970). For the most part, the major concentration of thermosensitive units has been found to be in the AH/POA. These observations have been made in several laboratory species. It is possible that the thermosensitive neurons are capable of picking up the small changes in local blood temperature or metabolic heat production. Hellon (1967, 1970) provided clear evidence that specific neurons within the anterior hypothalamic area of the rabbit changed their firing rate in a similar fashion when either their local temperature was changed or the temperature of the skin of the animal was altered.

With respect to temperature regulation, therefore, the AH/POA is essential and when this area is removed warm-blooded animals become quite poikilothermic. Not only can the AH/POA sense its own temperature but also it would appear that information flows directly to it from thermosensors in other regions of the body, primarily the skin. This region is therefore central for the integration of the inputs which provide the basis for outflow from the hypothalamus to produce a well-coordinated, appropriate thermoregulatory response.

B. Fever

1. Steps in the Production of Fever

Pyrogens are generally distinguished by their origin. "Bacterial pyrogen" refers to the pyrogenic material extracted from killed bacteria. In our discussion, any pyrogenic factor from outside the body will be referred to as "exogenous pyrogen." Exogenous pyrogen injected into the AH/POA causes fever (Villablanca and Myers, 1965) and, similarly, the pyrogenic material that is liberated from cells within the body (endogenous pyrogen) has been microinjected into the brain(s) of several species, and the general finding is that the area from which it best elicits fever is also the AH/POA (Cooper, 1965; Cooper *et al.*, 1965, 1967; Repin and Kratskin, 1967; Jackson, 1967). In these experiments, fever was caused by injections of pyrogen into the AH/POA in far smaller quantities than those needed to induce fever by intravenous injections of the same substance. For a complete discussion of the pathogenesis of fever, the reader is directed to a recent paper by Cooper *et al.* (1976).

Figure 1 is a diagrammatic flowchart for the events occurring in the production of fever. An exogenous pyrogen (bacterial pyrogen or endotoxin) interacts with a variety of cells within the body, in this particular example either polymorphonuclear neutrophils, monocytes, or Kupffer cells. From the interaction of the exogenous pyrogenic material with these cells an endogenous pyrogen is produced, which is assumed to find its way into the central nervous system and acts within the AH/POA to produce fever. Rather than acting directly, there is evidence to suggest that the endogenous pyrogen induces the synthesis and release of prostaglandins of the E series within the AH/POA, and this material in turn acts directly on the cells of the brain stem to initiate the febrile response. The role of these prostaglandins in the genesis of fever is the primary topic of this chapter.

2. Temperature Setpoint and Fever

It is generally agreed that the effect of a pyrogen is not to disorganize thermoregulation but to cause an upward shift in the level at which temperature is regulated (Fox and McPherson, 1954; Atkins, 1964; Cooper *et al.*, 1964). This idea is supported by the finding that pyrogen causes fever in dogs and the level of fever is not influenced by external temperatures (Thompson *et al.*, 1959), suggesting that thermoregulation occurs around a new elevated reference temperature. Hammel (1965) supported this concept that fever results in an elevated reference temperature with no interference in the ability to thermore-

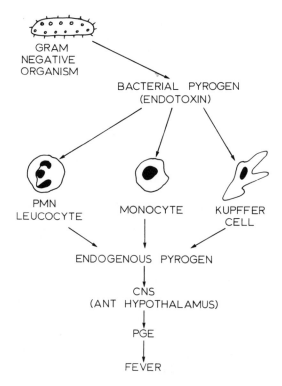

Fig. 1. Schema of the current concept of the mechanism of action of exogenous pyrogens in producing fever in mammals.

gulate by demonstrating that when the hypothalamus of the dog was cooled during a pyrogen fever a "hyperfever" developed and during heating of the hypothalamus a "hypofever" developed. In addition, Cabanac *et al.*, (1970) demonstrated that dogs will behaviorally thermoregulate around a higher body temperature following the injection of a pyrogen.

Of course, one can as yet only speculate about the physiological basis of a reference temperature. Nevertheless, a theory has been put forward by Myers and Veale (1970, 1971), which suggests that the reference temperature for the body is determined by the ratio of calcium and sodium ions within the posterior hypothalamus.

It would seem, therefore, that thermoregulation can be thought of as a "normal" process which involves the thermosensitive cells within the body, and in particular the anterior preoptic region of the brain stem, as well as the controlled release of the monoamines norepinephrine and 5-hydroxytryptamine. On the other hand, fever produced by pyrogens can be thought of as a "pathological" process which involves the production of an endogenous pyrogen, and perhaps the release of a prostaglandin of the E series in the AH/POA.

II. ROLE OF PGs IN FEVER

Since Milton and Wendlandt (1970) first demonstrated that prostaglandins of the E series, when injected directly in the cerebral ventricular system of unanesthetized cats, produced an increase in body temperature, evidence has been accumulating to support the concept that prostaglandins are involved in the febrile response. This observation stimulated a new phase of research on fever. In nearly all of the species examined so far (and there have been several), PGE produces a sharp fever. The cyclic, oxygenated, C_{20} fatty acids are present in hypothalamic tissue as well as in cerebrospinal fluid (Holmes, 1970; Holmes and Horton, 1968; Horton and Main, 1967; Samuelsson, 1964). Further, it has been known for several years that brain tissue from several different species is capable of the biosynthesis of prostaglandins (Wolfe *et al.,* 1967; Kataoka *et al.,* 1967). According to Coceani (1974), we may assume from the available evidence that the brain has all components of the prostaglandin system. With respect to the removal of prostaglandins from the brain tissue, this matter seems somewhat less clear. Several investigators have observed prostaglandins in the superfusion fluid of the cortex, spinal cord, and cerebral ventricular cavities and in perfusions of the substance of the brain itself (Ramwell and Shaw, 1966; Feldberg and Myers, 1966; Ramwell *et al.,* 1966; Beleslin and Myers, 1971). In fact, we have been able to demonstrate that labeled PGE_1 placed into the hypothalamic tissue appears intact in the cerebral spinal fluid after a delay of approximately 10 min (Cooper and Veale, 1973).

When injected directly into the substance of the brains of unanaesthetized animals, prostaglandins of the E series produce a profound hyperthermia even when the quantities injected are as low as 100 pg (Veale and Cooper, 1974*a;* Veale and Whishaw, 1976). So far as we are aware, this is the smallest amount of any substance which when injected directly into the brain stem can produce fever.

In our laboratory, we have conducted an extensive series of microinjection "mapping" studies in the brains of several species including rat, rabbit, cat, and monkey. One of the questions we attempted to answer in this line of investigation was that of correspondence of sites of action of endogenous pyrogen and prostaglandins of the E series. The endogenous pyrogen was produced by incubating leukocytes with exogenous pyrogen *in vitro*. This substance was then utilized in the mapping studies. In this work, two significant observations were made. First, the only region of the brain which we were able to find to be sensitive to the direct application of either PGE_1, PGE_2, or endogenous pyrogen was the AH/POA. Microinjection of these materials into other regions of the brain, including the midbrain, did not cause temperature changes. The second significant observation was that we were unable to find any locus within the brain which responded to prostaglandins of the E series by producing a temperature response and not to endogenous pyrogen or *vice versa*. From this work, we concluded that both endogenous pyrogen and prostaglandins of the E series produced fever only when they were injected into the AH/POA. If temperature responses were observed from injections in other loci within the brain, histological examination revealed that these injection sites were in communication with the cerebral ventricular system or the subarachnoid space. This would permit the tissue of AH/POA to be reached either from its inside or outside surface.

The important discovery by Vane (1971) and Flower and Vane (1972) that antipyretics inhibit the synthesis of prostaglandins provided powerful support for the theory which suggested that prostaglandins were released by endogenous pyrogen and in turn produced fever. These important observations made initially in homogenates of guinea pig lung (Vane, 1971) and more recently in homogenates of rat and dog brains (Flower and Vane, 1972), as well as other tissues, certainly are compatible with the idea that a rise in temperature and fever is induced by increased synthesis and the release of prostaglandins. Further, Milton and Wendlandt (1971) made the important observation that antipyretics which reduce pyrogen fever did not lower prostaglandin fever. They suggested that antipyretics did not interfere with the action of prostaglandins but did with their release.

An additional piece of evidence which further supports the theory for prostaglandin involvement in fever comes from the work of Stitt and Hardy (1972) and Stitt (1973). These workers were able to observe that in the rabbit the latency for the febrile response following injection of PGE into the AH/POA was shorter than that seen following local injection of an endogenous pyrogen into the exact same site. Other lines of evidence also tend to support the theory, and the reader is directed to a review by Veale and Cooper (1974a) in which additional lines of evidence are discussed.

The theory proposed by Milton and Feldberg and their colleagues that endogenous pyrogen causes the release of prostaglandins of the E series (probably E_2) to produce a fever appears, then, to have a solid base of experimental evidence. In the remainder of this chapter, we will present very recent evidence related to the prostaglandin theory and evidence which challenges it.

III. PG SYNTHESIS IN BRAIN DURING FEVER

A. Appearance of PGs in Cerebrospinal Fluid

Support for the theory that prostaglandins are mediators of pyrogen fever came from observations that CSF collected from the third cerebral ventricle of febrile cats contained 2.5–4.0 times as much prostaglandinlike activity as do samples collected from these same animals when nonfebrile (Feldberg and Gupta, 1973). Following the fall in temperature induced by paracetamol, an antipyretic, prostaglandinlike activity again decreased to the low levels observed prior to injection of pyrogen. The increase in prostaglandinlike activity during fever could also be detected in cisternal CSF, thus suggesting that the prostaglandin release was widespread throughout the neural structures bordering the ventricular system (Feldberg *et al.*, 1973; Dey *et al.*, 1974). Prostaglandinlike activity has, in fact, been detected in many areas of the brain (Feldberg and Myers, 1966; Holmes and Horton, 1968; Beleslin and Myers, 1971). Experiments similar to those reported in cats have also been carried out in rabbits (Philipp-Dormston and Siegert, 1974; Harvey *et al.*, 1975) and in these animals, also, fever is accompanied by high levels of prostaglandins in the ventricular and cisternal CSF. If PGE release is a response of all brain tissue to endogenous pyrogen, then we could envisage PGE release and its suppression by antipyretics to be coincidental with another process involving the action of endogenous pyrogen locally within the brain stem and not yet known.

B. Presence of PGs in the AH/POA

The presence of increased prostaglandin levels in CSF of febrile animals is largely circumstantial evidence for a role of prostaglandins in fever. Since the AH/POA is the only area of the brain shown to be sensitive to the hyperthermic effects of directly applied prostaglandins, the demonstration of prostaglandin release from this area would strengthen the theory. We carried out some experiments which did suggest that during fever some substance was released in the brain which was inactivated by efflux into the cerebral ventricles (Cooper and Veale, 1972). When the cerebral ventricles of a rabbit were filled with sterile, pyrogen-free mineral oil, no change was seen in the rabbit's body temperature; however, this procedure greatly potentiated a fever caused by pyrogen administered intravenously. Furthermore, in animals maintained in the febrile state by a continuous infusion of endogenous pyrogen, injection of the oil led to an abrupt increase in temperature to a much higher fever. Thus it appeared that the oil prevented the efflux of some substance that was present during fever but was not a normal product of thermoregulation.

In a further series of experiments, we showed that it was possible for prostaglandins released within the tissue of the AH/POA to diffuse into the ventricular system (Veale and Cooper, 1974*b*). Using a push-pull cannula apparatus (Myers, 1970; Veale, 1972), we perfused a small area of tissue within the AH/POA with a physiological solution containing [^{3}H]PGE$_1$, and simultaneously perfused the ventricular system from lateral ventricle to cisterna magna. When the perfusate from the ventricular system was analyzed for the presence of [^{3}H]PGE$_1$, the presence of the labeled prostaglandin was evident within 10 min of the time it was first perfused in the AH/POA. This work, therefore, extends the observation of Feldberg and co-workers (Feldberg and Gupta, 1973; Feldberg *et al.*, 1973) by showing that prostaglandins within the AH/POA can make their way to the ventricular system. Interestingly, experiments in which labeled acetylcholine, norepinephrine, and 5-hydroxytryptamine were perfused in the AH/POA in a similar manner did not demonstrate diffusion of the labeled compounds into the ventricles (Myers *et al.*, 1971). These substances, unlike prostaglandins are believed to be neurotransmitters involved in normal thermoregulation (Hellon, 1975).

We have also attempted to detect prostaglandin release within the AH/POA during pyrogen-induced fever. In the nonfebrile rabbit and on other occasions in the same animal during a fever induced by intravenous bacterial pyrogen, we utilized the push-pull cannula apparatus to wash the tissue of the AH/POA with sterile, pyrogen-free Krebs solution. The collected perfusates were then assayed for prostaglandin using a radioimmunoassay specific for prostaglandins of the E, A, and B series. Since it would appear that rabbit brain contains little, if any, PGA or PGB, we interpreted our results in terms of PGE (Coceani, 1974). We have carried out these experiments on 11 different animals and to date we have not seen consistent changes in the amounts of PGE between perfusates taken from febrile or nonfebrile animals. Upon histological examination, the perfusion sites were within or close to the AH/POA and all sites were tested for sensitivity to the hyperthermic action of directly applied PGE$_1$. It may be that tissue disruption caused by the push-pull perfusion may have caused rapid prostaglandin biosynthesis in itself which masked the effect of the pyrogen-induced synthesis. However, it is of interest that even in animals that were not febrile, and whose body temperature remained constant following the perfusion, prostaglandin was detected in the perfusate. Thus, under these conditions, endogenous release of prostaglandin within the tissue of the AH/POA is not accompanied by fever. In an earlier study, Beleslin and Myers (1971) had observed the presence of prostaglandin-like substance in perfusate from various brain regions including the hypothalamus, and they could not correlate its presence with any thermoregulatory response. The evidence from tissue perfusion studies therefore does not favor the theory of prostaglandin release during fever, but this may be due to methodological difficulties.

C. Source of PGs

The site of prostaglandin biosynthesis and release within the brain is still unclear. Kataoka *et al.* (1967) examined a subcellular fraction of rat cerebral cortex and observed significant ability to synthesize PGE_1 and $PGF_{2\alpha}$ only in the light microsomal fraction. They concluded, however, that the nerve endings were the storage sites of the prosta-glandinlike substance. This study was not confirmed by Hopkin *et al.* (1968), who found the majority of the prostaglandinlike activity in rabbit brain homogenate to be in the supernatant fraction. Quite possi-bly this represents prostaglandin released from a labile particulate fraction. Coceani (1974) has suggested that published values of prosta-glandin concentration in brain are largely artifactual and represent prostaglandin biosynthesis during homogenation.

Prostaglandin release has been correlated with neuronal activity both in the autonomic nervous system (Brody and Kadowitz, 1974) and within the central nervous system (Coceani, 1974). Since prostaglan-dins appear to have a modulating rather than classical neurotransmitter type of action on neuronal function (Horton and Main, 1967; Coceani *et al.*, 1971), it is possible that the site of synthesis and storage may be different from that of putative neurotransmitters. There is evidence that a rat glioma clone is capable of releasing PGE into the medium following stimulating with dibutyral cyclic AMP (Hamprecht *et al.*, 1973). Whether a similar synthesis could occur in glia cells *in vivo* remains to be determined. Perhaps with the use of neuronal and glial isolation technique now available (Norton and Poduslo, 1970), it will be possible to stimulate relatively pure preparations of neurons and glial cells with pyrogens and observe them for prostaglandin synthesis activity.

We have recently speculated that PGE_1, if indeed a mediator in fever, may come, in part, from white blood cells (Cooper *et al.*, 1975).

During fever, white cells interact with bacterial pyrogen and become sequestered in capillary beds throughout the body. Furthermore, polymorphonuclear leukocytes are capable of releasing prostaglandins following phagocytosis (Higgs and Youltin, 1972; Higgs *et al.*, 1975) and following mitogen and antigen stimulation (Ferraris *et al.*, 1974). Thus, white cells sequestered in the capillaries of the AH/POA could release PGE_1 which could diffuse into the tissue and have an effect on the neurons involved in elevating body temperature.

IV. DIFFICULTIES WITH THE PG HYPOTHESIS

A. Lesions of the AH/POA and Fever

It is generally accepted that the AH/POA is the only area of the brain into which localized injections of prostaglandins cause fever. However, it has been known for some years that even in animals in which the AH/POA had been lesioned, it was possible to produce a fever with intravenous pyrogen (Ranson *et al.*, 1939; Chambers *et al.*, 1949). This makes it likely that there may be a site in the brain other than the AH/POA where pyrogens exert their effect. We have carried out experiments in which large lesions were made in the AH/POA of rabbits (Veale and Cooper, 1975; Cooper *et al.*, 1976). This procedure destroyed the animal's ability to thermoregulate against a hot or cold environment, and on histological examination of the brains of these animals the tissue of the AH/POA appeared to have been fully destroyed. In these lesioned animals, injections of prostaglandins either into the lateral cerebral ventricle or directly into the area of the lesion in amounts which had been effective in elevating temperature prior to the time of the lesion were now without effect on body temperature. Nevertheless, intravenous injections of endogenous pyrogen caused fever which was slower in onset and took longer to reach maximum but was of approximately the same magnitude as that observed prior to when the lesioning occurred. Thus even though these animals had lost their ability to respond with fever to prostaglandins, they retained at least a component of the febrile response to intravenous pyrogen. This would suggest that there may a region of the brain sensitive to the hyperthermic action of pyrogens where fever is produced without the involvement of prostaglandins. However, both components of the febrile response would appear to have a common outflow, for lesions in the posterior hypothalamus abolish both prostaglandin fever and pyrogen fever.

B. Lack of "Prostaglandin Fever" in Lambs

We have recently studied the ability of the newborn lamb to develop fever in response to an intravenous injection of bacterial or leukocyte pyrogen (Pittman *et al.*, 1973, 1974; Cooper *et al.*, 1975). In a continuation of this work, we examined the temperature response of the newborn lamb to prostaglandins injected into the lateral cerebral ventricle (Pittman *et al.*, 1975*a*). In these experiments, 1.0–200 μg of PGE_1 or PGE_2 was injected under local anesthesia into a lateral ventricle. On 15 of 40 occasions, this procedure was followed by a temperature rise, although the fever was generally slow in onset and rose gradually to a maximum of less than 1°C. In the remaining experiments, rectal temperature either remained constant or fell. Although the lambs ranged in age from 4 to 7 days postnatal, there did not appear to be any consistent response with respect to age, amount of prostaglandin injected, or the ability of the lamb to develop a fever in response to intravenous pyrogen. We are quite certain that the injected material entered the ventricular system, for we have obtained thermoregulatory responses to newborn lambs to norepinephrine and 5-hydroxytryptamine that are consistent and agree with published data from adult sheep (Bligh *et al.*, 1971). If the injected PGE were to diffuse into the hypothalamus in a similar manner, our data would suggest that lambs may be capable of developing fever independently of the central involvement of prostaglandins. In further support of this concept is our more recent finding that prostaglandins microinjected directly into the tissue of the hypothalamus in newborn lambs are ineffective in causing fever, even in sites at which injections of norepinephrine cause temperature changes (Pittman *et al.*, 1975*b*).

Although prostaglandins cause fever when injected into the cerebral ventricles of adult sheep (Bligh and Milton, 1973; Hales *et al.*, 1973; our unpublished observations), it is of interest that Martin and Baile (1973) did not observe fever in the adult sheep following injection of PGE_1 or PGE_2 into the anterior hypothalamus. We also have made microinjections of PGE_1 and PGE_2 into over 80 sites within the anterior, lateral, or posterior hypothalamus of the adult sheep and on no occasion have we observed fever. Thus, at least within the area of the hypothalamus, prostaglandins do not have a febrile action in sheep. The echidna, a monotreme which also becomes febrile after intravenous pyrogen, responds to intraventricular PGE_1 or PGE_2 with a fall in body temperature (Baird *et al.*, 1974). Recently, Artunkal and Marley (1974) found that in young chicks PGE_1 given by various routes elevated body temperature when ambient temperature was thermoneutral, but lowered it when ambient temperature was below thermoneutrality. As experiments are carried out on more species, we may see more instances when prostaglandins do not cause fever.

C. PG Antagonists

Antipyretics such as salicylate, indomethacin, and paracetamol are believed to exert their action by the inhibition of prostaglandin synthesis. Recently compounds have become available which appear to competitively antagonize the effects of prostaglandins (Sanner, 1974). If such compounds were to lower fever, this could be interpreted as evidence in favor of a role for prostaglandins in fever. However, brief reports in the literature indicate that the prostaglandin antagonist SC-19220 administered intraperitoneally to rabbits (Sanner, 1974) and to cats (Clark and Cumby, 1975) did not significantly reduce a pyrogen-induced fever.

In cooperation with Dr. E. Preston, we have injected 2 mg quantities of SC-19220 directly into the lateral ventricles of four rabbits which had been previously fitted with Collison cannulas. The effect on body temperature was identical to that which would have occurred if we had injected prostaglandins rather than a prostaglandin antagonist. The animals became vasoconstricted and a rapid fever developed with a latency of onset of 1–2 min. Control injections of the vehicle (1% Tween 80 in 0.9% NaCl) were without effect on body temperature. Thus neither published data nor our preliminary results with use of a specific prostaglandin antagonist appear to provide any confirmatory evidence for a role for prostaglandins in fever. Perhaps with the use of a broader range of doses, or even with different types of prostaglandin antagonists, it may be possible to antagonize a PGE or pyrogen fever.

D. Action of PGs on Single Hypothalamic Neurons

It has been possible to characterize the action of certain neurons involved in temperature regulation by recording directly from them with fine extracellular electrodes. Thus in urethane-anesthetized cats (Cabanac *et al.,* 1968; Wit and Wang, 1968; Eisenman, 1969), neurons have been identified within the AH/POA that were either thermally insensitive, responsive to warming with an increase in firing rate (warm sensitive), or responsive to cooling with an increase in activity (cold sensitive).

Following intravenous administration of bacterial pyrogen, the warm-sensitive neurons showed a reduction in activity while the cold-sensitive types showed an activation and no effect was seen on thermally insensitive neurons. Acetylsalicylate appeared to reverse the pyrogen effects (Wit and Wang, 1968). Similar effects were seen in urethane-anesthetized cats in which leukocyte pyrogen and Na aspirin were microinjected directly into the AH/POA (Schoener and Wang, 1974). If pyrogens do exert their action through the mediation of prostaglandins, it should be possible to obtain similar results from direct application of prostaglandins to the AH/POA. Indeed, a number of cold-sensitive neurons tested by Ford (1974) did show excitation following microiontophoresis of PGE_1. Thermally insensitive or warm-sensitive neurons were unresponsive. However, in a more extensive study on rabbits by Stitt and Hardy (1975), in which 138 PO/AH units were studied following microiontophoresis of PGE_1, less than 9.0% of the total population tested showed any effect due to the prostaglandin. This effect, when present, was a mild facilitation and could not be correlated with thermal sensitivity or insensitivity. Poulain and Carette (1974) also observed facilitation in 57 of 77 cell bodies located in preoptic, septal, and arcuate areas of the guinea pig diencephalon. However, they did not make any attempt to correlate these effects with thermal sensitivity of the various cells.

In view of the findings of Stitt, it would appear that there is little basis for postulating prostaglandin mediation of pyrogen action. However, it may be that the site of action of pyrogen and prostaglandin is presynaptic to the thermosensitive neurons; therefore, iontophoresis of PGE would have no effect on these neurons (Stitt and Hardy, 1975). Indeed, there is evidence that both PGE_1 (Coceani *et al.*, 1971; Bergström *et al.*, 1973; Brody and Kadowitz, 1974) and bacterial pyrogen (Parnvas *et al.*, 1971) can modulate the release of transmitter substances at nerve terminals. If, as has been postulated (Cooper and Veale, 1974), fever is a pathological drive different from normal thermoregulation, it is not surprising that there is no effect of PG on neurons involved in sensing and regulating normal body temperature.

V. PGs IN NORMAL TEMPERATURE REGULATION

A. Antipyretics and Normal Temperature Regulation

If prostaglandins are released, within the brain, as a direct result of the action of pyrogens, it should be possible to show that they are not involved in thermoregulation in the nonfebrile animal. Circumstantial evidence for this concept is the fact that antipyretics which have been shown to inhibit prostaglandin synthesis in *in vitro* systems

(Vane, 1971; Flower and Vane, 1972) do not lower normal body temperature in afebrile rabbits (Cranston *et al.*, 1970), cats (Clark and Cumby, 1975), and humans (Rosendorff and Cranston, 1968) at room temperature.

Even when animals were engaged in heat production and conservation similar to that which occurs during fever, intravenous antipyretics were without effect on body temperature. This has been shown for rabbits (Pittman *et al.*, 1976) and cats (Cranston *et al.*, 1975) exposed to cold temperatures and for rabbits with an elevated body temperature following hypothalamic cooling (Cranston *et al.*, 1970). Recently, however, evidence has appeared that rats, given intraperitoneal injections of sodium salicylate, would lower their temperature when placed in the cold (Satinoff, 1972; Francesconi and Mager, 1975) and if given the opportunity would lower their temperatures by increasing the amount of time spent escaping the heat (Polk and Lipton, 1975). A hypothermic effect of sodium acetylsalicylate has also been observed following injection either into the third ventricle or intraperitoneally into afebrile monkeys (Chai and Lin, 1975). In disagreement with our findings (Pittman *et al.*, 1976) and those of Cranston *et al.* (1970) is the observation of a hypothermic effect of sodium salicylate in nonfebrile rabbits placed in a cold environment or when they were cooled locally in the hypothalamus (Murakami and Sakata, 1975). If the mode of action of antipyretics is indeed via the inhibition of prostaglandin synthesis, these results would suggest that prostaglandins may play a role in thermoregulation in the nonfebrile animal, in particular, the rat. There is a possibility, however, that some of the animals used in these experiments had subclinical infections which made them slightly febrile. Thus salicylate could lower their temperatures. It is also possible that the action of salicylate may be, in part, via a mechanism other than that just mentioned. Interestingly, Beckman and Rozkowska-Ruttiman (1974) have shown an excitatory effect of salicylate iontophoresed onto warm-sensitive cells in the hypothalamus and septal area of the rat. Barker and Levitan (1971) exposed cells in the buccal ganglion of the marine mollusk *Nowana* to salicylate and observed a membrane hyperpolarization and a decrease in total membrane resistance. Thus the effect of salicylate in the afebrile animal may be independent of its effect on the prostaglandin synthetase system.

B. PG Efflux in Afebrile Animals

Further evidence for the lack for a role for prostaglandins in thermoregulatory reactions to cold is the report that the amount of prostaglandin-like activity in the CSF collected from the cisterna magna did not change in conscious cats exposed to a hot environment and in the same animals exposed to a cold environment (Cranston *et al.*, 1975*a*). Our own experiments (unpublished data) show no consistent changes in the amount immunoreactive PGE in a perfusate collected by push-pull perfusion of the AH/POA of unanesthetized rabbits exposed to cold over that collected when they were resting at room temperature. Thus the evidence would not favor a role for prostaglandin in cats and rabbits during normal thermoregulation.

VI. SUMMARY OF THE EXISTING EVIDENCE

With the exception of the rat, there is no evidence to support the view that prostaglandins released in the substance of the brain play a part in normal thermoregulation. It may be that the techniques have been, as yet, too imprecise, or that the wrong measurements have been made. The assumption that the antipyretic dosage in experiments demonstrating the lack of action of antipyretics on normal thermoregulation, while adequate to reduce prostaglandin synthesis within the brain at levels at which it would operate in febrile situations, may not be adequate to prevent the synthesis and release of very minute amounts which themselves or acting through their intermediary metabolites could have an effect on body temperature regulation. This aspect of the action of prostaglandins is still open for further experimental work.

The role of prostaglandins of the E series in the genesis and perpetuation of fever also remains an interesting enigma. There is a considerable bulk of circumstantial evidence suggesting that prostaglandins of the E series, synthesized and released in the tissue of the AH/POA are involved in the fever produced by intravenous pyrogens, and by inference from the action of antipyretics in naturally occurring fevers. The cell structures from which the prostaglandin is released are not yet known, and the mechanism whereby released prostaglandins drive heat production and heat conservation pathways is equally unresolved at present. A considerable amount of evidence has also been adduced, in this chapter, to suggest that there may be a component of fever which can occur independently of the action of prostaglandins within the anterior hypothalamus and preoptic areas. On the basis of the characteristics of the types of fever produced by intravenous endogenous pyrogens or microinjections of prostaglandins before major lesions of the AH/POA, and the fevers produced by intravenous endogenous pyrogen following such lesions, it would be interesting to speculate that endogenous pyrogen might produce fever by two mechanisms. The first would be a fever of rapid onset and short duration due to the release of prostaglandin in the AH/POA and a second phase of fever of more gradual onset and greater duration produced by the action of endogenous pyrogen at a site other than the AH/POA and as yet unknown. This latter speculation would suggest a further mapping study of the brain with microinjections of endogenous pyrogen.

Although most studies have concluded that the AH/POA is the only site of action of endogenous pyrogen, the work of Rosendorff and Mooney (1971) also suggests a site of action within the midbrain. A number of brain loci have not yet been investigated by microinjections of either endogenous pyrogen or prostaglandins, despite the fact that they are known to be sensitive to changes in body temperature. The frontal cortex overlying the orbital area of the skull is known to have an input into the hypothalamus and it is possible that there may be a thermoregulatory function (Newman, 1974). Newman and Wolstencraft (1960) also described an area of the medullary reticular substance which could evoke peripheral vasodilatation in the cat in response to very marked heating. If there is an action of endogenous pyrogen or other substances on cells of the reticular activating system which project rostrally to the AH/POA thermoregulatory neurons, then one would expect that it would be necessary to apply the substances to be tested somewhere in the region of the cell bodies of these projections. To date, this has not been done, probably in the main for technical reasons, and one would suggest that this could be an area worth investigating.

The biggest stumbling block to date in the acceptance of the hypothesis of the role of prostaglandins in the genesis of fever, or at least in its having the sole role of production of fever, lies in our inability to detect the release of this substance during artificially induced fevers from the actual tissues of the region where it is known to have its singular action. Again, it may be that the substance is released in such a way as to reach the AH/POA rapidly through the ventral surface of the brain or from the cells lining the third ventricle, and thus it would not be possible to detect on push–pull perfusion of the tissue itself if the effect or mechanism lay close to these two surfaces and if the mechanism for its removal was very rapid.

Much then remains unexplained, and what was a few years ago thought to be the ultimate answer to the central mechanism of fever has now opened up once again to the need for exciting and critical experimental reevaluation.

VII. ADDENDUM

Two additional pieces of evidence have been published recently (Cranston *et al.*, 1975*b,c*) which cast further doubt on the hypothesis that PGE release within the AH/POA is the sole final common path in the genesis of the febrile response to pyrogens. In the first article, the authors demonstrated that a level of intravenous salicylate infusion could be achieved which would prevent the appearance of PGE in the cerebrospinal fluid of the cisterna magna but which was without effect on the fever due to simultaneous intravenous pyrogen infusion. It could be argued that if the blood vessels of the AH/POA form an alternative route of PGE excretion and that if the whole brain contributes to the cisternal cerebrospinal fluid level of PGE during the febrile reaction then changes in cisternal cerebrospinal fluid PGE levels might not reflect the minute increases in AH/POA tissue levels of PGE such as occur in fever. Nevertheless, the evidence remains damaging to the PGE theory of fever.

In the second article (Cranston and Mitchell, 1975) it was shown that intraventricular administration of a specific PGE antagonist SC-19220 effectively prevented fever produced by topical administration of PGE, but did not abolish fever induced by intravenous pyrogen. Again, although the diffusion volume of the SC-19220 from the ventricular system is not known, the evidence points powerfully to a central action of pyrogen, separate from the involvement of PGE release in the AH/POA.

ACKNOWLEDGMENTS

This work was supported by the Medical Research Council of Canada. Q. J. Pittman was supported by a Medical Research Council Studentship. The prostaglandins were kindly supplied by J. Pike, Upjohn Co., Kalamazoo, Michigan. The SC-19220 was kindly supplied by R. L. Bergstrom, Searle Laboratories, Chicago. The authors are indebted to G. Mears for carrying out the prostaglandin assays.

VIII. REFERENCES

Artunkal, A. A., and Marley, E., 1974, Hyper- and hypothermic effects of prostaglandin E_1 (PGE$_1$) and their potentiation by indomethacin, in chicks, *J. Physiol.* **242**:141.

Atkins, E., 1964, Elevation of body temperature in disease, *Ann. N.Y. Acad. Sci.* **121**:26.

Baird, J. A., Hales, J. R. S., and Lang, W. J., 1974, Thermoregulatory responses to the injection of monoamines, acetylcholine and prostaglandins into a lateral cerebral ventricle of the echidna, *J. Physiol.* **236**:539.

Barker, J. L., and Levitan, H., 1971, Salicylate: Effect on membrane permeability of molluscan neurons, *Science* **172**:1245.

Beckman, A. L., and Roskowska-Ruttimann, E., 1974, Hypothalamic and septal neuronal responses to iontophoretic application of salicylate in rats, *Neuropharmacology* **13**:393.

Beleslin, D. B., and Myers, R. D., 1971, Release of an unknown substance from brain structures of unanaesthetized monkeys and cats, *Neuropharmacology* **10**:121.

Bergström, S., Farnebo, L. A., and Fuxe, K., 1973, Effect of prostaglandin E_2 on central and peripheral catecholamine neurons, *Eur. J. Pharmacol.* **21**:362.

Bligh, J., and Milton, A. S., 1973, The thermoregulatory effects of prostaglandin E_1 when infused into a lateral cerebral ventricle of the Welsh mountain sheep at different ambient temperatures, *J. Physiol.* **229**:30P.

Bligh, J., Cottle, W. H., and Maskrey, M., 1971, Influence of ambient temperature on the thermoregulatory responses to 5-hydroxytryptamine, noradrenaline and acetylcholine injected into the lateral cerebral ventricles of sheep, goats and rabbits, *J. Physiol.* **212**:377.

Brody, M. J., and Kadowitz, P. J., 1974, Prostaglandins as modulators of the autonomic nervous system, *Fed. Proc.* **33**:48.

Cabanac, M., 1970, Interaction of cold and warm temperature signals in the brain stem, in: *Physiological and Behavioral Temperature Regulation* (J. D. Hardy, A. P. Gagge, and J. A. J. Stolwijk, eds.), p. 549, Charles C. Thomas, Springfield, Ill.

Cabanac, M., Stolwijk, J. A. J., and Hardy, J. D., 1968, Effect of temperature and pyrogens on single-unit activity in the rabbit's brain stem, *J. Appl. Physiol.* **24**:645.

Cabanac, M., Duclaux, R., and Gillet, A., 1970, Thermoregulation comportementale chez le chien: Effets de la fièvre et de la thyroxine, *Physiol. Behav.* **5**:697.

Chai, C. Y., and Lin, M. T., 1975, Hypothermic effect of sodium acetysalicylate on afebrile monkeys, *Br. J. Pharmacol.* **54**:475.

Chambers, W. W., Koenig, H., Koenig, R., and Windue, W. F., 1949, Site of action in the central nervous system of bacterial pyrogen, *Am. J. Physiol.* **159**:209.

Clark, W. G., and Cumby, H. R., 1975, Effects of prostaglandin antagonist SC-19220 on body temperature and on hyperthermic responses to prostaglandin E_1 and leucocytic pyrogen in the cat, *Prostaglandins* **9**:361.

Coceani, F., 1974, Prostaglandins and the Central Nervous System, *Arch. Intern. Med.* **133**:119.

Coceani, F., Puglisi, L., and Lavers, B., 1971, Prostaglandins and neuronal activity in spinal cord and cuneate nucleus, *Ann. N.Y. Acad. Sci.* **180**:289.

Cooper, K. E., 1965, The role of the hypothalamus in the genesis of fever, *Proc. R. Soc. Med.* **58**:740.

Cooper, K. E., and Veale, W. L., 1972, The effect of injecting an inert oil into the cerebral ventricular system upon fever produced by intravenous leucocyte pyrogen, *Can. J. Physiol. Pharmacol.* **50**:1066.

Cooper, K. E., and Veale, W. L., 1973, Exchange between the blood brain and cerebrospinal fluid of substances which can induce or modify febrile responses, in: *The Pharmacology of Thermoregulation* (Symp. San Francisco, 1972), p. 278, Karger, Basel.

Cooper, K. E., and Veale, W. L., 1974, Fever, an abnormal drive to the heat-conserving and -producing mechanisms? in: *Recent Studies of Hypothalamic Function* (Int. Symp. Calgary, 1973) (K. E. Cooper and K. Lederis, eds.), p. 391, Karger, Basel.

Cooper, K. E., Cranston, W. I., and Snell, E. S., 1964, Temperature regulation during fever in man, *Clin. Sci.* **27**:345.

Cooper, K. E., Cranston, W. I., and Honour, A. J., 1965, Effects of intraventricular and intra hypothalamic injection of noradrenaline and 5-HT on body temperature in conscious rabbits, *J. Physiol. (London)* **181**:852.

Cooper, K. E., Cranston, W. I., and Honour, A. J., 1967, Observations on the site and mode of action of pyrogens in the rabbit brain, *J. Physiol.* **191**:325.

Cooper, K. E., Pittman, Q. J., and Veale, W. L., 1975, Observations on the development of the "fever" mechanism in the fetus and newborn, in: *Temperature Regulation and Drug Action* (P. Lomax, E. Schonbaum, and J. Jacob, eds.), p. 43, Karger, Basel.

Cooper, K. E., Preston, E., and Veale, W. L., 1976*a*, Effects of intraventricular reserpine and atropine on fever produced by intravenous pyrogens, *J. Physiol.* **254**:729–741.

Cooper, K. E., Veale, W. L., and Pittman, Q. J., 1976*b*, The pathogenesis of fever, in: *Brain*

Dysfunction in Infantile Febrile Convulsions (M. A. B. Brazier and F. Coceani, eds.), pp. 107–115, Raven Press, New York.

Cranston, W. I., Luff, R. H., Rawlins, M. D., and Rosendorff, C., 1970, The effects of salicylate on temperature regulation in the rabbit, *J. Physiol.* **208**:251.

Cranston, W. I., Hellon, R. F., and Mitchell, D., 1975a, Is brain prostaglandin synthesis involved in responses to cold? *J. Physiol.* **249**:425.

Cranston, W. I., Hellon, R. F., and Mitchell, D., 1975b, Fever and brain prostaglandin release, *J. Physiol.* **248**:27.

Cranston, W. I., Hellon, R. F., and Mitchell, D., 1975c, A dissociation between fever and prostaglandin concentration in cerebrospinal fluid, *J. Physiol.* **253**:583.

Dey, P. K., Feldberg, W., Gupta, K. P., Milton, A. S., and Wendlandt, S., 1974, Further studies on the role of prostaglandin in fever, *J. Physiol.* **241**:629.

Eisenman, J. S., 1969, Pyrogen-induced changes in thermosensitivity of septal and preoptic neurons, *Am. J. Physiol.* **216**:330.

Feldberg, W., 1975, Body temperature and fever: Changes in our views during the last decade, *Proc. Roy. Soc. Lond. B.* **191**:199.

Feldberg, W., and Myers, R. D., 1963, A new concept of temperature regulation by amines in the hypothalamus, *Nature (London)* **200**:1325.

Feldberg, W., and Myers, R. D., 1964, Effects of temperature of amines injected into the cerebral ventricles: A new concept of temperature regulation, *J. Physiol. (London)* **173**:226.

Feldberg, W., and Myers, R. D., 1966, Appearance of 5-hydroxytryptamine and an unidentified pharmacologically active lipid acid in effluent from perfused cerebral ventricles, *J. Physiol.* **184**:837.

Feldberg, W., and Gupta, K. P., 1973, Pyrogen fever and prostaglandinlike activity in cerebrospinal fluid, *J. Physiol.* **228**:41.

Feldberg, W., Gupta, K. P., Milton, A. S., and Wendlandt, S., 1973, Effect of pyrogen and antipyretics on prostaglandin activity in central C.S.F. of unanaesthetized cats, *J. Physiol.* **234**:279.

Ferraris, V. A., Derubertis, F. R., Thomas, T. H., and Wolfe, L., 1974, Release of prostaglandin by mitogen- and antigen-stimulated leucocytes in culture, *J. Clin. Invest.* **54**:378.

Flower, R. J., and Vane, J. R., 1972, Inhibition of prostaglandin synthetase in brain explains the anti-pyretic activity of paracetamol (4-acetamido-phenol), *Nature (London)* **240**:410.

Ford, D. M., 1974, A selective action of prostaglandin E_1 on hypothalamic neurons in the cat which respond to brain cooling, *J. Physiol.* **242**:142P.

Fox, R. H., and MacPherson, R. K., 1954, The regulation of body temperature during fever, *J. Physiol.* **125**:21.

Francesconi, R. P., and Mager, M., 1975, Salicylate, tryptophan, and tyrosine hypothermia, *Am. J. Physiol.* **228**:1431.

Hales, J. R. S., Bennett, J. W., Baird, J. A., and Fawcett, A. A., 1973, Thermoregulatory effects of prostaglandins E_1, E_2, $F_{1\alpha}$, and $F_{2\alpha}$ in the sheep, *Pflügers Arch.* **339**:125.

Hammel, H. T., 1965, Neurons and temperature regulation, in: *Physiological Controls and Regulations* (W. S. Yamamoto and J. R. Brobeck, eds.), p. 71, Saunders, Philadelphia.

Hamprecht, B., Jaffe, B. M., and Philpott, G. W., 1973, Prostaglandin production by neuroblastoma, glioma, and fibroblast cell lines; stimulation by N6-O2'-dibutyryl adenosine 3':5'-cyclic monophosphate, *FEBS Lett.* **36**:193.

Harvey, C. A., Milton, A. S., and Straughan, D. W., 1975, Prostaglandin E levels in cerebral spinal fluid of rabbits and the effects of bacterial pyrogen and antipyretic drugs, *J. Physiol.* **248**:26P.

Hellon, R. F., 1967, Thermal stimulation of hypothalamic neurons in unanaesthetized rabbits, *J. Physiol.* **193**:381.

Hellon, R. F., 1970, Hypothalamic neurons responding to changes in hypothalamic and ambient temperatures, in: *Physiological and Behavioral Temperature Regulation* (J. D. Hardy, A. P. Gagge, and J. A. J. Stolwijk, eds.), p. 463, Charles C Thomas, Springfield, Ill.

Hellon, R. F., 1975, Monoamines, pyrogens and cations: Their actions on central control of body temperature, *Pharmacol. Rev.* **26**:289.

Higgs, G. A., and Youlten, L. J., 1972, Prostaglandin production by rabbit peritoneal polymorpho-nuclear leucocytes *in vitro, Br. J. Pharmacol.* **44**:330P.

Higgs, G. A., McCall, E., and Youlten, L. J. F., 1975, A chemotactic role for prostaglandins released from polymorphonuclear leucocytes during phagocytosis, *Br. J. Pharmacol.* **53**:539.

Holmes, S. W., 1970, The spontaneous release of prostaglandins into the cerebral ventricles of the dog and the effect of external factors on this release, *Br. J. Pharmacol.* **38**:653.

Holmes, S. W., and Horton, E. W., 1968, The identification of four prostaglandins in dog brain and their regional distribution in the central nervous system, *J. Physiol.* **195**:731.

Hopkin, J. M., Horton, E. W., and Whittaker, V. P., 1968, Prostaglandin content of a particulate and supernatent fraction of rabbit brain homogenates, *Nature (London)* **217**:71.

Horton, E. W., and Main, I. H. M., 1967, Identification of prostaglandins in central nervous tissue of the cat and chicken, *Br. J. Pharmacol.* **30**:582.

Jackson, D. L., 1967, A hypothalamic region responsive to localized injection of pyrogens, *J. Neurophysiol.* **30**:586.

Kataoka, K., Ramwell, P. N., and Jessop, S., 1967, Prostaglandins: Localization in subcellular particles of rat cerebral cortex, *Science* **157**:1187.

Lomax, P., 1970, Drugs and body temperature, *Int. Rev. Neurobiol.* **12**:1.

Martin, F. H., and Baile, C. A., 1973, Feeding elicited in sheep by intrahypothalamic injections of PGE$_1$, *Experientia* **29**:306.

Milton, A. S., and Wendlandt, S., 1970, A possible role for prostaglandin E$_1$ as a modulator for temperature regulation in the central nervous system of the cat, *J. Physiol. (London)* **207**:76.

Milton, A. S., and Wendlandt, S., 1971, Effects on body temperature of prostaglandins of the A, E, and F series on injection into the third ventricle of unanaesthetized cats and rabbits, *J. Physiol. (London)* **218**:325.

Murakami, N., and Sakata, Y., 1975, Effects of antipyretics on normothermic rabbits, *Jpn. J. Physiol.* **25**:29.

Myers, R. D., 1970, An improved push–pull cannula system for perfusing an isolated region of the brain, *Physiol. Behav.* **5**:243.

Myers, R. D., 1974, *Handbook of Drug and Chemical Stimulation of the Brain,* Van Nostrand Reinhold, New York.

Myers, R. D., and Veale, W. L., 1970, Body temperature: Possible ionic mechanism in the hypothalamus controlling the set point, *Science* **170**:95.

Myers, R. D., and Veale, W. L., 1971, The role of sodium and calcium ions in the hypothalamus in the control of body temperature of the unanaesthetized cat, *J. Physiol.* **212**:411.

Myers, R. D., and Waller, M. B., 1973, Differential release of acetylcholine from the hypothalamus and mesencephalon of the monkey during thermoregulation, *J. Physiol.* **230**:273.

Myers, R. D., Tytell, M., Kawa, A., and Rudy, T., 1971, Micro-injection of ^{3}H-norepinephrine into the hypothalamus of the rat: Diffusion into tissue and ventricles, *Physiol. Behav.* **7**:743.

Nakayama, T., and Hardy, J. D., 1969, Unit responses in the rabbit's brain stem to changes in brain and cutaneous temperature, *J. Appl. Physiol.* **27**:848.

Newman, P. P., 1964, *Visceral Afferent Functions of the Nervous System,* Edward Arnold Publishers Ltd., London.

Newman, P. P., and Wolstencroft, J. H., 1960, Cardiovascular and respiratory responses to heating the carotid blood, *J. Physiol.* **152**:87.

Norton, W. T., and Poduslo, S. E., 1970, Neuronal soma and whole neuroglia of rat brain: A new isolation technique, *Science* **167**:1144.

Parnas, I., Reinhold, R., and Fine, J., 1971, Synaptic transmission in the crayfish: Increased release of transmitter substance by bacterial endotoxin, *Science* **171**:1153.

Philipp-Dormston, W. K., and Siegert, R., 1974, Prostaglandins of the E and F series in rabbit cerebrospinal fluid during fever induced by Newcastle disease virus, *E. coli*-endotoxin, or endogenous pyrogen, *Med. Microbiol. Immunol.* **159**:279.

Pittman, Q. J., Cooper, K. E., Veale, W. L., and Van Petten, G. R., 1973, Fever in newborn lambs, *Can. J. Physiol. Pharmacol.* **51**:868.

Pittman, Q. J., Cooper, K. E., Veale, W. L., and Van Petten, G. R., 1974, Observations on the development of the febrile response to pyrogens in sheep, *Clin. Sci. Mol. Med.* **46**:591.

Pittman, Q. J., Veale, W. L., and Cooper, K. E., 1975*a*, Temperature responses of lambs after centrally injected prostaglandins and pyrogens, *Am. J. Physiol.* **228**:1034.

Pittman, Q. J., Veale, W. L., and Cooper, K. E., 1975*b*, Effect of prostaglandin, bacterial pyrogen and norepinephrine, injected into the hypothalamus, on thermoregulation in the newborn lamb, *Neurosci, Abst.* **1**:301.

Pittman, Q. J., Veale, W. L., and Cooper, K. E., 1976, Observations on the effect of salicylate in fever and the regulation of body temperature against cold, *Can. J. Physiol. Pharmacol.* **54**:101–106.

Polk, D. L., and Lipton, J. M., 1975, Effects of sodium salicylate, aminopyrine and chlorpromazine on behavioral temperature regulation, *Pharmacol. Biochem. Behav.* **3**:167.

Poulain, P., and Carette, B., 1974, Iontophoresis of prostaglandins on hypothalamic neurons, *Brain Res.* **79**:311.

Ramwell, P. W., and Shaw, J. E., 1966, Spontaneous and evoked release of prostaglandins from the cerebral cortex of anaesthetized cats, *Am. J. Physiol.* **211**:125.

Ramwell, P. W., Shaw, J. E., and Jessop, R., 1966, Spontaneous and evoked release of prostaglandins from frog spinal cord, *Am. J. Physiol.* **211**:998.

Ranson, S. W. Jr., Clarke, G., and Magoun, H. W., 1939, Effect of hypothalamic lesions on fever induced by intravenous injection of typhoid–paratyphoid vaccine, *J. Lab. Clin. Med.* **25**:160.

Repin, I. S., and Kratskin, I. L., 1967, An analysis of hypothalamic mechanisms of fever, *Fiziol. Zh. SSSR* **53**:1206.

Rosendorff, C., and Cranston, W. I., 1968, Effects of salicylate on human temperature regulation, *Clin. Sci.* **35**:81.

Rosendorff, C., and Mooney, J. J., 1971, Central nervous system sites of action of a purified leucocyte pyrogen, *Am. J. Physiol.* **220**:597.

Samuelsson, B., 1964, Identification of a smooth muscle-stimulating factor in bovine brain, *Biochim. Biophys. Acta* **84**:218.

Sanner, J. H., 1974, Substances that inhibit the actions of prostaglandins, *Arch. Intern. Med.* **133**:133.

Satinoff, E., 1972, Salicylate: Action on normal body temperature in rats, *Science* **176**:532.

Schoener, E. P., and Wang, S. C., 1974, Effects of leucocytic pyrogen and Na aspirin on PO/AH neurons in cats, *Fed. Proc.* **33**:458.

Stitt, J. T., 1973, Prostaglandin E$_1$ fever induced in rabbits, *J. Physiol.* **232**:163.

Stitt, J. T., and Hardy, J. D., 1972, Evidence that prostaglandin E$_1$ may be a mediator in pyrogenic fever in rabbits, *Biometeorology* **5**:112.

Stitt, J. T., and Hardy, J. D., 1975, Microelectrophoresis of PGE$_1$ onto single units in the rabbit hypothalamus, *Am. J. Physiol.* **229**:240.

Tanner, J. M., 1951, The relation between serum cholesterol and physique in healthy young men, *J. Physiol.* **115**:371.

Thauer, R., 1970, Thermosensitivity of the spinal cord, in: *Physiological and Behavioral Temperature Regulation* (J. D. Hardy, A. P. Gagge, and J. A. I. Stolwijk, eds.), p. 472, Charles C Thomas, Publisher, Springfield, Ill.

Thompson, R. H., Hammel, H. T., and Hardy, J. D., 1959, Calorimetric studies in temperature regulation: The influence of cold, neutral, and warm environment upon pyrogenic fever in normal and hypothalectomized dogs, *Fed. Proc.* **18**:159.

Vane, J. R., 1971, Inhibition of prostaglandin synthesis as a mechanism of action for aspirin-like drugs, *Nature (London) New Biol.* **231**:232.

Veale, W. L., 1972, A stereotaxic method for the push-pull perfusion of discrete regions of brain tissue of the unanaesthetized rabbit, *Brain Res.* **42**:479.

Veale, W. L., and Cooper, K. E., 1973, Species differences in the pharmacology of temperature regulation, in: *The Pharmacology of Thermoregulation* (E. Schönbaum and P. Lomax, eds.), p. 289, Karger, Basel.

Veale, W. L., and Cooper, K. E., 1974*a*, Evidence for the involvement of prostaglandins in fever, in: *Recent Studies of Hypothalamic Function* (K. Lederis and K. E. Cooper, eds.), p. 359, Karger, Basel.

Veale, W. L., and Cooper, K. E., 1974*b*, Prostaglandin in cerebrospinal fluid following perfusion of hypothalamic tissue, *J. Appl. Physiol.* **37**:942.

Veale, W. L., and Cooper, K. E., 1975, Comparison of sites of action of prostaglandin E and leucocyte pyrogen in brain, in: *Temperature Regulation and Drug Action* (P. Lomax, E. Schönbaum and J. Jacob, eds.), p. 218, Karger, Basel.

Veale, W. L., and Whishaw, I. Q., 1976, Body temperature responses at different ambient temperatures following injections of prostaglandin E_1 and noradrenaline into the brain, *Pharmacol. Biochem. Behav.* **4**:143–150.

Villablanca, J., and Myers, R. D., 1965, Fever produced by microinjection of typhoid vaccine into hypothalamus of cats, *Am. J. Physiol.* **208**:703.

Vogt, M., 1954, The concentration of sympathin in different parts of the central nervous system under normal conditions and after administration of drugs, *J. Physiol.* **123**:451.

Wit, A., and Wang, S. C., 1968, Temperature-sensitive neurons in preoptic/anterior hypothalamic region: Actions of pyrogen and acetylsalicylate, *Am. J. Physiol.* **215**:1160.

Wolfe, L. S., Coceani, F., and Pace-Asciak, C., 1967, Brain prostaglandins on the isolated rat stomach, in: *Prostaglandins, Nobel Symposium 2* (S. Bergström and B. Samuelsson, eds.), p. 265, Almquist and Wiksell, Stockholm.

Prostaglandins and the Lung

Aleksander A. Mathé

Laboratory of Biogenic Amines and Allergy,
and Departments of Psychiatry and Pharmacology
Boston University School of Medicine
Boston, Massachusetts 02118

I. INTRODUCTION

Although the discovery of the pulmonary capillaries dates back three centuries (Malpighius, 1661) and the true nature of the interchange of gases in the lung was elucidated 200 years ago (Lavoisier, 1778), the first observation that lung may have functions other than respiratory was made in this century (Starling and Verney, 1925). Subsequent work has established the many and variegated nonrespiratory functions of this organ. Several reviews have dealt with the metabolic and endocrinological aspects of lung function, including the release of the chemical mediators of anaphylaxis (Heinemann and Fishman, 1969; Said, 1973; Bakhle and Vane, 1974; Fishman and Pietra, 1974; Kaliner and Austen, 1975).

The discovery of $PGF_{2\alpha}$ in the lung and demonstration of the effects of prostaglandins on the airways (Bergström *et al.*, 1962; Änggård and Bergström, 1963) have stimulated investigation of the action of prostaglandins on the respiratory smooth muscle and pulmonary vasculature (Cuthbert, 1973; Smith, 1973*a*). The aim of this chapter is to review the current knowledge and speculations on the role of prostaglandins in both respiratory and nonrespiratory lung function.

II. PROSTAGLANDIN CONTENT AND SYNTHESIS IN THE LUNG

A. Content

Prostaglandins of the E type were first obtained from the sheep vesicular gland and of the F type from the sheep and swine lung (Bergström *et al.*, 1962). Later, it was shown that incubation of guinea pig lung homogenates with arachidonic acid produced predominantly $PGF_{2\alpha}$ and relatively little PGE_2 (Bergström *et al.*, 1963). Further experiments by Änggård (1965) demonstrated the presence of small amounts of PGE_2 in sheep and a much higher concentration range of $PGF_{2\alpha}$ in human, monkey, guinea pig, and sheep lungs. Very small amounts of $PGF_{1\alpha}$ and $PGF_{3\alpha}$ were also detected in sheep lung. Karim *et al.* (1967) investigated the distribution of prostaglandins in human tissues. Using a bioassay, 1.3 and 2.4 ng PGE_2 and 2.4 and 50.0 ng $PGF_{2\alpha}$ per gram wet weight lung parenchyma were found in two samples. The levels of PGE_2 and $PGF_{2\alpha}$ for the bronchi were reported as 4.5 and 7.8 and 1.0 and 2.5, respectively. Prostaglandin content was further studied in chicken, rabbit, guinea pig, rat, cat, and dog lung (Karim *et al.*, 1968). Except for cat lung, higher levels of $PGF_{2\alpha}$ than PGE_2 were reported in lungs of these species. In view of the subsequent realization that prostaglandins are not stored in the tissue but synthesized *de novo* and released upon stimulation by factors such as mechanical damage or simple tissue handling, the interpretation of these results is difficult. Furthermore, the postmortem period, during which there can occur a substantial release and catabolism of prostaglandins, varied—some tissues being obtained up to 24 hr after death.

Spontaneous release of prostaglandins from human and guinea pig lungs, using a radioimmunoassay to measure PGE_2, $PGF_{2\alpha}$, and 15-keto-13,14-dihydro-$PGF_{2\alpha}$,* has been recently studied (Mathé and Levine, unpublished results; Yen *et al.*, 1976). Macroscopically healthy parts of human lungs obtained at surgery (usually for cancer) were freed of pleura, large bronchi, and large blood vessels, and the tissue and bronchi were sliced into approximately 1–2 mm^3 fragments. After washing, the fragments were incubated in Tyrode's solution (37°C, pH 7.4, bubbled with 5% CO_2 in O_2) and the amounts of prostaglandins and histamine spontaneously released after 10-min incubation were measured. Healthy guinea pigs were sacrificed, and

---

*The following notations are used for our data presented in this chapter: PGE_2, immunoreactive PGE_1 and/or PGE_2; $PGF_{2\alpha}$, immunoreactive $PGF_{2\alpha}$; 15-keto-13,14-dihydro-$PGF_{2\alpha}$ (metabolite), immunoreactive 15-keto-13,14-dihydro-$PGF_{2\alpha}/E_2$. For the discussion of radioimmunoassay specificities, see Mathé *et al.* (1976*f*).

the lungs were rinsed via the pulmonary artery with Tyrode's solution and excised together with trachea. Lung fragments and pieces of trachea were then prepared and treated as human lungs. The results are presented in Table I.

The results show that more $PGF_{2\alpha}$ than PGE_2 was found in media incubated with parenchymal tissue. In contrast, the PGE_2-to-$PGF_{2\alpha}$ ratio was markedly higher in the media from the tracheobronchial tree than lung parenchyma. This indicates either that more $PGF_{2\alpha}$ than PGE_2 was synthesized, or that the PGE_2 9-ketoreductase (9K-PGR) was more active, or that PGE_2 was removed faster than $PGF_{2\alpha}$ in lung parenchyma than was the case for the trachea and bronchi. The data would also indicate that the catabolism of prostaglandins is slower in tracheal tissue than in lung parenchyma. A less likely possibility is that further oxidation of the measured metabolite is considerably faster in the former than in the latter. Thus, although the concentrations of the parent compounds released from the airways and lung parenchyma are different, the total synthesizing capacity (assuming that PGE_2 plus $PGF_{2\alpha}$ plus metabolite adequately reflect the amount of PGs synthesized) may be quite similar. Larger PGE_2 than $PGF_{2\alpha}$ synthesis in the airways with a seemingly slow prostaglandin degradation rate is consistent with the importance for the organism of insuring the patency of the tracheobronchial tree. On the other hand, larger concentrations of $PGF_{2\alpha}$ than PGE_2 in lung parenchyma, which is associated with an apparently fast catabolic rate, would serve the useful purpose of maintaining adequate circulation. Lung circulation may, at least in part, depend on the capacity of $PGF_{2\alpha}$ to constrict segments of pulmonary vasculature, and thereby direct the blood flow from areas of lesser to those of higher aeration.

Table I. Spontaneous Release of Prostaglandins from Human and Guinea Pig Lung Fragments[a]

	PGE_2	$PGF_{2\alpha}$	Metabolite
Human:			
Parenchyma	1.2 ± 0.1	2.3 ± 0.1	2.7 ± 0.4
Bronchi	1.9 ± 0.1	1.7 ± 0.2	0.4 ± 0.0
Guinea pig:			
Parenchyma	1.5 ± 0.1	3.8 ± 0.3	8.6 ± 1.0
Trachea	7.3 ± 1.1	5.3 ± 0.5	trace

[a]Lung fragments were incubated in Tyrode's solution (37°C, pH 7.4, bubbled with 5% CO_2 in O_2), and spontaneously released prostaglandins were measured in the medium. Results expressed as mean ± SE ng prostaglandin released/g wet weight/min. $N = 9$ lungs/group.

One of the main conditions for normal respiratory lung function is maintenance of adequate ventilation–perfusion ratios. Larger synthesis of the bronchodilatory PGE_2 in the airways and of the vasoconstrictive $PGF_{2\alpha}$ in the lung parenchyma seems to be consistent with this function.

In addition to measuring the content or release of prostaglandins, in some experiments tissues were incubated or perfused with the labeled precursors of prostaglandins and the prostaglandin products were determined. This approach has led Samuelsson (1965) to postulate the existence of an endoperoxide intermediate in prostaglandin synthesis. Several years later, two endoperoxides named PGH_2 and PGG_2 (PGG_2 differs from PGH_2 by the attachment of the 15-L-hydroperoxy group instead of the 15-L-hydroxy group) were isolated from sheep vesicular gland preparations incubated with [^{14}C]arachidonic acid and from platelets during aggregation induced by thrombin (Hamberg and Samuelsson, 1973; Nugteren and Hazelhof, 1973; Hamberg *et al.*, 1974). Endoperoxides and their products were also found when the effluents from guinea pig lungs, perfused via the pulmonary artery, were analyzed for prostaglandins and hydroxy acids before and after injection of 30 μg arachidonic acid (Hamberg and Samuelsson, 1974). Basal outflow of these compounds was low. However, following arachidonic acid perfusion there was a marked increase in the release of thromboxane B_2 (TxB_2, previously called PHD (Hamberg *et al.*, 1976) [8-(1-hydroxy-3-oxopropyl)-9, 12-L-dihydroxy-5, 10-heptadecadienoic acid]), HHT (12-hydroxy-5,8,10-heptadecatrienoic acid), PGE_2, $PGF_{2\alpha}$, and also HETE (12-L-hydroxy-5,8,10,14-eicosatetraenoic acid) which is a product of the lipoxygenase and not the synthetase system (cyclo-oxygenase) activity. Structures of these compounds are shown in Fig. 1. Similar results were obtained after guinea pig lung homogenates were incubated with [^{14}C]arachidonic acid. It is noteworthy that several times more TxB_2 and HHT (both PGG_2 metabolites) than $PGF_{2\alpha}$ and PGE_2 were measured. This indicates that an important pathway of endoperoxide catabolism, in addition to prostaglandins (and their metabolites), is production of thromboxanes and other hydroxy acids. In summary, lungs of all species investigated to date possess the capacity to synthesize and release ample amounts of prostaglandins, their intermediates, and their metabolites.

B. Synthesis

The synthetase system has been most extensively studied in sheep and bovine seminal vesicles, and ample literature on the subject is available (e.g., Samuelsson, 1972; Sih and Takeguchi, 1973; Samuelsson and Paoletti, 1976). Since, as already discussed, prostaglandin

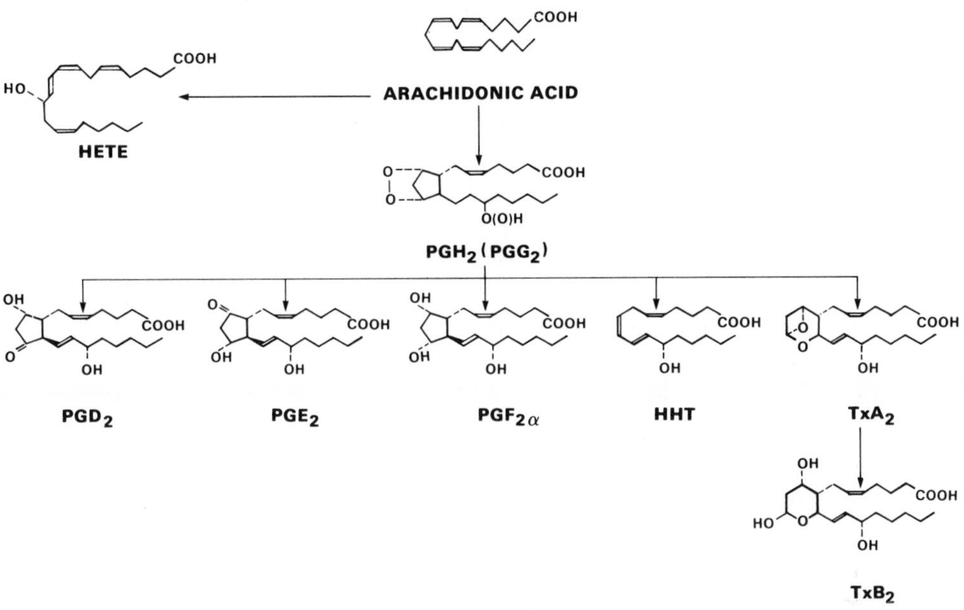

Fig. 1. Transformation of arachidonic acid in guinea pig lung.

liberation can be equated with *de novo* synthesis, the ability of lungs from several species to release significant amounts of prostaglandins in response to a variety of stimuli indicates the presence of an active prostaglandin synthetase system. Änggård and Samuelsson (1965) have shown that guinea pig lung homogenates convert arachidonic acid into $PGF_{2\alpha}$ and PGE_2 (thus demonstrating that the biosynthesis of prostaglandins is not restricted to the reproductive system). However, when compared with the extensive exploration of factors influencing the release of prostaglandins from the lung, there have been few studies on prostaglandin lung enzymes.

As in other organs, the synthetase system of lung is present in the microsomal fractions of tissue homogenates. Parkes and Eling (1974) showed that MI, a compound less polar than PGE_2, was the major material produced by guinea pig lung microsomes and that MII ($PGF_{2\alpha}$) constituted only a minor fraction. Concomitantly, Hamberg and Samuelsson (1974) isolated TxB_2 in guinea pig lung homogenates and perfusates. This MI material has since been shown to be the same substance as TxB_2 (Hamberg and Samuelsson, 1974; Eling and Anderson, 1975). The ratio of TxB_2 to $PGF_{2\alpha}$ produced by nonstimulated washed microsomes was 30:1. Epinephrine, norepinephrine, and 5-hydroxytryptamine (5-HT) stimulated synthesis of both TxB_2 and $PGF_{2\alpha}$. An additional interesting finding was the existence of a heat-stable factor in the 100,000g supernatant fraction, which appeared to be essential for

$PGF_{2\alpha}$ but not TxB_2 synthesis (Parkes and Eling, 1974; Eling and Anderson, 1975). As in other organs, aspirin-like drugs inhibited guinea pig lung synthetase activity (Vane, 1971).

In an initial study of the guinea pig lung microsomal prostaglandin synthetase enzyme (Leslie and Mathé, unpublished) the enzyme preparations alone incubated with the substrate (arachidonic acid) synthesized little prostaglandin. To produce significant amounts of $PGF_{2\alpha}$ in both antigen-sensitized and control preparations two cofactors had to be added, namely hydroquinone (HQ) and reduced glutathione (GSH). Alone HQ had little or no effect. Both preparations synthesized from $1\frac{1}{2}$ to 4 times more $PGF_{2\alpha}$ than PGE_2. Furthermore, the antigen-sensitized preparations produced more $PGF_{2\alpha}$ than the controls. Thus in a typical experiment the sensitized preparation would produce 175 ng $PGF_{2\alpha}$/mg protein/8 min whereas the control one would synthesize 109 ng $PGF_{2\alpha}$ under identical conditions. These results are consistent with the greater outflow of $PGF_{2\alpha}$ than PGE_2 observed in both sensitized and control perfused lungs, and the larger spontaneous outflow of $PGF_{2\alpha}$ and metabolite from sensitized lungs (Mathé *et al.*, 1976*f*). Benzie *et al.* (1975) also reported that microsomal preparations from sensitized guinea pig lungs incubated with labeled arachidonic acid synthesized more of the labeled products $PGF_{2\alpha}$ and PGE_2 as compared to controls.

Neither in the absence nor in the presence of added exogenous cofactors did histamine increase prostaglandin production. Thus the ability of histamine to increase prostaglandin release from both sensitized and control guinea pig lungs (Yen *et al.*, 1976) may not be a direct effect on the activity of prostaglandin synthetase.

A cytoplasmic factor was found to stimulate the synthetase preparations to produce $PGF_{2\alpha}$ in the absence of exogenous cofactors (Leslie and Mathé, unpublished). This stimulatory factor was present in both sensitized and control lungs. Addition of this factor to the homologous lung produced more $PGF_{2\alpha}$ in the sensitized lung than in the control lung. In the sensitized lungs, twice as much $PGF_{2\alpha}$ was produced by the cytoplasmic factor as was by the two exogenously added cofactors. In contrast, in the control lung the stimulation of synthesis was equal to that obtained in the presence of cofactors. Alteration in the prostaglandin synthetase activity in the state of sensitization may be secondary to changes in cofactors. It could be speculated that this might be due to either removal of an inhibitory factor or increased availability of a stimulatory compound.

In conclusion, as in other organs the synthetase system in the lung is contained in the microsomal fraction. Either a cytoplasmic factor or HQ and GSH are necessary for its activation in *in vitro* experiments.

More $PGF_{2\alpha}$ than PGE_2 is synthesized under such conditions. The enzyme activity appears to be higher in antigen-sensitized than in non-sensitized control preparations.

III. CATABOLISM AND UPTAKE OF PROSTAGLANDINS IN THE LUNG

A. Catabolic Enzymes

1. 15-Hydroxyprostaglandin Dehydrogenase (15-PGDH)

The sequence of prostaglandin catabolism, i.e., 15-hydroxyl group oxidation, $\Delta^{13,14}$ double bond reduction, and β and ω oxidation has been extensively studied and reviewed (Samuelsson *et al.*, 1971; Samuelsson, 1972; Oesterling *et al.*, 1972; Pong and Levine, 1976). The isolation and properties of 15-PGDH have been reviewed by Marrazzi and Andersen (1974). Consequently, only those aspects of 15-PGDH which relate to lung will be reported here. The first step in the biological inactivation of prostaglandins is catalyzed by 15-PGDH (Änggård and Samuelsson, 1966). From the available data, the 15-PGDH activity may be the rate-limiting step in prostaglandin catabolism and also an important one, since the oxidized prostaglandins generally possess much less biological activity than the parent compound (see Section VD for further discussion). While this enzyme activity is distributed in most tissues (Änggård *et al.*, 1971), many of the initial observations on the catabolizing enzymes and the identification of the metabolites were made in guinea pig and swine lungs (Änggård and Samuelsson, 1966). Considering the marked capacity of the lung to synthesize prostaglandins and to take up (and inactivate) the circulating prostaglandins, the presence of an active 15-PGDH in lung is consistent with the view that prostaglandins are primarily local hormones. The enzyme is localized in the cytoplasmic fraction of lung homogenates, and it is NAD^+ dependent and specific for the C-15(S) alcohol group (Änggård *et al.*, 1971; Samuelsson, 1972). A second 15-PGDH which is $NADP^+$ dependent (type II) has been recently isolated (Lee and Levine, 1975). The significance of this enzyme type for prostaglandin inactivation in lung is not clear at present. It has been suggested, however, that, in general, 15-PGDH type II activity coupled with 9K-PGR might regulate the relative levels of PGE_2 and $PGF_{2\alpha}$ (Lee and Levine 1974*a,b*, 1975). Currently, little is known about factors regulating the lung 15-PGDH activity. Sun *et al.* (1976) reported that 15-PGDH activity in monkey lung is not affected by steroids, calcium, and cyclic AMP, but that it is inhibited by prostanoic acid and furosemide. In human placenta 15-PGDH is influenced by estrogen and progestesterone (Schlegel et al., 1974), and in thyroid gland it is inhibited by thyroid hormones (Tai et al., 1974). Eling and Anderson (1975) have shown that exposure to 100% oxygen does not change prostaglandin synthesis but decreases the catabolism (15-PGDH activity) in guinea pig lungs. By analogy, bradykinin, angiotensin, and other substances associated with lung function might modify the enzyme. Nakano and Prancan (1973) found that degradation of $[^3H]PGE_1$ in rat lung was reduced after an intraperitoneal injection of endotoxin 8 hr preceding the experiment. If prostaglandins are the result of or are of significance (either primarily or secondarily) in the development of certain pathological conditions (toxemia, fever, migraine, hypertension, renal disease), a locally decreased ability of an organ to catabolize these compounds or an impaired lung clearance could play an important etiological role.

2. Prostaglandin Δ^{13}-Reductase (13-PGR)

Existence of a second catabolic enzyme was indicated when it was shown that in addition to oxidation of the secondary alcohol at the C-15 position there occurred a reduction at the $\Delta^{13,14}$ double bond when labeled PGE_1, PGE_2, and $PGF_{2\alpha}$ were incubated with the cytoplasmic fraction of lung homogenates (Änggård and Samuelsson, 1964, 1965, 1966; Änggård et al., 1965, 1971). The properties of this enzyme were reviewed by Samuelsson (1972), and it was subsequently purified by Lee and Levine (1974c). 13-PGR from the lungs of several species used NADPH as a cofactor and not NADP or NAD. It reduced the $\Delta^{13,14}$ double bond only when the 15-keto group was there. This last finding confirmed previous investigations (Samuelsson, 1972) and pointed to the fact that the action of 15-PGDH precedes that of 13-PGR.

Studies on the variations of enzyme activities in lung with age (Pace-Asciak and Miller, 1973; Sun and Armour, 1974; Bedwani and Marley, 1975) have revealed increased 13-PGR activity in neonatal rats and high 15-PGDH activity in maternal rabbits near term. This indicates that an important function of the lung might be to protect the organism from excessive prostaglandins. In this context it is noteworthy that indomethacin, at a concentration of 15 μM, inhibited the NAD^+-dependent catabolism of PGE_1 and PGE_2 by the cytoplasmic fraction of dog spleen (unpublished data quoted by Flower, 1974). Similarly, purified 15-PGDH from bovine lung was inhibited by indomethacin and aspirin (Hansen, 1974). The I_{50} values for 15-PGDH were the same order of magnitude as the I_{50} values reported in literature for the synthetase. Thus it is possible to speculate that the known hypersensitivity of a segment, of the asthmatic population to aspirin and aspirin-like drugs might be due to an altered susceptibility to the inhibition of the 15-PGDH, and not of the synthetase system, as is currently proposed (Szczeklik *et al.*, 1975).

3. PGE₂ 9-Ketoreductase (9K-PGR)

The finding of 5β,7-L-dihydroxy-11-ketotetranorprostanoic acid, the formation of which involves reduction of the 9-keto group of the five-membered ring, in guinea pig urine following injection of [³H]PGE₂ (Hamberg and Samuelsson, 1969) indicated the existence of a prostaglandin interconverting enzyme. In another experiment, PGF₂α,9,11 L-15-trihydroxyprost-5-enoic acid, and 9,11-L-dihydroxy-15-ketoprost-5-enoic acid were isolated after incubation of guinea pig liver homogenates with labeled PGE₂ (Hamberg and Israelsson, 1970). Subsequently, Leslie and Levine (1973) demonstrated the presence of a 9K-PGR in several organs of rat and five other species tested. Enzyme activities were high in the cytoplasmic fractions of heart, kidney, and brain, but enzyme was also found in the lung. In further experiments on monkey, pigeon, and chicken tissues (Lee and Levine 1974*a,b*, 1975) this enzyme was partially characterized. Cytoplasmic fractions from lung homogenates possessed enzymatic activity. As in other organs, the cytoplasmic 9K-PGR was mainly NADPH dependent and to much smaller extent NADH dependent. Further, a natural factor stimulating the 9K-PGR was discovered. Conversion of PGE₂ to PGF₂α was inhibited by NADP⁺ and NAD⁺. Since 15-PGDH is inhibited by NADH and NADPH (but is NAD⁺- and NADP⁺-dependent, see p. 175), it was proposed that the ratio of reduced to oxidized pyridine nucleotides could control both the interconversion and the first step in the catabolism of prostaglandins. Thus metabolic events regulating the availability of pyridine nucleotides could exert profound effects on lung. Since PGE₂ and PGF₂α frequently have opposite effects on smooth muscle contraction, adenylate cyclase activity, release of mediators of the antigen–antibody reaction, and release of the autonomic nervous system transmitters, it is apparent that the enzymes regulating the conversion of PGE to PGF and the catabolism of these two compounds might be of crucial importance in the control of both the respiratory and nonrespiratory lung function.

B. Uptake and Inactivation

Early studies on the distribution of radioactivity after injection of [³H]PGE₁ or PGF₂α in rats and mice (Samuelsson, 1964; Hansson and Samuelsson 1965; Gréen *et al.*, 1967) demonstrated that significant amounts of label were found in the lung, although less than in kidney and liver. At the same time, the lung was established as an important organ in the synthesis and catabolism of prostaglandins and its capacity to take up and inactivate these compounds was investigated. In dog, cat, and rabbit, prostaglandins released from the spleen by nerve stimulation and intravenously infused $PGF_{2\alpha}$, PGE_2, and PGE_1 were 90% or more inactivated after one single passage through the pulmonary circulation (Ferreira and Vane, 1967). In contrast, the biological effects of infused PGA_1 and PGA_2 were not lost after one passage through the lungs of dog and cat (McGiff et al., 1969; Horton and Jones, 1969). This differential uptake and/or catabolism of prostaglandins is in accord with the general ability of lung to discriminatively take up and inactivate, or in some instances activate, certain hormones and autacoids (Bakhle and Vane, 1974; Fishman and Pietra, 1974). Piper et al. (1970) reported that guinea pig lung inactivated all three classes, E, F, and A, of prostaglandins. Since the disappearance of prostaglandin activity was not followed by an increased level of these compounds in the lung, it was concluded that they were metabolized and not stored. Using tritiated PGE_1 and PGA_1, Gillis and co-workers demonstrated that both prostaglandins are rapidly metabolized (E > A) during one passage through isolated perfused rabbit lung (Hook and Gillis, 1975; Gross and Gillis, 1975).

Little is known about removal of prostaglandins from human lung circulation. Golub et al. (1974) determined PGA_1 in blood samples from the pulmonary artery and the left ventricle of four patients undergoing cardiac catheterization. The percent obtained of the original prostaglandin injected ranged from 0 to 24%. Stjärne et al. (unpublished results) rapidly injected a mixture of $[^3H]PGE_2$, $[^{14}C]$inulin, and $[^{125}I]$albumin into the right ventricle of human subjects and sampled the blood at 15-sec intervals from the brachial artery. Removal was calculated from changes in $^3H/^{14}C$, $^3H/^{125}I$, and $^{14}C/^{125}I$ ratios, according to the method of Stjärne et al. (1975). Significant disappearance (close to 50%) of tritium was found only in the first period. It was suggested that in human lung PGE_2 is rapidly removed and inactivated, and the metabolite is then released back into the pulmonary circulation. Gillis et al. (personal communication) injected $[^3H]PGE_1$ into the cubital vein and withdrew simultaneous samples from the pulmonary artery and left atrium. Over 90% of tritium in blood sampled at the pulmonary artery, immediately after injection, was associated with unchanged PGE_1. In contrast, in the left atrial blood unchanged PGE_1 accounted for only 20% of the radioactivity. The remaining 80% was in the form of a more polar metabolite, possibly 15-keto-PGE_1. These data represent direct evidence of extensive metabolic conversion of PGE_1 in a single passage through the pulmonary vasculature of human subjects.

Further analysis of the uptake process was carried out in *in vitro* and *in vivo* experiments by Bito and co-workers (Bito, 1972, 1975; Bito *et al.*, 1976). Adult and fetal lungs of cat, rat, and rabbit incubated for 60 min with [^{3}H]PGE_1, $PGF_{1\alpha}$, $PGF_{2\alpha}$, and PGA_1 accumulated substantial amounts of the label. Moreover, the concentration ratio (^{3}H concentration in tissue/medium) for all four prostaglandins in rat and rabbit lung was above 1. The accumulation of radioactivity was associated with tissues which are known to inactivate prostaglandins. Tritium was readily eluted from the organs tested. The conclusion was drawn that there exists either a reversible absorption of prostaglandins onto tissue components or a carrier-mediated active transport. The possible existence of such a transport system in the lung was further supported by experiments showing that after 2 min of perfusion with tritiated prostaglandins the ^{3}H concentration was approximately 5 times higher in the lung tissue than in the perfusate. Simultaneously, the [^{14}C]sucrose concentration ratio was less than 1 (Bito, 1972, 1975; Bito *et al.*, 1976). However, it is conceivable that the major part of the radioactivity in the above experiments represented metabolites and that the concentration ratio of unchanged prostaglandins was 1 or less. Indeed, Piper *et al.* (1970) did not find a higher content of prostaglandins in the lung after pulmonary perfusion of large amounts of these compounds. Ryan *et al.* (1975) perfused rat lungs with [^{3}H]$PGF_{1\alpha}$ and blue dextran. The initial delay in appearance of ^{3}H in the outflows compared to blue dextran indicated $PGF_{1\alpha}$ uptake in the lung. Since blue dextran did not leave the pulmonary bed, it can be argued that this finding did not conclusively distinguish between an active uptake and passive diffusion. About 30% of the radioactivity was retained by the lung in the form of two metabolites and approximately 25% of ^{3}H in the effluents represented $PGF_{1\alpha}$. Others (Anderson, 1975; Eling and Anderson, 1975) have also studied the uptake and metabolism of prostaglandins in the isolated perfused rat lung. Both [^{3}H]$PGF_{2\alpha}$ and [^{3}H]PGE_1 were removed from the pulmonary circulation and subsequently catabolized to 15-keto-$PGF_{2\alpha}$/E and 15-keto-13,14-dihydro-$PGF_{2\alpha}$/E. The appearance of the metabolites in the venous effluent was delayed relative to simultaneously infused [^{14}C]dextran. In contrast, appearance of [^{3}H]PGA_1 and PGB_1 was not delayed, nor were these two prostaglandins metabolized. These investigators suggested that a transport system was involved in the passage of $PGF_{2\alpha}$ and PGE_1 across pulmonary cell membranes. Accumulation against a concentration gradient, storage in a pool, a requirement of energy, and a carrier have been generally associated with an active transport of a compound. None of these conditions has, as yet, been convincingly demonstrated for prostaglandins. One of the difficulties is that the uptake of prostaglandins in the lung has not been dissociated from its rapid metabolism. With a specific inhibitor of 15-PGDH, the existence of an active transport system would be quite convincing if concentration ratios above 1 can be found. The histological localization of uptake is not known. However, it has been speculated that the endothelial cells represent the site of catabolism (Hook and Gillis, 1975). If correct, this would be similar to the catecholamine uptake processes, where a substantial percent of radioactivity is found in endothelial cells after perfusing the lung with labeled norepinephrine (Hughes *et al.*, 1969).

Recovery of tritium from rabbit lungs perfused with [^{3}H]PGE$_1$ or PGA$_1$ was the same at 37°C and 8°C. However, thin-layer chromatograms showed that while at 37°C most of the ^{3}H was recovered as a metabolite, at lower temperature the peak of tritium recovery corresponded to PGE$_1$ or PGA$_1$ (Hook and Gillis, 1975; Gross and Gillis, 1975). Polyphloretin phosphate (PPP), a known antagonist of prostaglandin action on smooth muscle preparations (Eakins *et al.*, 1970; Mathé *et al.*, 1971), and its analogues increased the passage of infused PGE$_2$ and PGF$_{2\alpha}$ through guinea pig and rabbit pulmonary circulation, as assayed biologically (Crutchley and Piper, 1974, 1975). This effect is probably due to the blocking of prostaglandin catabolism, as PPP inhibits 15-PGDH activity (Marrazzi and Matschinsky, 1972).

In an *in vivo* experiment on anesthetized dogs (Jackson *et al.*, 1973), effects of intravenous and intraarterial PGF$_{2\alpha}$ infusion on cardiac output and central venous pressure were measured before and after aspirin administration. The investigators calculated that following aspirin the mean amounts of intravenously infused PGF$_{2\alpha}$ reaching the arterial circulation were increased by over 50% (from 7.3% to 12.8%), and suggested that aspirin inhibited uptake of prostaglandins by lung. Other nonsteroidal antiinflammatory drugs (NSAIDs) also inhibited the pulmonary removal of prostaglandins. In guinea pig lung, a mean of the 2% of intravenously infused PGE$_2$ was reaching the arterial circulation in control experiments. After infusion of phenylbutazone (50 μg/ml) and indomethacin (20 μg/ml), this was increased to 4% and 20%, respectively (Crutchley and Piper, 1974). Indomethacin and aspirin have been reported to inhibit 15-PGDH in homogenates of bovine lung (Hansen, 1974), and this can probably account for these results. Pregnancy is one of the conditions known to alter the inactivation of prostaglandins by lung. Bedwani and Marley (1975) injected PGE$_2$ into the vena cava and the ascending aorta and found that the difference in blood pressure changes between the vein and artery was significantly smaller in pregnant than in nonpregnant rabbits. In lung homogenates from the respective animals, the metabolism of PGE$_2$ was faster in the pregnant animals. In nonpregnant rabbits, a 12-day treatment period with progesterone (5–10 mg/kg) also increased the *in vivo* inactivation of PGE$_2$, whereas pretreatment with cortisol (25 mg/kg) or estradiol (5–10 μg/kg) had no effect.

The material discussed in this section points to the close interdependence between prostaglandin removal from circulation, inactivation of prostaglandins, and catabolism, although each phenomenon is often studied *per se* and using different techniques. Clearly, the lung very efficiently metabolizes circulating prostaglandins. No 15-PGDH activity has been detected in plasma, and there is no loss of PGF and probably only a very slow loss of PGE in noncirculating blood (Ferreira and Vane, 1967; Ramwell and Shaw, 1967; Holmes *et al.,* 1968; Horton, 1972; Gutierrez-Cernosek *et al.,* 1972; Levine *et al.,* 1973). It would seem, therefore, that the lung plays a cardinal role in regulating the amounts of prostaglandins, either synthesized in the lung or derived from other organs, reaching the arterial circulation.

IV. RELEASE OF PROSTAGLANDINS FROM THE LUNG

A. Mechanical and Chemical Stimuli

1. Mechanical Stimuli

Release of prostaglandins, rabbit aorta contracting substance [RCS, now identified as a mixture of thromboxane A$_2$ (TxA$_2$) and small amounts of endoperoxides (Hamberg *et al.,* 1975; Svensson *et al.,* 1975)], and RCS-releasing factor from the lung was discovered by Piper and Vane (1969) when studying the chemical mediators in the perfusates from anaphylactic guinea pig lungs. This finding has stimulated many investigations on the role of prostaglandins in the lung and the conditions for release. It was found that a variety of mechanical stimuli, e.g., embolization with a colloidal suspension of various particles (1– 120 μm in diameter), agitation, stroking, and scratching of the tissue, and ventilation, released prostaglandins and RCS from chopped or perfused guinea pig and rat lungs, tracheas, and also sliced human lungs (Lindsey and Wyllie, 1970; Berry *et al.,* 1971; Piper and Vane, 1971; Piper and Walker, 1973; Palmer *et al.,* 1973; Orehek *et al.,* 1973). Mincing the lung in itself released large amounts of prostaglandins from both guinea pig and human lung (Yen *et al.,* 1976; Mathé *et al.,* 1976*b*).

2. Chemical Stimuli, Autacoids

Release of prostaglandins has been demonstrated in allergic phenomena and inflammation, two prevalent conditions. It is therefore natural that the effects of the mediators of these conditions on the release of prostaglandins have been extensively explored in several species. Thus phospholipase A, arachidonic acid, slow reacting substance C, slow reacting substance of anaphylaxis (SRS-A), 5-HT, bradykinin, histamine, and acetylcholine (ACh) release prostaglandins (and also RCS activity) when injected into the pulmonary artery or added to media containing lung fragments or to organ baths with suspended tracheal preparations (Piper and Vane, 1969, 1971; Alabaster and Bakhle, 1970; Vargaftig and Dao Hai, 1971; Bakhle and Smith, 1972; Piper and Walker, 1973; Palmer *et al.*, 1973; Orehek *et al.*, 1973, 1975; Grodzinska *et al.*, 1975).

In the works cited, bioassay, with its inherent difficulties of distinguishing between E and F prostaglandins and their biologically active metabolites, as well as between prostaglandins and SRS-A, and possibly other unidentified mediators, was used. However, generally similar data were obtained in subsequent experiments employing a sensitive radioimmunoassay for PGE_2, $PGF_{2\alpha}$, and the 15-keto-13,14-dihydro-$PGF_{2\alpha}$. Liebig *et al.* (1974) reported that the mean ± SE release from whole perfused guinea pig lung by histamine (10 μg/ml for 10 min) was 35.2 ± 7.3, 13,6 ± 3.1, and 147.2 ± 30.0 ng of $PGF_{2\alpha}$, PGE_2, and the metabolite, respectively. The effects of autacoids and vasoactive substances have been further studied by others (Mathé and Levine, 1973; Yen *et al.*, 1976; Mathé *et al.*, 1976*b;* Mathé, unpublished). Infusion of histamine (10 μg/ml) into the pulmonary artery of control guinea pigs caused a mean ± SE increase in the outflow of PGE_2, $PGF_{2\alpha}$, and the metabolite from 0.5 ± 0.2, 0.4 ± 0.1, and 0.8 ± 0.3 ng/min to 1.7 ± 0.2, 2.4 ± 0.6, and 3.4 ± 1.3 ng/min, respectively. In the sensitized lungs the amounts were elevated from 0.7 ± 0.1, 1.0 ± 0.2, and 0.7 ± 0.1 ng/min to 2.2 ± 0.4, 3.2 ± 0.6, and 10.9 ± 2.7 ng/min, respectively. Although the increase in prostaglandin release was larger from the sensitized than from the control lungs, only the outflow of metabolite was statistically different in this experiment. In others, however, significantly more prostaglandins during histamine infusion were measured in the outflows from the sensitized lungs. A rapid return to the control values was obtained within 4 min after termination of histamine infusion. Preinfusion with pyrilamine (histamine receptor-1 blocker) significantly diminished the subsequent response to histamine, and predominantly decreased the outflow of $PGF_{2\alpha}$. In contrast, pretreatment with metiamide (histamine receptor-2 blocker) decreased the subsequent liberation of PGE_2 by histamine but not that of $PGF_{2\alpha}$. The data suggest that the release of $PGF_{2\alpha}$ by histamine is predominantly mediated via the receptor 1, whereas the release of PGE_2 is mediated by the receptor 2. Since in whole perfused guinea pig lung, histamine raises cyclic GMP levels by stimulation of the receptor 1 and cyclic AMP levels by an action on receptor 2 (Mathé *et al.*, 1974*b*), cyclic GMP could be linked to $PGF_{2\alpha}$ and cyclic AMP to PGE_2.

Infusion of cat paw SRS into the pulmonary circulation of guinea pig caused a marked release of prostaglandins; the sensitized lungs released more prostaglandins than the healthy control lungs (Mathé *et al.*, 1976*b*). This phenomenon was similar to the response of lung to histamine. In control lungs the mean ± SE outflows of PGE_2, $PGF_{2\alpha}$, and 15-keto-13,14-dihydro-$PGF_{2\alpha}$ increased in the presence of SRS from 0.3 ± 0.1, 1.5 ± 0.2, and 2.0 ± 0.2 ng/min to 0.8 ± 0.1, 3.8 ± 0.5, and 7.5 ± 1.1 ng/min, respectively. The comparable changes from the sensitized lungs were from 0.5 ± 0.1, 2.3 ± 0.3, and 2.7 ± 0.4 ng/min to 2.7 ± 0.3, 8.4 ± 1.0 and 22.1 ± 3.1 ng/min. Histamine was not released by SRS in these experiments. SRS did not liberate prostaglandins as effectively from lung fragments. This result was again parallel to that obtained with histamine. Both compounds may be more rapidly metabolized by the fragments than by the intact tissues; however, a more likely explanation is that the prostaglandin release induced by the trauma of preparing the lung fragments masks effects of other stimuli. In the same series of experiments, guinea pig lungs were also perfused with bradykinin, ACh (with and without eserine), and 5-HT (1–50 μg/ml) and 8-bromo-cyclic GMP (50–300 μg/ml), and in all cases prostaglandins were determined in the outflows. Bradykinin released PGE_2 and $PGF_{2\alpha}$, which confirmed the results obtained by others. In addition, large amounts of metabolite were found. Release of prostaglandins by ACh was inconsistent; in the majority of the experiments, little prostaglandin increase could be detected. The effects of 5-HT and 8-bromo-cyclic GMP were also small and not uniform.

Many experiments have supported the hypothesis that smooth muscle contraction and distortion of the cell membrane cause release of prostaglandins (Piper and Vane, 1971). Isolated guinea pig trachea contracted by histamine or ACh released prostaglandins of the E and F types (Orehek *et al.*, 1973; Grodzinska *et al.*, 1975). Abolishment of the contractile response by atropine or by removal of Ca^{2+} from the media was followed by the disappearance of prostaglandins release (Orehek *et al.*, 1973; Bouhuys, 1975). On the other hand, Dunlop and Smith (1975) reported that a dose of ACh which induced a maximal contraction of isolated human bronchial strips did not significantly increase the amount of $PGF_{2\alpha}$ released. Our data agree with this finding. Ferreira *et al.* (1976) could not detect a release of prostaglandins from isolated rabbit jejunum contracted by ACh and eserine, although the same preparation, both spontaneously and when manipulated physically, released PGE_2 and $PGF_{2\alpha}$. Investigation of prostaglandin release from isolated rabbit heart demonstrated that mechanical performance *per se* was not decisive for the liberation of prostaglandins (Junstad and Wennmalm, 1973; Wennmalm, 1975). Consequently, although mechanical manipulation and trauma of tissue are regularly associated with the release of prostaglandins, smooth muscle contraction does not consistently lead to prostaglandin release. The initial tone and the metabolic state of the smooth muscle might partially account for the variability in prostaglandin release by contraction. An additional source of differences could be the amount of other cells (possibly liberating more prostaglandins than the smooth muscle cells) present in the smooth muscle preparation.

B. Antigen–Antibody Reaction and Other Physiological and Pathological Stimuli

1. Anaphylaxis and Sensitization with Antigen

Following the discovery of Piper and Vane (1969, 1971) that anaphylaxis liberates prostaglandins, this phenomenon has been studied in several species. Lungs of actively sensitized rats and guinea pigs released prostaglandins, SRS-A, and histamine when challenged with appropriate antigen (Piper and Walker, 1973). In the same series of experiments, a mixture of PGE_2 and $PGF_{2\alpha}$ (2–4 ng/ml) was found in perfusates from challenged, passively sensitized human lungs. Histamine and SRS-A, but not RCS, were also released. No prostaglandins were detected in perfusion fluid that had passed through passively sensitized lung tissue which had not been challenged.

Using a radioimmunoassay, Mathé and Levine (1973, and unpublished data) found that during anaphylaxis, in addition to the parent prostaglandins, 15-keto metabolites and 15-keto-13,14-dihydro metabolites were released from perfused guinea pig lungs. As no antibody to the E_2 metabolites was available, 15-keto-13,14-dihydro-$PGF_{2\alpha}$ was assayed before and after alkali treatment (pH 12.5, 100°C for 5 min). Since the serological activity of the E_2 but not the $F_{2\alpha}$ metabolite is destroyed by such procedure, an attempt was made to assess the amount of each metabolite. Quantitatively, the release of the 15-keto-13,14-dihydro-PGE_2 and $PGF_{2\alpha}$ was several times larger than that of unchanged prostaglandins and 15-keto metabolites. It also appeared that more E_2 than $F_{2\alpha}$ metabolite was present. Since neither the 15-PGDH nor 13-PGR activity could be detected in the perfusates, it was concluded that prostaglandins released were metabolized before they reached the pulmonary circulation. These results were confirmed and extended in another study (Liebig *et al.*, 1974). Prostaglandins appeared in the outflows during the first minute after antigen challenge. The release of $PGF_{2\alpha}$ lasted longer and the overall concentration in the perfusate was higher than that of PGE_2. Large amounts of 15-keto-13,14-dihydro-$PGF_{2\alpha}$ were also released. At its peak, 30 ng/min was being released which was approximately 3 times more than $PGF_{2\alpha}$ and

PGE_2. Moreover, larger amounts of the metabolite than the parent compound were found in the perfusates throughout the whole experiment. Similarly, anaphylactic release of $PGF_{2\alpha}$ and histamine was also demonstrated from guinea pig lung fragments (Hitchcock, 1974, 1975).

Mathé *et al.* (1976*f*) measured prostaglandins and histamine in guinea pig lung perfusates at 60-sec intervals, preceding, during, and after the antigen–antibody reaction. The mean maximal outflows of $PGF_{2\alpha}$ and PGE_2 were simultaneous and occurred 2 min after the start of challenge. The respective values were 5.4 ± 1.6 and 3.1 ± 1.1. ng/ml. The increased outflow of $PGF_{2\alpha}$ by anaphylaxis lasted longer than the enhanced release of PGE_2. Peak histamine concentration in perfusates preceded that of prostaglandins. Changes in the time course of the outflow of histamine, PGE_2, and $PGF_{2\alpha}$ would indicate that the release of histamine precedes that of prostaglandins. This is consistent with the notion that histamine is a "primary" mediator of the antigen–antibody reaction (Kaliner and Austen, 1974, 1975) and that it releases prostaglandins (Bakhle and Smith, 1972; Mathé and Levine, 1973; Liebig *et al.*, 1974; Yen *et al.*, 1976), which are the "secondary" mediators. However, the converse may also be true; under some circumstances, prostaglandins may release histamine (Crunkhorn and Willis, 1971; Søndergaard and Greaves, 1971; Tauber *et al.*, 1973; Hitchcock, 1974). Another possibility is that the initial release is simultaneous, reflecting an identical underlying mechanism, but the time required to reach the pulmonary circulation and be washed out is different for histamine and prostaglandins.

A basal outflow of mediators from guinea pig lungs perfused with Tyrode's solution via the pulmonary artery was also demonstrated; it was significantly larger in the antigen-sensitized animals than in the healthy control animals (Mathé *et al.*, 1976*f*). In the sensitized group, the mean $\pm$ SE release of histamine (expressed as percentage released) was 3.8 ± 0.3 and of PGE_2, $PGF_{2\alpha}$ and 15-keto-13,14-dihydro-$PGF_{2\alpha}$ (expressed as total ng) was 3.4 ± 0.8, 5.5 ± 1.7, and 12.5 ± 3.5, respectively. In control group, the percentage released was 0.4 ± 0.03 for histamine and that for each prostaglandin was 1.5 ± 0.3, 1.1 ± 0.5, and 5.3 ± 0.3, respectively. The levels of prostaglandins were not measured in the lung tissue since they are not stored but are synthesized *de novo*. However, the activity of the prostaglandin synthetase system was found to be higher in the sensitized than control lung microsomal preparations (Leslie and Mathé, unpublished). The level of histamine, which is stored, was determined in the lungs. No differences were found between the control and sensitized guinea pigs (Mathé *et al.*, 1976*f*). Thus the increased percentage of histamine released was not a reflection of a change in the lung content of histamine. These results suggested an increased turnover of both prostaglandins and histamine in the sensitized lung, and it was proposed that antigen sensitization in itself may change the biological properties of the organism. In line with this suggestion is the observation that the PGE_2-to-$PGF_{2\alpha}$ ratio was less in the outflows from the sensitized compared to the control lungs, and similarly in the anaphylactic as compared to the sensitized lungs; there was a larger increase of $PGF_{2\alpha}$ than PGE_2 (Mathé et al., 1976f). Whether this is due to a larger increase in synthesis of $PGF_{2\alpha}$ than PGE, or conversion of PGE_2 to $PGF_{2\alpha}$ by 9K-PGR, or possibly a faster catabolism of PGE_2 than $PGF_{2\alpha}$ is not known.

Dunlop and Smith (1975) passively sensitized pieces of human bronchi and measured by radioimmunoassay the release of $PGF_{2\alpha}$ during antigen challenge. The mean release of $PGF_{2\alpha}$ increased from 25.5 ± 7.3 to 148.1 ± 57.5 pg/mg dry weight of bronchial tissue. A significant correlation existed between the amounts of prostaglandin liberated and the degree of contraction obtained, suggesting that the bronchoconstriction observed after antigen–antibody reaction may, in part, be due to the release of prostaglandins. In another series of investigations (Strandberg et al., 1976; Mathé, unpublished) the anaphylactic liberation of prostaglandins and histamine from passively sensitized human lung fragments and pieces of bronchi, obtained at surgery, was studied. After a 10-min incubation with antigen, mean $\pm$ SE ng PGE_2, $PGF_{2\alpha}$, and 15-keto-13,14-dihydro-$PGF_{2\alpha}$ per gram wet weight tissue increased from 13.5 ± 1.9, 18.6 ± 2.2, and 7.9 ± 1.0 to 20.1 ± 1.9, 35.5 ± 4.0, and 19.4 ± 2.1 for lung parenchyma, and from 18.1 ± 1.5, 20.3 ± 2.1, and 4.3 ± 0.6 to 24.0 ± 2.2, 27.2 ± 3.0, and 5.1 ± 0.8 for the isolated bronchi. Doubling the time of incubation did not affect the amount of unchanged prostaglandins, but elevated by close to 100% the amount of metabolite.

The anaphylactic release of prostaglandins was also demonstrated in vivo. Thus Palmer et al. (1973) showed release of an RCS-like substance into the carotid arterial blood after intravenous challenge of sensitized guinea pigs. Burka and Eyre (1974) have studied anaphylaxis in calves by measuring cardiovascular and respiratory changes in anesthetized animals, and also used the carotid blood for tissue superfusion to assess the liberation of mediators. The results indicated that in vivo the antigen–antibody reaction released prostaglandins of both E and F types from the lungs of calves. Strandberg and Hamberg (1974) determined the major urinary metabolite of F-class prostaglandins, 5,7-L-dihydroxy-11-ketotetranorprostanoic acid, by gas–liquid chromatography/mass spectrometry in guinea pig urine before and during anaphylaxis. The basal excretion ranged from 0.5 to 1.1 $\mu g/24$ hr. In three out of four animals tested, approximately 100% elevation in the excretion of metabolite was found. Using a radioimmunoassay, Ruff *et al.* (1975) measured high levels of $PGF_{2\alpha}$ and histamine in blood of anaphylactic guinea pigs. Interestingly, no PGE_2 was detected in these samples.

Plasma levels of 15-keto-13,14-dihydro-$PGF_{2\alpha}$ in peripheral venous blood of five asthmatic patients were measured by gas–liquid chromatography/mass spectrometry (Gréen *et al.*, 1974). Inhalation of a specific allergen leading to signs and symptoms of an asthmatic attack was associated with an up to eightfold increase in the plasma metabolite within 5 min of the asthmatic attack. The range of values was approximately 10–50 pg/ml plasma before and 50–200 pg/ml after the inhalation. In another study (Mathé *et al.*, unpublished), levels of 15-keto-13,14-dihydro-$PGF_{2\alpha}$ were determined in blood samples drawn simultaneously from cubital vein and brachial artery of healthy controls and mild-to-moderate asthmatics in remission. The mean concentrations of the metabolite, expressed as ng/ml, were 3.3 and 2.9 in the arterial and venous blood, respectively, of the controls and 2.8 and 2.5 in the corresponding asthmatic samples. It would seem that, in contrast to the rise during an acute, allergen-provoked asthmatic attack, plasma levels of 15-keto-13,14-dihydro-$PGF_{2\alpha}$ are not higher in asthmatics in remission compared to healthy control subjects. The results from the two studies cited are not directly comparable, since both the design and the method of prostaglandin determination were different.

In addition to prostaglandins and their metabolites, endoperoxide intermediates and their metabolic products are also released during anaphylaxis. Hamberg *et al.* (1976) demonstrated that following injection of antigen into perfused sensitized guinea pig lungs approximately 4–10 times more TxB_2 than during basal conditions was found in the outflows; the basal values were approximately 100 ng, while the range of values during anaphylaxis was 330–843 ng. Finally, it is noteworthy that the lipid content (such as cholesterol, triglycerides, choline, and ethanolamine phospholipids) of guinea pig lung is decreased following anaphylaxis *in vivo* (Smith, 1962; Goadby and Smith, 1962; Goadby, 1975). Since these lipids can serve as the source of arachidonic and dihomo-γ-linoleic acids, which are substrates for the synthetase system, this finding is an additional indication of increased prostaglandin synthesis during the antigen–antibody reaction and also points to the local derivation of the substrates for the lung synthetase system.

2. Hypoxia, Lung Edema, Oxygen, and Ascorbic Acid

Other physiological or pathological stimuli, in addition to the anaphylactic reaction, liberate prostaglandins. Thus hypoxia (ischemia) has been shown to release prostaglandins in several organs (McGiff *et al.*, 1970; Jaffe *et al.*, 1972; Wennmalm *et al.*, 1974). In isolated perfused cat lungs ventilated with 2% or 21% O_2 in 5% CO_2 in N_2, a degree of pulmonary hypertension with concomitant release of PGE- and PGF-like material was observed by Said *et al.* (1974). These airway and vascular responses *in vivo* were decreased by infusion of aspirin (>50 mg/kg body weight). In contrast, Even *et al.* (1975) using a radioimmunoassay did not find increases in PGE_1, PGE_2, or $PGF_{2\alpha}$ in human arterial blood, in spite of a marked elevation of pulmonary arterial pressure induced by hypoxia. Moreover, neither indomethacin (3 mg/kg) nor aspirin (60 mg/kg) influenced this response. If prostaglandins indeed are released by hypoxia (an assumption which the accumulated results on organs like heart and kidney would seem to favor), a possible explanation for the apparent contradiction between the two reports is that the prostaglandin metabolites which show smooth muscle activity could be detected by the bioassay but not by the radioimmunoassay. However, the failure of indomethacin and aspirin to inhibit the cardiovascular and respiratory changes indicates either insufficient drug concentration, release of other, unknown mediators, or species differences. The possible significance of the prostaglandin liberation by hypoxia is that the released compounds might contribute to the regulation of blood flow and cause reactive hyperemia.

Isolated, perfused cat lungs released prostaglandin-like substances into the alveolar space and circulation when pulmonary edema was induced (Said and Yoshida, 1976). No other work on the role of prostaglandins in lung edema would appear to have been published.

Lung 15-PGDH was inhibited *in vitro*, after exposure of guinea pigs to 100% O_2 for 48 hr. The degree of inhibition was dependent on the duration of exposure (Parkes and Eling, 1975). However, the synthetase system was not influenced. Exposure of antigen-sensitized guinea pigs to ozone, NO_2, and SO_2 increased the subsequent symptoms of anaphylaxis (Matsumura *et al.*, 1972). It is conceivable that this was due to inhibition of prostaglandin catabolism. If applicable to humans, these phenomena could have significant clinical and environmental health implications.

Claims have been made that the prevalence of asthma is higher in populations on ascorbic-acid-deficient diets. The finding that tracheas from guinea pigs maintained on a vitamin-C-free diet released more $PGF_{2\alpha}$ and less PGE_2 than the control animals (Puglisi *et al.*, 1976) is thus of potential clinical interest. After a 30-min incubation, the following mean $\pm$ SE amounts of prostaglandins (expressed as ng/g tissue per hr) were found in the media from ascorbic acid deprived and control animals: 248 ± 88 vs. 130 ± 48 $PGF_{2\alpha}$ and 508 ± 182 vs. 926 ± 176 PGE_2. Puglisi *et al.* (1976) also observed that ascorbic acid ($50-100$ $\mu g/$ ml) added to the organ bath increased the mean PGE_2 concentration in the media from 150 to 442 ng/g tissue/hr. On the other hand, Hitchcock (1975) found that preincubation of lung fragments from ovalbumin-sensitized guinea pigs with ascorbic acid markedly increased the anaphylactic release of both histamine and $PGF_{2\alpha}$. The mean $PGF_{2\alpha}$ liberated during anaphylaxis in control and ascorbic acid (10^{-6} M) treated lungs was 40 and 180 ng/g tissue, respectively. Further experiments are necessary to clarify the role of ascorbic acid in prostaglandin metabolism and in allergic phenomena.

C. Inhibition of Prostaglandin Release

A variety of compounds block the release of prostaglandins from the lung. Liberation of prostaglandins from guinea pig, rat, and human lungs by anaphylaxis, autacoids, vasoactive substances, and mechanical stimuli is antagonized by NSAIDs and some thiol derivatives (Piper and Vane 1969, 1971; Vargaftig and DaoHai, 1971, 1972; Bakhle and Smith, 1972; Walker, 1973; Piper and Walker, 1973; Palmer *et al.*, 1973; Grodzinska *et al.*, 1975). Further studies employing radioimmunotechniques for prostaglandin assay have extended these results (Liebig *et al.*, 1974; Dunlop and Smith, 1975; Mathé *et al.*, 1976*f*). It is of interest that Walker (1973) and Mathé *et al.* 1976*f*) found that indomethacin inhibited the release of both prostaglandins and histamine from perfused guinea pig lung, while Liebig *et al.* (1974) obtained an increase in histamine release. NSAIDs were found to also decrease the antigen-induced *in vivo* release of prostaglandins in guinea pigs and calves (Strandberg and Hamberg, 1974; Burka and Eyre, 1974; Ruff *et al.*, 1975).

Catecholamines ($1-10$ $\mu g/ml$) inhibit the anaphylactic release of prostaglandins from whole perfused guinea pig lung (Mathé and Levine, 1973; Mathé *et al.*, 1976*f*; Liebig *et al.*, 1974). Both dibutyryl and 8-bromo analogues of cyclic AMP, in concentrations of $50-300$ $\mu g/ml$, inhibited the anaphylactic release of prostaglandins from whole perfused and minced guinea pig lung. Disodium chromoglycate (DSCG) ($0.3-20$ $\mu g/ml$) diminished the antigen-induced release of prostaglandins from minced human lung tissue only when the release of histamine and SRS-A had been almost completely suppressed (Piper and

Walker, 1973). Diisopropylfluorophosphate (5 mM) abolished the release of prostaglandins and histamine from anaphylactic lungs and reduced the spontaneous outflow from sensitized guinea pig lungs (Mathé *et al.*, 1976*f*). Pyrilamine, metiamide, and methysergide (specific receptor antagonists of histamine and 5-HT) inhibit the release elicited by these agonists (Bakhle and Smith, 1972; Yen *et al.*, 1976).

Hydrocortisone can also decrease the anaphylactic release of prostaglandins, RCS, and histamine from perfused guinea pig lungs, as measured by bioassay (Gryglewski *et al.*, 1975). When arachidonic acid (1 μg/ml) was infused into the hydrocortisone-treated lungs and the allergen challenge repeated, prostaglandins partially reappeared in the effluent.

Incubation in medium kept at 4°C very markedly decreased both the spontaneous and antigen-induced release of prostaglandins from minced guinea pig lung (Mathé *et al.*, unpublished results). In the same series of experiments, Tyrode's solution without Ca^{2+} and Mg^{2+} had little effect on the liberation of prostaglandins. Furthermore, absence of Ca^{2+} in the fluid superfusing spirally cut guinea pig trachea did not prevent release of prostaglandins elicited by mechanical irritation of the mucosal surface (Orehek *et al.*, 1975).

Inhibitors of the prostaglandin synthetase system in several tissues and organs have been extensively reviewed (Flower, 1974; Ferreira and Vane, 1974; Robinson and Vane, 1974). Data currently available would indicate that properties of the synthetase system in the lung are similar to those in the other organs. Since prostaglandins released from the lung by a variety of stimuli can originate from tracheal and bronchial mucosa, endothelial cells of the vasculature, smooth muscle cells, mast cells, thrombocytes/megakaryocytes possibly trapped in the capillaries, and probably other cells (Piper and Vane, 1971; Coburn *et al.*, 1974; Orehek *et al.*, 1975; Kadowitz *et al.*, 1975*a*; Strandberg *et al.*, 1976), the mode of action of prostaglandin inhibitors is difficult to assess. The antagonism of prostaglandin release from the lung by NSAIDs is by suppression of the prostaglandin synthetase system, as discovered by Vane (1971) in guinea pig lung. Other compounds and manipulation of the physicochemical environment, which are known to inhibit release of prostaglandins, may act similarly. A second possible mechanism of action is on the availability of the substrate, without affecting the synthetase. Thus Gryglewski *et al.* (1975) have suggested (based on the results mentioned above) that corticosteroids act by stabilizing membrane phospholipids, thereby decreasing the supply of the substrate available to the synthetase. Kantrowicz *et al.* (1975) and Tashjian *et al.* (1975) have found that corticosteroids inhibit the release of PGE_2, $PGF_{2\alpha}$, and the metabolites from tissue and cell cultures, and similarly proposed that the mode of action is by decreasing the availability of the substrate. That seems to be a plausible explanation, as Vane (1971) found that hydrocortisone (50 μg/ml) did not affect the synthetase activity in guinea pig lung. Phospholipase A increases prostaglandin synthesis by cleavage of the precursor acids from the cell membrane phospholipids (Kunze and Vogt, 1971; Kunze, 1972; Vargaftig, 1974). The action of compounds which antagonize prostaglandin release induced by the broncho- and vasoactive substances, anaphylaxis, or mechanical irritation may be by a direct inhibition of phospholipase A or by antagonism of the events leading to activation of this enzyme and to cell distortion.

D. Possible Role of Prostaglandins in Bronchial Asthma and Other Lung Diseases

The role, harmful or beneficial, of prostaglandins in the etiology and pathogenesis of bronchial asthma has not yet been clearly established. The antigen–antibody reaction substantially increases the synthesis of endoperoxides, $PGF_{2\alpha}$, and PGE_2, a large proportion of which will be further metabolized to TxB_2 and the 15-keto-13,14-dihydro metabolites before reaching the circulation. An increased prostaglandin synthesis may also occur in asthma in remission or during an acute attack which is induced by antigen, or without a detectable allergic component, or one which is aspirin-provoked. A useful approach to the problem of this disease is to consider the tracheobronchial smooth muscle of the asthmatic as predisposed, genetically or secondary to an acquired lesion, to hyperreact to a variety of stimuli. Tone of the airways could be viewed as an equilibrium between the dilatory effects of epinephrine and PGE_2 and the constrictive action of ACh, $PGF_{2\alpha}$, histamine, SRS-A, possibly other, unidentified mediators, and conceivably the α-adrenoceptor effect of norepinephrine. The influence of each can not be solely assessed by the levels found in the blood or perfusates, since these do not take into account the actual concentration at the receptor site (Goldstein, 1949), the sensitivity of a receptor to its specific agonist, and the interaction with other compounds having similar effects.

The elevated tone of the airways, found in asthmatics even in remission, might be caused by one or all of the following: increased spontaneous release of histamine and other bronchoconstrictors, liberation of $PGF_{2\alpha}$, and/or reduced stimulation of or damage to the adenylate cyclase system. Results recently obtained in our laboratory are consistent with these possibilities (Mathé *et al.*, 1974a, 1975, 1976a,e,f). Reflex parasympathetic constriction and lesion of the smooth muscle might also play a role. We would like to hypothesize that in some cases of asthma there is production of antibodies to $PGF_{2\alpha}$ (Mathé and Levine, unpublished results), increased synthesis and increased catabolism of $PGF_{2\alpha}$ (Mathé *et al.*, 1973*a,b*, 1976*d*), and, possibly, diminished availability of epinephrine (Mathé, 1971; Mathé *et al.*, 1974*a;* Mathé and Volicer, 1976*a,b*).

The symptoms of acute allergen-provoked asthma are probably elicited by the liberation of histamine, SRS-A, and other mediators, including prostaglandins. To speculate, when the $PGF_{2\alpha}$-specific antibody or nonspecific protein-binding capacity becomes saturated, the levels of $PGF_{2\alpha}$ and the metabolites would be elevated in tissues and possibly plasma. In an analogous manner to the protective effect binding to protein has on hormones, $PGF_{2\alpha}$ bound to $PGF_{2\alpha}$-antibody would not be susceptible to degradation by 15-PGDH. After an acute attack, there could be a gradual dissociation of $PGF_{2\alpha}$ from its binding sites. This could partially explain the increased tone in remission of the asthmatic airways.

In the heart and the gastrointestinal tract of several species, ACh inhibits the liberation of catecholamines and, conversely, epinephrine and norepinephrine antagonize the release of ACh (reviews by Kosterlitz and Lees, 1972; Muscholl, 1973). Similar mechanisms appear to also exist in rabbit lung (Mathé *et al.*, 1976*c,d*). If such a mechanism is found in human lung, occurrence of asthma in the absence of a known allergy could be explained by an alteration of the presynaptic parasympathetic–sympathetic balance in the direction of the parasympathetic system. This would lead directly to increased tone (constriction) of airways and possibly to release of prostaglandins. Whether histamine and other mediators would be liberated by such a mechanism is not known. However, it has been demonstrated that the anaphylactic release of mediators is modulated by the sympathetic and parasympathetic system (reviews by Austen, 1974; Kaliner and Austen, 1974, 1975).

The suggestion has been put forward that aspirin and other NSAIDs induce asthma by inhibition of prostaglandin synthesis in the lung, i.e., in aspirin-sensitive asthmatics patency of the airways is primarily dependent on PGE synthesis, and to a lesser extent on β-adrenergic stimuli (Szczeklik *et al.*, 1975). An attack would develop since the bronchoconstrictive action of histamine and other mediators would no longer be counteracted by the dilatory effect of PGE. If correct, the effects of aspirin should be prevented by histamine receptor-1 blockade or by pretreatment with PGE. Moreover, the ratio of PGE (and its metabolites) to epinephrine excreted in urine should also be different in the aspirin-sensitive and non-sensitive asthmatics.

The massive anaphylactic release of prostaglandins and precursors will exert local effects on the airways and pulmonary vasculature, as well as influence cyclic nucleotides and further release of mediators. If the parent compounds or active metabolites reach the arterial circulation, systemic signs, due to altered capillary permeability and vascular tone, and other cardiovascular, immunological, and general metabolic effects of prostaglandins might ensue. Other lung diseases, perhaps best exemplified by emphysema, sarcoidosis, and chronic interstitial pneumonia, could alter uptake and metabolism of prostaglandins and other active substances normally processed by the lung. This could have local, intrapulmonary effects and also contribute to systemic symptomatology. These aspects of pulmonary diseases appear not to have been investigated as yet. However, it has been suggested that prostaglandin release from lung, not necessarily triggered by a pulmonary disease, elicited pathophysiological phenomena in other organs, e.g., cutaneous vasodilatation in carcinoid syndrome and migraine (Bakhle and Vane, 1974; Sandler, 1972).

V. BRONCHOPULMONARY EFFECTS OF PROSTAGLANDINS

A. F Prostaglandins

The first bronchopulmonary effects of crude prostaglandin preparations were observed by von Euler (1939) on isolated perfused lungs of cats and rabbits. Following isolation and identification of $PGF_{2\alpha}$, Änggård and Bergström (1963) found that $PGF_{2\alpha}$ caused contraction of isolated tracheal and bronchial chain preparations from cats, rabbits, and guinea pigs. Using a Konzett–Rössler lung preparation, they also reported increased bronchial resistance in cat following injection of 15–30 μg $PGF_{2\alpha}$ into a jugular vein. The actions of prostaglandins were further tested on spirally cut human bronchi suspended in an organ bath (Collier and Sweatman, 1968; Sweatman and Collier, 1968; Mathé et al., 1971). $PGF_{2\alpha}$ (0.08–0.8 μg/ml) caused contraction which was antagonized by PGE_1 and PGE_2 in comparable concentrations. Neither atropine nor mepyramine (both 1 μg/ml) altered the response to $PGF_{2\alpha}$. The effect of $PGF_{2\alpha}$ on isolated human bronchi was specifically antagonized by fenamates and PPP. It was found that neither α- nor β-adrenoceptor blockade modified the action of $PGF_{2\alpha}$ on human bronchial strips. Methysergide did not influence the response either, whereas epinephrine counteracted the bronchial response to $PGF_{2\alpha}$. Polyphloretin phosphate blocked the constriction caused by $PGF_{2\alpha}$, but did not alter the bronchodilation elicited by PGE_1 and PGE_2. The responses of the same strips to ACh and histamine were also unaltered by PPP (Mathé et al., 1971).

In unanesthetized guinea pig, intravenous administration of $PGF_{2\alpha}$

resulted in a marked increase in respiratory resistance associated with a decrease in pulmonary compliance. The immediate effects of $PGF_{2\alpha}$ were the same in atropine-treated and control animals. On the other hand, 3–30 min after termination of $PGF_{2\alpha}$ infusion the observed changes were smaller in atropinized animals. It was suggested that the effect of $PGF_{2\alpha}$ is both a direct one on the smooth muscle and an indirect one that is cholinergically mediated (Drazen and Austen, 1974, 1975). Such an hypothesis is in agreement with the other work reviewed here. In two other investigations, bronchopulmonary consequences of intravenous $PGF_{2\alpha}$ administration to cats and guinea pigs were not modified by bilateral cervical vagotomy (Frey and Schäfer, 1974; Strandberg and Hedqvist, 1975). It should be noted, however, that in these experiments the animals were anesthetized and changes in airway insufflation pressure measured by the Konzett–Rösller method, a technique which does not allow for differentiation between changes in bronchial tone and changes in pulmonary circulation (Widdicombe, 1963).

In an *in vivo* investigation on cats and guinea pigs, $PGF_{2\beta}$ was shown to antagonize the bronchoconstricting action of aerosolized $PGF_{2\alpha}$. Alone, $PGF_{2\beta}$ was a potent bronchodilator (Rosenthale *et al.*, 1973). *In vitro*, this stereoisomer also antagonized the contraction of guinea pig tracheal strips by $PGF_{2\alpha}$ and histamine, and reduced the spontaneous tone of the preparations (Baum *et al.*, 1974).

The pulmonary action of $PGF_{2\alpha}$ on human subjects was originally examined by Mathé and co-workers. $PGF_{2\alpha}$ was aerosolized with an ultrasonic nebulizer and the aerosol mist was inhaled by healthy volunteers. Changes in airway resistance and thoracic gas volume were measured in a whole-body plethysmograph and the results were conventionally expressed as specific airway conductance (SG_{aw}, the reciprocal of airway resistance at a given thoracic gas volume). Sodium chloride did not affect the SG_{aw}, whereas $PGF_{2\alpha}$ produced a decrease. One minute after inhalation of 64,256, 521, and 1024 μg $PGF_{2\alpha}$, the mean decrease of SG_{aw} was 14, 21, 31, and 49%, respectively. The effect of $PGF_{2\alpha}$ was relatively short-lasting and complete recovery occurred within 10 min. Approximately 3 times the dose, on a molar basis, of histamine was required to induce a comparable decrease (Hedqvist *et al.*, 1971). The effects of inhalation of $PGF_{2\alpha}$ were then tested on asthmatic subjects (Mathé *et al.*, 1973a,b). The largest dose required to reduce the SG_{aw} by 50% was 0.71 nmol and the mean dose was 0.275 nmol. Thus a remarkable difference in the sensitivity of asthmatic and control subjects to inhalation of $PGF_{2\alpha}$ was established. Furthermore, the maximal effect of $PGF_{2\alpha}$ in the controls was observed 1 min after the inhalation; in the asthmatics, it was seen 5–15 min later. The effect of $PGF_{2\alpha}$ in the patients was longer lasting and the recovery was more gradual. Thus in five out of ten asthmatics, SG_{aw} remained more than 50% below the baseline 60 min after the inhalation. These patients also had wheezing, dyspnea, and the usual sensation of an asthmatic attack. Inhalation of isoproterenol or intravenous injection of aminophylline relieved the distress and increased SG_{aw} considerably in these subjects. The asthmatics responded with bronchoconstriction to lower doses of histamine than did the healthy subjects, confirming the previous observations (Curry, 1946; Tiffeneau, 1958). However, there seemed to be a significant difference in the pattern of response to the two drugs. The ratio of histamine to $PGF_{2\alpha}$ doses, each causing about 50% fall of SG_{aw}, was approximately 3:1 in the controls and 2000:1 in the asthmatic patients.

In the final study (Mathé and Hedqvist, 1975, and unpublished observations), a considerable variability in the response to $PGF_{2\alpha}$ between subjects was observed. The dose eliciting a 50% decrease in SG_{aw} ranged from 4 to 1024 ng in asthmatics, whereas in healthy subjects it was many hundred times larger, from 128 to 1024 μg. Moreover, in a few control subjects, a 50% reduction of SG_{aw} could not be achieved, even with the highest dose given. Repeated $PGF_{2\alpha}$ administration elicited a diminished response in some asthmatics; in healthy subjects, this generally did not occur; two typical cases are presented in Fig. 2 (top). The *in vitro* response of human bronchi to repeated administration of $PGF_{2\alpha}$ did not change. The contractile effect of $PGF_{2\alpha}$ on an isolated bronchial strip suspended in an organ bath is shown in Fig. 2 (center). No signs of tachyphylaxis were noted *in vivo* when $PGF_{2\alpha}$ was administered intravenously to atropinized and β-blocked guinea pigs (Fig. 2, bottom).

Some subjects were retested over a 2-year period. Although the mean responses to $PGF_{2\alpha}$ were not significantly different, the variability in response to $PGF_{2\alpha}$ was more pronounced in asthmatics than in healthy subjects. Test–retest correlation coefficients were 0.08 and 0.49 for the two groups, respectively. Results from an asthmatic subject and a healthy subject are shown in Fig. 3.

The effect of the parasympathetic tone on the bronchoconstrictive action of $PGF_{2\alpha}$ was investigated by measuring SG_{aw} changes elicited by $PGF_{2\alpha}$ inhalation before and after 1 mg atropine given parenterally. In asthmatics, the mean SG_{aw} decreases 1, 5, and 10 minutes after $PGF_{2\alpha}$ inhalation were 46%, 49%, and 42% respectively, from basal control, and 24%, 27%, and 28%, respectively, from atropine control. In healthy subjects, the corresponding decreases were 43%, 49%, and 34% before atropine, and 32%, 45%, and 33% after. The bronchoconstricting action of $PGF_{2\alpha}$ was diminished in asthmatics by atropine, 1

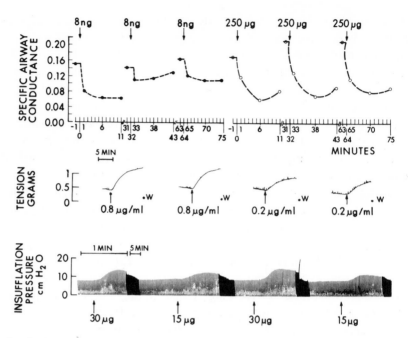

Fig. 2. *In vivo* and *in vitro* effects of repeated administration of PGF$_{2\alpha}$. Top: Changes in specific airway conductance after inhalation of 8 ng and 250 μg of PGF$_{2\alpha}$ by an asthmatic patient and a healthy subject, respectively. Closed circles denote asthmatic patient and open circles healthy subject. Center: Strip of isolated human bronchus in an organ bath. PGF$_{2\alpha}$ was added to final concentration of 0.2 and 0.8 μg/ml. "W" denotes wash. Changes in tension were recorded on a Grass polygraph; Bottom: Effect of intravenous PGF$_{2\alpha}$, 15 and 30 μg, on airway (insufflation) pressure in an anesthetized, atropinized, and propranolol-pretreated guinea pig, recorded on a Grass polygraph. Reprinted in modified form from Mathé and Hedqvist (1975) with the permission of the editors.

and 5 but not 10 min after inhalation. In contrast, atropine had no significant protective effect in healthy subjects.

The presumed interaction between the effects of PGF$_{2\alpha}$ and PGE$_2$ was also investigated. In three cases in which PGE$_2$ alone caused bronchodilation, it also diminished the constricting action of PGF$_{2\alpha}$ when administered simultaneously. Conversely, in five cases in which PGE$_2$ alone caused bronchoconstriction, it enhanced the effect of PGF$_{2\alpha}$ when given simultaneously.

Pulmonary effects of PGF$_{2\alpha}$ on humans were further investigated by Smith and Cuthbert (1972). In four healthy subjects, inhalation of PGF$_{2\alpha}$ aerosol caused bronchoconstriction. Inhalation of PGE$_2$ in similar doses (55 μg), 8 min after PGF$_{2\alpha}$, antagonized the bronchoconstriction induced by PGF$_{2\alpha}$. Bronchoconstriction was also observed after intravenous administration of PGF$_{2\alpha}$ (Smith, 1973b). In patients undergoing termination of pregnancy, continuous infusion of PGF$_{2\alpha}$ (5–

200 µg/min) increased lung resistance. This was in agreement with a reported case of respiratory distress developing in a pregnant patient after intravenous infusion of $PGF_{2\alpha}$ (Fishbourne *et al.*, 1972). The fact that much larger doses of the intravenous $PGF_{2\alpha}$ than the aerosolized $PGF_{2\alpha}$ are required to induce bronchoconstriction underlines the capacity of the lung to inactivate the circulating prostaglandins. In a further experiment (Smith *et al.*, 1975), the hyperreactivity of asthmatic subjects to inhaled $PGF_{2\alpha}$ and the generally greater hyperreactivity to this compound than to histamine were confirmed. No correlations could be made between the increased reactivity to $PGF_{2\alpha}$ and age, type of asthma, or treatment. Premedication with DSCG or flufenamic acid or inhalation of atropine (1 mg) did not affect the bronchoconstriction caused by $PGF_{2\alpha}$.

The effects of atropine and DSCG on $PGF_{2\alpha}$-induced bronchoconstriction have also been studied by other investigators. Thus Newball *et al.* (1974) found that atropine (2–2.5 mg intravenously) or DSCG (40 mg by inhalation) administered 10 and 30 min, respectively, before the test had no protective effect on bronchoconstriction caused by inhalation of $PGF_{2\alpha}$. The hyperreactivity of the asthmatic bronchial tree to $PGF_{2\alpha}$ was corroborated by Pasargiklian *et al.* (1976). The investiga-

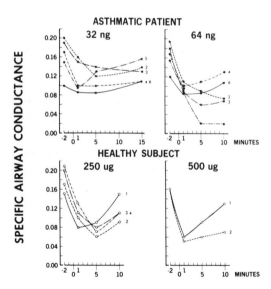

Fig. 3.   Effect of repeated $PGF_{2\alpha}$ inhalation on specific airway conductance. During a 17-month period, 32 and 64 ng of $PGF_{2\alpha}$ were administered 6 times to an asthmatic patient (top, closed circles) and 250 and 500 µg 4 times to a healthy subject (bottom, open circles). Numbers in parentheses denote the temporal sequence of experiments. The asthmatic patient did not inhale both doses in experiments 1 and 5, and the healthy subject did not inhale both doses in experiments 3 and 4. Reprinted from Mathé and Hedqvist (1975) with permission of the editors.

tors also reported that one female asthmatic patient tolerated large doses of $PGF_{2\alpha}$ by inhalation and, in contrast, that one control subject exhibited a pathological bronchial hyperreactivity to $PGF_{2\alpha}$ after influenzal bronchitis. The partial protective effect of atropine (Mathé and Hedqvist, 1975) was confirmed by these investigators. Moreover, pretreatment with DSCG diminished the bronchoconstrictive action of $PGF_{2\alpha}$. The constriction caused by $PGF_{2\alpha}$ was promptly counteracted by β-receptor stimulation, whereas premedication with an L-adrenergic blocker did not alleviate it. Generally similar results, i.e., hyperreactivity to inhaled $PGF_{2\alpha}$, partial protection by atropine, and no protection by L-adrenergic blockade and DSCG, were reported by Patel (1975).

The results presented show that $PGF_{2\alpha}$ is a potent bronchoconstrictor and that asthmatic subjects exhibit a marked hyperreactivity to its inhalation. Although part of this effect is mediated by the muscarinic receptors, the substantial part is apparently exerted via a specific receptor. Several mechanisms, alone or in combination, could account for the hyperreactivity: (1) Release by $PGF_{2\alpha}$ of other mediators, such as histamine or ACh, to which the asthmatic tracheobronchial tree has been sensitized; indeed, some data are consistent with this possibility (Tauber *et al.*, 1973; Hitchcock, 1974). (2) Sensitization by $PGF_{2\alpha}$ to the contractile action of other mediators. Potentiation of a variety of responses to ACh, angiotensin, bradykinin, vasopressin, oxytocin, Ca^{2+}, and nerve stimulation after treatment with PGE_1, PGE_2, $PGF_{2\alpha}$, or NSAIDs in several guinea pig and rat organs points to that possibility (Eliasson and Risley, 1966; Clegg *et al.*, 1966; Eagling *et al.*, 1971; Denko *et al.*, 1972; Ferreira *et al.*, 1973; Thomas and West, 1973; Baudouin-Legros *et al.*, 1974; Sjöstrand and Swedin, 1974; Chan *et al.*, 1974). Changes in the response of isolated guinea pig trachea to $PGF_{2\alpha}$ and its metabolites, histamine, ACh, other constricting agents, and nerve stimulation, after pretreatment with NSAIDs, also support such a notion (Orehek *et al.*, 1973; Finch *et al.*, 1974; Farmer *et al.*, 1974; Grodzinska *et al.*, 1975; Boot *et al.*, 1976). (3) Sensitization of the tracheobronchial smooth muscle to $PGF_{2\alpha}$, possibly comparable to the transient exaggerated response to histamine and ACh which is observed in sensitized guinea pigs *in vivo* within 2 hr after antigen challenge (Popa *et al.*, 1973). This could be due to receptor change, alteration of the cell membrane with concomitant changes in electrical impulse propagation and movement of ions, or altered response of the cyclic nucleotide system [e.g., elevation of cyclic AMP levels in response to histamine stimulation is smaller in antigen-sensitized than in control guinea pig lungs (Mathé *et al.*, 1976*a*); a similar change in cyclic AMP or a large increase in cyclic GMP in response to $PGF_{2\alpha}$ could exist in asthmatic patients]. (4)

Smooth muscle sensitization secondary to diminished availability of catecholamines with the hypothesized inverse relationship between the action and release of catecholamines and prostaglandins (Hedqvist, 1970; Mathé, 1971). (5) Altered catabolism of $PGF_{2\alpha}$. Either less $PGF_{2\alpha}$ may be catabolized by 15-PGDH or the catabolic pathway could be altered so that predominantly 13,14-dihydro-$PGF_{2\alpha}$, a more potent constrictor than the parent compound, is formed instead of 15-keto-13,14-dihydro-$PGF_{2\alpha}$. Lastly, (6) hypersensitization of $PGF_{2\alpha}$ receptors, either due to decreased bioavailability of $PGF_{2\alpha}$ or due to only a relative deficiency, i.e., increased PGE_2-to-$PGF_{2\alpha}$ ratio. Both mechanical and biochemical responses are markedly enhanced when the receptor has been previously deprived of its specific agonist. The converse has also been shown (Kakiuchi and Rall, 1968; Makman, 1971; Trendelenburg, 1972; Deguchi and Axelrod, 1973; Mathé *et al.*, 1974a). The enhanced bronchoconstriction elicited by administration of $PGF_{2\alpha}$ after pretreatment with indomethacin (Puglisi, 1973; Frey and Shäfer, 1974; Farmer *et al.*, 1974; Grodzinska *et al.*, 1975) is consistent with such a possibility. The hypothesized decreased bioavailability of $PGF_{2\alpha}$ could be secondary to a diminished synthesis of $PGF_{2\alpha}$, which, however, appears unlikely. It could also be due to an unchanged or increased production of $PGF_{2\alpha}$ (a more probable hypothesis), but with altered disposition. This could be a consequence of changed activity of the 9K-PGR due to a change in the ratio of oxidized to reduced pyridine nucleotides; increased activity of 15-PGDH, conceivably catabolizing more effectively $PGF_{2\alpha}$ than PGE_2; nonspecific binding of $PGF_{2\alpha}$; or, more speculatively, production of antibodies to $PGF_{2\alpha}$. While the mechanisms described under (1) and (2) would appear more plausible than the others, and the existing experimental results support those hypotheses, not enough knowledge has been accumulated to evaluate any of the above possibilities. However, they can have a useful purpose in further research.

B. E Prostaglandins

The actions of the E class of prostaglandins were first examined on preparations of isolated trachea from five species and on insufflation pressure in three different animals *in vivo* (Main, 1964). PGE_1 relaxed the smooth muscle and inhibited the bronchoconstrictory action of histamine and ACh. Subsequently, PGE_1 and PGE_2 were found to relax strips of human bronchi suspended in an organ bath. Neither α- nor β-adrenoceptor blockade affected the dilatory properties of PGE_1 and PGE_2 (Sweatman and Collier, 1968; Sheard, 1968; Mathé *et al.*, 1971). In the latter study, PGE_2 caused slight bronchoconstriction in some human bronchial preparations, consistent with the variable effect of PGE_2 on the guinea pig trachea (James, 1969). Frey and Schäfer (1974) injected PGE_2 (0.5–10 μg/kg) into anesthetized, ventilated cats which were vagotomized. Although the compound *per se* produced slight elevation of airway pressure, it counteracted the bronchoconstriction caused by 5-HT or carbachol in the majority of cases. In some instances, however, PGE_2 provoked additional constriction.

The bronchodilating effects of PGE_1 and PGE_2 have been extensively investigated *in vivo* by Rosenthale *et al.* (1968, 1970, 1971, 1975). In guinea pigs, dogs, and other species, both inhalation and intravenous administration of PGE_1 and PGE_2 effectively antagonized the bronchoconstriction elicited by histamine, ACh, 5-HT, and bradykinin. Changes in lung resistance caused by anaphylaxis were also suppressed by these prostaglandins. Adrenalectomy, double pithing, and pretreatment with reserpine or β-adrenoceptor blockers did not modify the dilatory activity of PGE_1 and PGE_2. Furthermore, tachyphylaxis was not observed. Administered as aerosols, both PGE_1 and PGE_2 were at least 10 times more potent as bronchodilators than isoproterenol. Similar results with a PGE_2 aerosol administered *in vivo* were reported by Large *et al.* (1969).

In the original study with humans, PGE_1 and PGE_2 (0.275–55 μg) were inhaled by healthy and asthmatic subjects and marked bronchodilatory effects were obtained with the highest doses (Cuthbert, 1969, 1971). Ten times more isoproterenol was required to achieve a similar degree of dilation. No systemic signs caused by the administered prostaglandins were observed. However, symptoms due to irritation of the upper respiratory tract were conspicuous in many experimental subjects. Other investigators who used up to 100 μg of PGE_1 have reported similar results (Herxheimer and Roetscher, 1971). In further studies, Smith (1973*b*) observed that continuous intravenous infusion of PGE_2 (2.5–20 μg/min) in pregnant women resulted in small but significant increases in lung resistance in five out of eight women. Intravenous administration of PGE_2 (5–20 μg/min) to asthmatic patients again demonstrated that, in some subjects, PGE_2 may act as a bronchoconstrictor (Smith, 1974*a*). In a subsequent paper (Smith *et al.*, 1975), inhalation of PGE_1 and PGE_2 was reported to cause bronchodilation. Symptoms of irritation of the respiratory tract, but no bronchoconstriction, were observed in both healthy and asthmatic participants. Mathé and Hedqvist (1975) found that in asthmatic patients the response to PGE_2 was variable; bronchodilation was obtained in five, no change in three, and bronchoconstriction in 18 tests. The highest PGE_2 dose inhaled (100 μg) resulted in the greatest mean decrease in SG_{aw}, -27.7%. With this dose, five out of six patients exhibited a decrease in SG_{aw}; they also complained of throat irritation and tightness in the chest. In healthy subjects, PGE_2 caused less irritation than in asthmatics. The healthy also reacted more uniformly, bronchodilation being obtained in 22 out of 23 tests. The highest dose of PGE_2 (100 μg) elicited the highest mean increase in SG_{aw}, 42%. Inhalation of PGE_2 caused bronchodilation also in chronic bronchitic, asthmatic, and healthy subjects (Kawakami *et al.*, 1973). However, in some subjects from all three groups bronchoconstriction was observed. In a recent study (Pasargiklian *et al.*, 1976), the bronchodilatory effect of inhalation of 48 μg PGE_1 or PGE_2 by asthmatic patients was very small. In the same patients, both prostaglandins had a significant protective effect on the bronchoconstriction induced by inhalation of water or specific allergens. While a cholinergically mediated reflex could possibly explain the *in vivo* bronchoconstrictive effect of PGE_2 (when obtained), this can not explain the *in vitro* results. Conversion of the administered PGE_2 to $PGF_{2\alpha}$ and release of $PGF_{2\alpha}$, histamine, or ACh by PGE_2 (Tauber *et al.*, 1973; Willis *et al.*, 1974) are possibilities yet to be explored.

The bronchodilatory effects of aerosolized PGE_1 and PGE_2 in asthmatic subjects, as described by Cuthbert (1969, 1971), have raised the possibility of prostaglandin usage as therapeutic agents in diseases with reversible airway obstruction. The ability of PGE to stimulate adenylate cyclase and hence cyclic AMP and to inhibit anaphylactic release of histamine and SRS-A in several models of allergy (Koopman *et al.*, 1971; Bourne *et al.*, 1972; Lichtenstein *et al.*, 1972; Tauber *et al.*, 1973) also points to their potential as therapeutic agents. However, the consistent irritation of the upper respiratory tract after inhalation of both PGE_1 and PGE_2 and the variability of the response to PGE_2 currently preclude their clinical application. Derivatives of PGE_1 and PGE_2 have been tested *in vitro* and *in vivo* on several species and had some dilatory effects on preparations with both normal bronchial tone and tone raised by constrictors (Greenberg and Beaulieu, 1974; Greenberg, 1976; Rosenthale *et al.*, 1976; Dessy and Weiss, 1976). However, certain 15-C and 16-C PGE-methylesters can elicit bronchoconstriction in anesthetized guinea pigs and on isolated human bronchi (Strandberg and Hedqvist, personal communication). Inhalation of 15(S)-methyl-PGE_2 and 15(S)-methyl-PGE_2 methylester aerosol by healthy and asthmatic subjects (Smith, 1974b; Mathé *et al.*, unpublished data) had only a weak bronchodilatory effect. Since these compounds were not irritative, it is conceivable that other, more potent analogs could be developed.

In conclusion, PGE_1 exerts a significant and consistent bronchodilatory action, while the bronchodilatory effects of PGE_2 are smaller and bronchoconstriction has been observed in several studies. When administered as aerosols, both prostaglandins are irritative to the upper respiratory tract, thereby limiting their therapeutic usefulness. Analogs devoid of these side effects but retaining other properties of the parent compounds hold therapeutic promise in the future.

C. A and B Prostaglandins

Although prostaglandins of the A series have significant effects in the cardiovascular and renal systems, their known actions on the tracheobronchial tree appear to be moderate. (Nakano, 1973; Lee, 1976; Kadowitz *et al.*, 1975*a*, 1976). In cat trachea chain preparations, PGA_1 inhibited the smooth muscle contraction elicited by ACh. PGA_1 was 30 times less effective than PGE_1 or isoproterenol in antagonizing the action of bronchoconstrictor substances (Horton and Jones, 1969). Given intravenously to anesthetized, artificially ventilated guinea pigs, both PGA_1 and PGA_2 diminished the bronchoconstriction elicited by histamine, ACh, and 5-HT. Again, A prostaglandins were less active than PGE_1 or isoproterenol against the spasmogens used (Dessy *et al.*, 1973). In anesthetized, ventilated dogs, infusion of PGA_2 had no effect on airway resistance.

The effects of PGBs on the airways were more marked and in the opposite direction than those of PGAs. Both PGB_2 and PGB_1 were constrictors, the former being more potent. Pretreatment with atropine, decamethonium, cocaine, tetrodotoxin, and reserpine did not modify the action of PGBs on the airways (Greenberg *et al.*, 1973, 1974*a,b*). Available data indicate that the effects of PGAs and PGBs are exerted by a direct stimulating and inhibitory action, respectively, on the smooth muscle. The effects of the prostaglandins A and B *in vivo* were consistent with those obtained *in vitro* and both were, in general, similar to the effects obtained on the pulmonary vasculature (Kadowitz *et al.*, 1975*a*, 1976).

D. Endoperoxide Intermediates and Metabolites

In addition to PGFs and PGEs, both the intermediates formed during synthesis and the metabolites exert significant biological effects on several organs, including the lung. The endoperoxide intermediates have recently been isolated (Hamberg and Samuelsson, 1973; Nugteren and Hazelhof, 1973), and their role in platelet aggregation and effects on a variety of smooth muscle preparations were investigated (Hamberg and Samuelsson, 1974; Hamberg *et al.*, 1975*a,b*, 1976). PGG_2, PGH_2, and PGD_2 (0.1 $\mu g/ml$) caused marked contraction of guinea pig tracheal ring preparations. When contractile action was compared with that of $PGF_{2\alpha}$, the mean effects of the same concentration of PGD_2, PGG_2, and PGH_2 were 5.3, 7.5, and 9.3 times larger. In experiments on anesthetized, artificially respirated guinea pigs, intravenously administered PGD_2, PGG_2, and PGH_2 increased the tracheal insufflation pressure and were 5–10 times more potent than $PGF_{2\alpha}$. When administered as an aerosol, however, PGG_2 and PGH_2 had a smaller effect than the equivalent dose of $PGF_{2\alpha}$. It was suggested that the relative inefficiency of nebulized endoperoxides is due to their rapid degradation ($t_{1/2} = 5$ min) and possible formation of PGE_2 while in the airways (Hamberg *et al.*, 1975*b*).

The effects of the prostaglandin metabolites have been more extensively tested. Änggård (1966) demonstrated in guinea pig and rabbit that 15-keto-PGE_1 and 15-keto-13,14-dihydro-PGE_1 were considerably less active than the parent compound on smooth muscle preparations *in vitro* and on blood pressure. Reduction of the $\Delta^{13,14}$ double bond only did not markedly reduce the biological effects compared to those of the parent compound. Very similar results were obtained with the corresponding $PGF_{2\alpha}$ metabolites on uterine contractility *in vivo* (Bygdeman *et al.*, 1974). Smooth muscle stimulating activity of 15-keto-$PGF_{2\alpha}$ has been reported (Dawson *et al.*, 1974). This metabolite was 1–3 times more potent as a bronchoconstrictor than $PGF_{2\alpha}$ on guinea pig trachea and human bronchial muscle. Since PGD_2 also elicited strong bronchoconstriction, it was suggested that the 9-hydroxyl group is necessary for contractile activity. In contrast, 15-keto-$PGF_{2\alpha}$ given as an aerosol to unanesthetized guinea pigs did not cause bronchoconstriction (Douglas, personal communication). In a recent study (Wasserman, 1975), lung resistance, dynamic lung compliance, and other pulmonary indices were measured after injection of $PGF_{2\alpha}$, 15-keto-$PGF_{2\alpha}$, 15-keto-13,14-dihydro-$PGF_{2\alpha}$ or 13,14-dihydro-$PGF_{2\alpha}$ into the jugular vein of anesthetized dogs. In agreement with Douglas, the 15-keto metabolites were only marginally effective, even in high doses. On the other hand, 13,14-dihydro-$PGF_{2\alpha}$ was approximately twice as potent as $PGF_{2\alpha}$. These data support the original suggestion of Änggård (1966) that oxidation of the OH group at C-15 abolishes the actions of prostaglandins on smooth muscle. However, other investigators (Crutchley and Piper, 1976; Boot *et al.*, 1976) found that 15-keto-PGE_2 was a potent bronchodilator *in vivo* and also in isolated guinea pig trachea. The effects of the other two metabolites tested, 15-keto-13,14-dihydro-PGE_2 and 13,14-dihydro-PGE_2, were considerably smaller. Results to date confirm that the 15-keto-13,14-dihydro metabolites are less biologically active than the 13,14-dihydro metabolites or the parent compounds. Since the above-mentioned metabolites exhibited less activity on other organs and the order of potency varied, it was suggested that the respiratory smooth muscle is specifically sensitive to the prostaglandin metabolites (Boot *et al.*, 1976).

E. Interaction of Prostaglandins and Indomethacin and Their Relation to the Tone of Airways

Reduction of the resting tone of several smooth muscle preparations by prostaglandin antagonists has led Bennett and Posner (1971) to suggest that intramural synthesis of prostaglandins maintains the tone of these preparations. This concept was further tested on several gastrointestinal preparations (Eckenfels and Vane, 1972; Bennett *et al.*, 1975; Ferreira *et al.*, 1976). Orehek *et al.* (1973) found that aspirin (50 μg/ml) and indomethacin (0.6–6 μg/ml) decreased the basal tone of superfused, spirally cut guinea pig trachea and suggested that prostaglandins regulated the airway smooth muscle tone. In the same preparation, pretreatment with aspirin or indomethacin diminished the contractile response to low doses (0.05–1.1 μg) of histamine and ACh, but enhanced the response to higher doses (0.5–2 μg) of the agonists. The proposed explanation was that the decreased response to low doses of histamine and ACh as well as the reduction of basal tone is due to inhibition of $PGF_{2\alpha}$ synthesis by NSAIDs. In contrast, enhanced effects of these two agonists at high doses were secondary to inhibition of PGE_2 production. This hypothesis was further developed, and the authors proposed that the sensitivity of the smooth muscle to prostaglandins varies according to the state of contraction. When relaxed, the muscle would be more sensitive to $PGF_{2\alpha}$ and when contracted it would respond more to PGE_2 (Orehek *et al.*, 1975).

Farmer *et al.* (1972, 1974) reported that indomethacin (0.05–1.6 μg/ml) decreased the intrinsic intraluminal pressure of guinea pig tracheal tube preparations. The decrease of basal tracheal tone was considered a consequence of inhibited synthesis of prostaglandins and not due to a diminished uptake of calcium into the smooth muscle (Northover, 1971, 1973), since 270 times more indomethacin was required to protect against methacholine-induced contractions than that which inhibited the intrinsic tone. It was also found that 1 and 10 μg/ml of indomethacin potentiated, while 100 μg/ml indomethacin decreased, the maximal response to administered $PGF_{2\alpha}$. Conceivably, indomethacin at low concentrations exerted an enhancing effect on $PGF_{2\alpha}$ contraction by the inhibition of PGE synthesis and/or increasing the sensitivity of the $PGF_{2\alpha}$ receptor to its agonist. The antagonizing action at high doses could be explained by a nonspecific membrane-stabilizing effect. On isolated guinea pig tracheal preparations, indomethacin (10

μg/ml), SC 19220 (100 ng/ml), and meclofenamate (10 ng/ml) potentiated the contractile response to $PGF_{2\alpha}$ over a wide range of concentrations (Puglisi, 1973). Effects of PGE_1 under the same conditions were reduced or reversed. The potentiation of the exogenous $PGF_{2\alpha}$ action could be explained, as above, by the inhibition of endogenous PGE and/or $PGF_{2\alpha}$ production, thereby making the receptor more sensitive to subsequent exposure to the same agonist. The reason for the decreased or reversed PGE_1 activity is not clear, although it could be due to the same factors.

Lambley and Smith (1975) also obtained a decrease in the resting tone of guinea pig tracheal chain preparations incubated with indomethacin (0.1–3.2 μg/ml) or SC 19220 (1–8 μg/ml). Incubation with arachidonic acid (0.5–4 μg/ml) restored the tracheal tone. PGE_2 caused contraction of tracheal muscle relaxed by indomethacin or SC 19220. It was proposed that the effects of prostaglandins might depend on the preexisting tone of the smooth muscle. In additional experiments (Dunlop and Smith, 1975), indomethacin (3.6 μg/ml) slowly relaxed passively sensitized human bronchial preparations. Frey and Schäfer (1974) measured changes in insufflation pressure (taken as an index of changes in the airway tone) in three cats which were vagotomized, completely relaxed by suxamethonium, and respirated. Intravenous indomethacin (2–10 mg/kg) had no effect on the intrinsic tone of the airways. However, in parallel to the results of others (Puglisi, 1973; Farmer *et al.*, 1972, 1974; Lambley and Smith, 1975), airway responses to prostaglandins were enhanced after indomethacin pretreatment. From the work presented, it would appear reasonable to conclude that endogenous prostaglandins are modulating both the intrinsic tone of the airways and the changes in tone induced by other bronchoactive compounds.

F. Effects of Prostaglandins on Pulmonary Circulation

Von Euler (1934, 1936) and Goldblatt (1933, 1935) discovered that crude extracts containing prostaglandins decreased the systemic arterial pressure and dilated blood vessels, effects which were not antagonized by atropine. Actions of prostaglandins in a variety of species on myocardial contractility and heart rate, on arterial, venous, and total lung resistance, and on other cardiovascular parameters have been extensively studied by many investigators, possibly most systematically by Bergström, Kadowitz, Nakano, Weeks, and their co-workers. Since there have been recent reviews on the cardiovascular and pulmonary vascular effects of prostaglandins (Nakano, 1973; Kadowitz *et al.*, 1975a), only a brief outline will be presented here. In intact animals and lung preparations of cat, calf, dog, swine, lamb, and guinea pig, $PGF_{2\alpha}$ was demonstrated to be one of the most active pressor substances. It contracts both arteries and veins, and in the dog responses could be obtained with concentrations as low as 10^{-12} M $PGF_{2\alpha}$ in lobar arterial blood (Änggård and Bergström, 1963; DuCharme *et al.,* 1968; Said, 1968; Nakano and Cole, 1969; Anderson *et al.,* 1972; Lewis and Eyre, 1972; Okpako, 1972; Kadowitz *et al.,* 1974*a,b*). Recently, it was also shown that administration of 500 and 1000 μg $PGF_{2\alpha}$ into the human pulmonary artery *in vivo* elicited a significant increase in the mean pressure in the same vessel and in total lung resistance (Pasargiklian *et al.,* 1976). *In vitro,* $PGF_{2\alpha}$ is a potent constrictor of isolated segments of arteries and veins (Kadowitz *et al.,* 1975*a*).

The 15-methyl analogue of $PGF_{2\alpha}$, administered intramuscularly or intravenously to dogs, produced a greater and longer-lasting elevation in pulmonary arterial pressure and a greater initial fall in systemic arterial oxygen tension than $PGF_{2\alpha}$ (Weir *et al.,* 1975). Both 15-methyl-$PGF_{2\alpha}$ and -$PGE_{2\alpha}$, infused into the dog lobar artery, markedly increased the pulmonary vascular pressure. They also contracted isolated segments of intrapulmonary artery and vein (Kadowitz *et al.,* 1975*d,* 1976). These findings would indicate the need for caution when using these analogues in human subjects as abortifacients or for other purposes.

Consistent with its general relaxing action on smooth muscle, PGE_1 was shown to decrease pulmonary vascular resistance by dilating veins and arteries in blood- or dextran-perfused isolated lungs and in intact lungs of rabbit, dog, calf, swine, and lamb (Hauge *et al.,* 1967; Nakano and Cole, 1969; Said, 1968; DuCharme *et al.,* 1968; Hyman, 1969; Lewis and Eyre, 1972; Kadowitz *et al.,* 1974*a,b*). β- and α-adrenoceptor blockade did not modify these effects. A fall in mean pulmonary arterial pressure by intravenous infusion of PGE_1 was also found in one study conducted in human (Carlson *et al.,* 1969). In contrast, PGE_2 increases the pulmonary vascular resistance in intact dog, swine, lamb, calf, and cat lung circulation (Said, 1968; Anderson *et al.,* 1972; Frey and Schäfer, 1974; Kadowitz *et al.,* 1975*b*). In the isolated, perfused calf and guinea pig lung, PGE_2 decreased the resistance to flow (Lewis and Eyre, 1972; Okpako, 1972). Results on isolated strips of vasculature, suspended in an organ bath, are generally in agreement with data obtained from whole organs and in intact animals *in vivo* (Kadowitz *et al.,* 1975*a,* 1976). PGE_1 dilated both arteries and veins, while PGE_2 had no significant effect on arteries but contracted the veins. The reasons for the difference in effects of PGE_1 and PGE_2 in the lung circulation and for the apparent opposite action of PGE_2 in the pulmonary compared to the systemic vascular beds are not known.

The effects of A and B prostaglandins in the pulmonary vascular bed have recently been extensively investigated. In dog, infusion of PGA$_1$ decreased lobar arterial and venous pressure, while PGA$_2$ had the opposite effects. Infusion of PGB$_1$ or PGB$_2$ increased lobar arterial and venous pressure. PGB$_2$ was found to be one of the most potent constrictors in the dog pulmonary vascular bed. Results on isolated pulmonary vessels were parallel to those obtained *in vivo* (Kadowitz and Hyman, 1975; Kadowitz *et al.*, 1975*c*, 1976; Greenberg *et al.*, 1973).

VI. INTERACTION OF PROSTAGLANDINS WITH CYCLIC NUCLEOTIDES IN THE LUNG

Since the first direct demonstration of the effects of prostaglandins on cyclic AMP (Butcher *et al.*, 1967) interaction of prostaglandins and cyclic nucleotides has been investigated in numerous tissues (reviews by Bourne, 1974; Kuehl, 1974; Kuehl *et al.*, 1974; Gorman, 1975; Sohn *et al.*, 1976). In general, PGEs stimulate adenylate cyclase to increase cyclic AMP levels, while PGFs affect the guanylate cyclase resulting in an enhanced cyclic GMP accumulation. The predominantly opposing effects of PGEs and PGFs on many systems would thus be consistent with the concept of two opposing "forces," cyclic AMP and cyclic GMP, regulating cell function (Goldberg *et al.*, 1975).

In the lung, an early investigation showed that PGE$_1$ (2.8 μM) increased cyclic AMP levels in rat lung fragments (Butcher and Baird, 1968). In contrast, PGE$_2$, PGA$_2$, and PGF$_{2\alpha}$ (5×10^{-5} M) did not elevate the cyclic AMP levels in rabbit lung fragments (Palmer, 1972). In subsequent experiments, adenylate cyclase from guinea pig lung was stimulated by PGE$_1$ and to a lesser extent by PGE$_2$ (Weinryb *et al.*, 1973). Increased enzyme activity was also obtained after incubation with PGF$_{2\alpha}$. The concentration of all drugs used was 10^{-4} M. PGE$_1$ (1 μg/ml) also potently stimulated adenylate cyclase activity in preparations of rat and monkey lung (White, 1974). The same concentrations of PGE$_2$, PGE$_3$, and PGA$_1$ were considerably less effective, while prostaglandins of the F series did not change the activity of this enzyme.

In rat lung slices (Kuo and Kuo, 1973), PGE$_1$ increased the level of cyclic AMP and had no effect on cyclic GMP. The increase in cyclic GMP accumulation elicited by ACh was antagonized by PGE$_1$. Stoner *et al.* (1974) found that cyclic AMP levels in guinea pig lung slices were significantly increased when incubated with PGE$_1$ (2 μM). On the other hand, PGE$_2$ had only a marginal effect. Incubation with PGF$_{2\alpha}$ (2 μM) did not change the cyclic AMP levels. In the same preparation, all three prostaglandins slightly reduced the cyclic GMP concentration. Bradykinin and ACh (1–100 μg/ml) also elevated the cyclic AMP levels in guinea pig slices (Stoner *et al.,* 1973). Since pretreatment with indomethacin (15μM) or aspirin (10 mM) abolished this effect, it can be surmised that the endogenously synthesized prostaglandins regulate the cyclic AMP levels in the lung. In agreement with these experiments was the finding that in antigen-sensitized guinea pig lungs cyclic AMP levels were lower in indomethacin-perfused (10 μM) than in NaCl-perfused (0.9%) preparations. The rise in cyclic AMP produced by anaphylaxis was also substantially reduced by the perfusion of 10 μM indomethacin through guinea pig lung (Mathé *et al.,* 1976*f*). In whole perfused guinea pig lungs, PGE_1 (1 and 10 μg/ml) markedly increased the cyclic AMP levels but had no effect on cyclic GMP. $PGF_{2\alpha}$, in the same concentrations, did not change the cyclic AMP levels and the mean cyclic GMP levels were increased by only 40% (Mathé and Volicer, unpublished data). Levels of cyclic AMP in the isolated guinea pig trachea were increased severalfold by incubation with 1–10 μg/ml PGE_1 (Murad and Kimura, 1974). As in other experiments, $PGF_{2\alpha}$ did not alter the cyclic AMP level. Neither PGE_1 nor $PGF_{2\alpha}$ affected the cyclic GMP levels in this preparation. In conclusion, except for the report of Palmer (1972), stimulation of adenylate cyclase–cyclic AMP system in a variety of lung preparations has always been obtained with PGE_1 and to a lesser extent by PGE_2, whereas $PGF_{2\alpha}$ generally had no effect. Prostaglandins of the E type do not affect the guanylate cyclase–cyclic GMP system and the reported data on the action of $PGF_{2\alpha}$ to alter this system have been inconclusive.

Increase in endogenous levels of cyclic AMP and administration of cyclic AMP derivatives inhibit the immunological release of mediators from whole lung or lung fragments, isolated mast cells, leukocytes, and other tissues of several species. In contrast, a decrease in cyclic AMP, with or without an increase in cyclic GMP levels, enhances the liberation of mediators (Lichtenstein and Margolis, 1968; Ishizaka *et al.,* 1971; Lichtenstein and DeBernardo, 1971; Koopman *et al.,* 1971; Orange *et al.,* 1971; Bourne *et al.,* 1972; Kaliner *et al.,* 1972, 1973; Lichtenstein *et al.,* 1972; Tauber *et al.,* 1973; Stechschulte and Austen, 1974; Mathé *et al.,* 1976*f*).

Cyclic nucleotides also play a role in smooth muscle contraction and dilation (Bär, 1974; Andersson *et al.,* 1975). Other areas of importance, although not specific to the lung, where the action of prostaglandins may be mediated by alterations of cyclic nucleotide levels are cell division, production of antibodies, and inflammation (Parker *et al.,* 1974; MacManus *et al.,* 1974; Sheppard, 1974; Bourne, 1974; Ferreira and Vane, 1974; Vargaftig, 1974; Zurier, 1974; Bekemeier *et al.,* 1974).

In conclusion, since it has been shown that the intracellular cyclic nucleotide levels will modulate the release of mediators, tone of the smooth muscle, production of antibodies, and the course of inflammatory processes, the ability of prostaglandins to alter cyclic nucleotide levels must have important consequences for lung function.

VII. SUMMARY AND CONCLUSIONS

Lungs of all species examined possess a marked capacity to synthesize but not store prostaglandins. While more $PGF_{2\alpha}$ than PGE_2 is spontaneously released from the lung parenchyma, the opposite is true for the airways. Parenchyma also appears to catabolize the prostaglandins to a larger extent than do the airways. Factors controlling this discriminative synthesis and catabolism are not known. The meaning of such a phenomenon is more apparent. It is suggested that the preponderance and apparently slow catabolism of PGE_2 in tracheobronchial tissue ensures the patency of the airways, whereas the preponderance in lung parenchyma of $PGF_{2\alpha}$, with faster catabolism, is important for shifting the blood flow from one location to another, hence achieving the appropriate ventilation–perfusion ratios.

The properties of prostaglandin enzymes obtained from a variety of tissues appear to be similar. Activities of the synthetase system, 15-PGDH, and 13-PGR are high in the lung as compared to most other organs, whereas that of 9K-PGR appears to be relatively low. High 15-PGDH activity, and apparent absence of intracellular compartments inaccessible to this enzyme, may explain the lack of lung capacity to retain unmetabolized prostaglandins. An active, carrier-mediated uptake process has been suggested for prostaglandins; the results hitherto obtained are consistent with such a hypothesis.

The lung very efficiently catabolizes prostaglandins, either synthesized locally or carried by the blood from other organs. PGEs and PGFs are almost completely inactivated by one passage through the pulmonary circulation, whereas the loss of PGAs is only partial. Prostaglandins can thus be added to the list of substances whose arterial concentrations, and thereby systemic effects, are controlled by the lung. Increased synthesis or decreased breakdown could influence the course of the pathophysiological processes within the lung as well as contribute to the systemic signs and symptoms of lung diseases, such as emphysema, sarcoidosis, and chronic interstitial pneumonia. No investigations appear to have been conducted in this particular area.

Endothelium, mucosa, smooth muscle cells, mast cells, and other cells and tissues have been implicated as the source of prostaglandins in the lung. Prostaglandins are released from the lung by mechanical stimuli and vasoactive and bronchoconstricting substances. Antigen-sensitized, nonchallenged lungs release spontaneously more $PGF_{2\alpha}$ and metabolites than do control lungs. Large quantities of thromboxanes, prostaglandins, and metabolites are released by the antigen–antibody reaction. A significant increase in 15-keto-13,14-dihydro-$PGF_{2\alpha}$ has been detected in venous blood of asthmatic patients after allergen challenge. However, elevated plasma levels of prostaglandins or metabolites have not been found in asthmatics in remission. A hypothesis has been put forth that development of antibodies to $PGF_{2\alpha}$ and increased synthesis of prostaglandins concomitantly with a faster breakdown of these compounds play a role in (some cases of) bronchial asthma.

Lung edema, emboli and hypoxia, and also lack of ascorbic acid are other conditions where some signs and symptoms are probably attributable to the increased generation of prostaglandins, predominantly of the F type.

Several groups of compounds inhibit prostaglandin release: NSAIDs by inhibiting the prostaglandin synthetase, corticosteroids probably by decreasing the availability of prostaglandin precursors, catecholamines by diminishing the anaphylactic reaction, and various compounds by blocking the receptor sites of specific agonists.

The potent broncho- and vasoconstrictive actions of $PGF_{2\alpha}$ have been demonstrated in many *in vitro* and *in vivo* preparations. Asthmatic subjects are hyperreactive to inhaled $PGF_{2\alpha}$, an effect which is only partially cholinergically mediated. The release by $PGF_{2\alpha}$ of antigen–antibody reaction mediators or potentiation of their effects and altered catabolism of $PGF_{2\alpha}$ are but a few possibilities that have been suggested to account for the hyperreactivity phenomenon.

In the lung, the actions of PGEs oppose those of $PGF_{2\alpha}$; PGE_1 is a broncho- and vasodilator, inhibits the release of histamine, and directly antagonizes the effects of $PGF_{2\alpha}$. Effects of PGE_2 are generally similar to those of PGE_1, although broncho- and vasoconstriction have been observed under some conditions. Therapeutic application of PGEs has been hampered because PGEs irritate the upper respiratory tract. Development of less irritative analogues holds therapeutic promise.

Prostaglandins also contribute to regulation of the basal tone of the airways, as demonstrated by changes in the tone after indomethacin treatment. Endoperoxides, TxA_2, and some prostaglandin metabolites exert potent effects on tracheobronchial and whole lung preparations. This emphasizes the fact that products at each step of prostaglandin synthesis and catabolism may have a biological function.

Prostaglandins of the E type stimulate adenylate cyclase in the lung. On the other hand, $PGF_{2\alpha}$ generally does not affect this system but may activate guanylate cyclase. The consequences of changes in the levels of cyclic nucleotides would be to alter the patency of the airways and the tone of the pulmonary vasculature as well as modulate the anaphylactic release of mediators.

The role of prostaglandins in the control of lung function has been discussed, and while some phenomena are well established, others have only been touched upon. The picture emerging is that prostaglandins modulate both the respiratory and nonrespiratory lung functions in the state of health and may account for some of the symptoms in pathophysiological states.

Future research might profitably be directed toward clarification of the primary vs. secondary mediators, of the interdependence of release and action of prostaglandins and other mediators, of the relationship between prostaglandins and the sympathetic–parasympathetic balance, of the relationship between the local and systemic symptoms of lung disease and altered prostaglandin synthesis and catabolism, and, lastly, of the possible role of prostaglandin antibodies in the development of allergy.

ACKNOWLEDGMENTS

This work was supported in part by NHLI, Grant HL-15677 and Pulmonary SCOR Grant HL-15063. The author wishes to express his gratitude to C. Leslie for helpful suggestions in the preparation of this chapter.

VIII. REFERENCES

Alabaster, V. A., and Bakhle, Y. S., 1970, The release of biologically active substances from isolated lungs by 5-hydroxytryptamine and tryptamine, *Br. J. Pharm. Chem.* **40**:582.

Anderson F. L., Krabos, A. C., Tsagaris, T. J., and Kuida, H., 1972, Effect of prostaglandins $F_{2\alpha}$ and E_2 on the bovine circulation, *Proc. Soc. Exp. Biol. Med.* **140**:1049.

Anderson, M. W., 1975, Studies on the uptake and metabolism of prostaglandins by the isolated perfused rat lung: A possible carrier-mediated transport system for prostaglandins in the lung, *Fed. Proc.* **33**:790.

Andersson, R., Nilsson, K., Wikberg, J., Johansson, S., Mohme-Lundholm, E., and Lundholm, L., 1975, Cyclic nucleotides and the contraction of smooth muscle, in: *Advances in Cyclic Nucleotide Research,* Vol. 5 (G. I. Drummond, P. Greengard, and G. A. Robison, eds.), pp. 491–532, Raven Press, New York.

Änggård, E., 1965, The isolation and determination of prostaglandins in lungs of sheep, guinea pig, monkey, and man, *Biochem. Pharmacol.* **14**:1507.

Änggård, E., 1966, The biological activities of three metabolites of prostaglandin E_1, *Acta Physiol. Scand.* **66**:509.

Änggård, E., and Bergström, S., 1963, Biological effects of an unsaturated trihydroxy acid $(PGF_{2\alpha})$ from normal swine lung, prostaglandin and related factors 13, *Acta Physiol. Scand.* **58**:1.

Änggård, E., and Samuelsson, B., 1964, Metabolism of prostaglandin E_1 in guinea pig lung: the structure of metabolites, *J. Biol. Chem.* **239**:4097.

Änggård, E., and Samuelsson, B., 1965, Biosynthesis of prostaglandins from arachidonic acid in guinea pig lung, *J. Biol. Chem.* **240**:3518.

Änggård, E., and Samuelsson, B., 1966, Purification and properties of a 15-hydroxyprostaglandin dehydrogenase from swine lung, *Ark. Kemi* **25**:293.

Änggård, E., Gréen, K., and Samuelsson, B., 1965, Synthesis of tritium-labeled prostaglandin E_2 and studies on its metabolism in guinea pig lung, *J. Biol. Chem.* **240**:1932.

Änggård, E., Larsson, C., and Samuelsson, B., 1971, The distribution of 15-hydroxy prostaglandin dehydrogenase and 13-reductase in tissues of the swine, *Acta Physiol. Scand.* **81**:396.

Austen, K. F., 1974, Systemic anaphylaxis in the human being, *N. Engl. J. Med.* **291**:661.

Bakhle, Y. S., and Smith, T. W., 1972, Release of spasmogenic substances induced by vasoactive amines from isolated lungs, *Br. J. Pharmacol.* **46**:543.

Bakhle, Y. S., and Vane, J. R., 1974, Pharmacokinetic function of the pulmonary circulation, *Physiol. Rev.* **54**:1007.

Bär, H.-P., 1974, Cyclic nucleotides and smooth muscle, in: *Advances in Cyclic Nucleotide Research*, Vol. 4 (P. Greengard and G. A. Robinson, eds.), pp. 195–237, Raven Press, New York.

Baudouin-Legros, M., Meyer, P., and Worcel, M., 1974, Effects of prostaglandin inhibitors on angiotensin, oxytocin and prostaglandin $F_{2\alpha}$ contractile effects on the rat uterus during the oestrous cycle, *Br. J. Pharmacol.* **52**:393.

Baum, T., Wendt, R. L., Peters, J. R., Butz, F., and Shropshire, A. T., 1974, Comparison of activities of two prostaglandin stereoisomers: $PGF_{2\alpha}$ and $PGF_{2\beta}$, *Eur. J. Pharm.* **25**:92.

Bedwani, J. J., and Marley, P. B., 1975, Enhanced inactivation of prostaglandin E_2 by the rabbit lung during pregnancy or progesterone treatment, *Br. J. Pharmacol.* **53**:547.

Bekemeier, H., Giessler, A. J., and Hirschleman, R., 1974, Prostaglandins and inflammation, *Pol. J. Pharmacol. Pharm.* **26**:5.

Bennett, A., and Posner, J., 1971, Studies on prostaglandin antagonists, *Br. J. Pharmacol.* **42**:584.

Bennett, A., Eley, K. G., and Stockley, H. L., 1975, The effects of prostaglandins on guinea pig isolated intestine and their possible contribution to muscle activity and tone, *Br. J. Pharmacol.* **54**:197.

Benzie, R., Boot, J. R., and Dawson, W., 1975, A preliminary investigation of prostaglandin synthetase activity in normal, sensitized and challenged sensitized guinea pig lung, *J. Physiol.* **246**:80P.

Bergström, S., Dressler, F., Krabisch, L., Ryhage, R., and Sjövall, J., 1962, The isolation and structure of a smooth muscle stimulating factor in normal sheep and pig lungs, *Ark. Kemi* **20**:63.

Bergström, S., Ryhage, R., Samuelsson, B., and Sjövall, J., 1963, The structure of prostaglandin E_1, $F_{1\alpha}$ and $F_{1\beta}$, *J. Biol. Chem.* **238**:3555.

Berry, E. M., Edmonds, J. F., and Wyllie, J. H., 1971, Release of prostaglandin E_2 and unidentified factors from ventilated lungs, *Br. J. Surg.* **58**:189.

Bito, L. Z., 1972, Accumulation and apparent active transport of prostaglandins by some rabbit tissues *in vitro*, *J. Physiol. (London)* **221**:371.

Bito, L. Z., 1975, Saturable energy-dependent, transmembrane transport of prostaglandins against concentration gradients, *Nature (London)* **256**:134.

Bito, L. Z., Wallenstein, M., and Baroody, R., 1976, The role of transport processes in the distribution and disposition on prostaglandins, in: *Advances in Prostaglandin and Thromboxane Research* (B. Samuelsson and R. Paoletti, eds), pp. 297–303, Raven Press, New York.

Boot, J. R., Dawson, W., and Harvey, J., 1976, Comparative biological activity of prostaglandin E_2 and its C_{20} metabolites on smooth muscle preparations, in: *Advances in Prostaglandin and Thromboxane Research* (B. Samuelsson and R. Paoletti, eds.), p. 958, Raven Press, New York.

Bouhuys, A., 1976, Action and interaction of pharmacologic agents on airway smooth muscle, in: *The Biochemistry of Smooth Muscle* (N. L. Stephens, ed.), University Park Press, Baltimore, in press.

Bourne, H. R., 1974, Immunology, in: *The Prostaglandins*, Vol. 2 (P. W. Ramwell, ed.), pp. 277–291, Plenum Press, New York.

Bourne, H. R., Lichtenstein, L. M., and Melmon, K. L., 1972, Pharmacologic control of allergic histamine release *in vitro:* Evidence of an inhibitory role of 3'-5'-adenosine monophosphate in human leukocytes, *J. Immunol.* **108**:695.

Burka, J. F., and Eyre, P., 1974, A study of prostaglandins and prostaglandin antagonists in relation to anaphylaxis in calves, *Can. J. Physiol. Pharmacol.* **52**:942.

Butcher, R. W., and Baird, C. E., 1968, Effects of prostaglandins on adenosine 3'-5'-monophosphate levels in fat and other tissues, *J. Biol. Chem.* **243**:1713.

Butcher, R. W., Pike, J. E., and Sutherland, E. W., 1967, The effect of prostaglandin E_1 on adenosine 3'-5'-monophosphate levels in adipose tissue, in: *Prostaglandins* (Proc. 2nd Nobel Symp., Stockholm) (S. Bergstrom and B. Samuelsson, eds.), p. 133, Almqvist and Wiksell, Uppsala.

Bygdeman, M., Gréen, K., Toppozada, M., Wiqvist, N., and Bergström, S., 1974, The influence of prostaglandin metabolites on the uterine response to $PGF_{2\alpha}$: A clinical and pharmacokinetic study, *Life Sci.* **14**:521.

Carlson, L. A., Ekelund, L. G., and Orö, L., 1969, Circulatory and respiratory effects of different doses of prostaglandin E_1 in man, *Acta Physiol. Scand.* **75**:161.

Chan, W. Y., Hruby, V. J., and du Vigneaud, V., 1974, Effects of magnesium ion and oxytocin inhibitors on the uterotonic activity of oxytocin and prostaglandins E_2 and $F_{2\alpha}$, *J. Pharmacol. Exp. Ther.* **190**:77.

Clegg, P. C., Hall, W. J., and Pickles, V. R., 1966, The action of ketonic prostaglandins on the guinea pig myometrium, *J. Physiol. (London)* **183**:123.

Coburn, R. F., Hitzig, B., and Yamaguchi, T., 1974, Prostaglandin (PG) synthesis in canine trachealis muscle (CTM) and guinea pig taenia coli (TC), *Fed. Proc.* **33**:451.

Collier, H. O. J., and Sweatman, W. J. F., 1968, Antagonism by fenamates of prostaglandin $F_{2\alpha}$ and slow reacting substance on human bronchial muscle, *Nature (London)* **219**:864.

Crunkhorn, P., and Willis, A. L., 1971, Cutaneous reaction to intradermal prostaglandins, *Br. J. Pharmacol.* **41**:49.

Crutchley, D. J., and Piper, P. J., 1974, Prostaglandin inactivation in guinea pig lung and its inhibition, *Br. J. Pharmacol.* **52**:197.

Crutchley, D. J., and Piper, P. J., 1975, Inhibition of the pulmonary inactivation of prostaglandins *in vivo* by DI-4-phloretin phosphate, *Br. J. Pharmacol.* **54**:301.

Crutchley, D. J., and Piper, P. J., 1976, Biological activity of the pulmonary metabolites of prostaglandin E_2, in: *Advances in Prostaglandin and Thromboxane Research* (B. Samuelsson and R. Paoletti, eds.), p. 959, Raven Press, New York.

Curry, J. J., 1946, The action of histamine on the respiratory tract in normal and asthmatic subjects, *J. Clin. Invest.* **25**:785.

Cuthbert, M. F., 1969, Effect on airway resistance of prostaglandin E_1 given by aerosol to healthy and asthmatic volunteers, *Br. Med. J.* **4**:723.

Cuthbert, M. F., 1971, Bronchodilator activity of aerosols of prostaglandins E_1 and E_2 in asthmatic subjects, *Proc. R. Soc. Med.* **64**:15.

Cuthbert, M. F., 1973, Prostaglandins and respiratory smooth muscle, in: *The Prostaglandins* (M. F. Cuthbert, ed.), pp. 253–285, Lippincott, Philadelphia.

Dawson, W., Lewis, R. L., McMahon, R. E., and Sweatman, W. J. F., 1974, Potent bronchoconstrictor activity of 15-keto prostaglandin $F_{2\alpha}$, *Nature (London)* **250**:331.

Deguchi, T., and Axelrod, J., 1973, Supersensitivity and subsensitivity of the β-adrenergic receptor in pineal gland regulated by catecholamine transmitter, *Proc. Natl. Acad. Sci. U.S.A.* **70**:2411.

Denko, C. W., Moskowitz, R. W., and Heinrich, G., 1972, Interrelated pharmacologic effects of prostaglandins and bradykinin, *Pharmacology* **8**:353.

Dessy, F., and Weiss, M. J., 1976, Broncholytic activity of some PGE_1 and PGE_2 derivatives in the guinea-pig, in: *Advances in Prostaglandin and Thromboxane Research* (B. Samuelsson and R. Paoletti, eds.), p. 960, Raven Press, New York.

Dessy, F., Maleux, M. R., and Cognioul, A., 1973, Bronchospasmolytic activity of some prostaglandins in the guinea pig, *Arch. Int. Pharmacodyn. Ther.* **206**:368.

Douglas, J. S., Dennis, M. W., and Ridgway, P., 1973, Airway constriction in guinea pigs: Interaction of histamine and autonomic drugs, *J. Pharmacol. Exp. Ther.* **184**:169.

Drazen, J. M., and Austen, K. F., 1974, Effects of intravenous administration of slow-reacting substance of anaphylaxis, histamine, bradykinin, and prostaglandin $F_{2\alpha}$ on pulmonary mechanics in the guinea pig, *J. Clin. Invest.* **53**:1679.

Drazen, J. M., and Austen, K. F., 1975, Atropine modification of the pulmonary effects of chemical mediators in the guinea pig, *J. Appl. Physiol.* **38**:834.

DuCharme, D. W., Weeks, J. R., and Montgomery, R. G., 1968, Studies on the mechanism of the hypertensive effect of prostaglandin $F_{2\alpha}$, *J. Pharmacol. Exp. Ther.* **160**:1.

Dunlop, L. S., and Smith, A. P., 1975, Reduction of antigen-induced contraction of sensitized human bronchus *in vitro* by indomethacin, *Br. J. Pharmacol.* **54**:495.

Eagling, E. M., Lovell, H., and Pickles, V. R., 1971, Prostaglandins, myometrial "enhancement" and calcium, *J. Physiol. (London)* **213**:53.

Eakins, K. E., Karim, S. M. M., and Miller, J. D., 1970, Antagonism of some smooth muscle actions of prostaglandins by polyphloretin phosphate, *Br. J. Pharmacol.* **39**:556.

Eckenfels, A., and Vane, J. R., 1972, Prostaglandins, oxygen tension, and smooth muscle tone, *Br. J. Pharmacol.* **54**:451.

Eliasson, R., and Risley, P. L., 1966, Potentiated response of isolated seminal vesicles to catecholamines and acetylcholine in the presence of PGE_1, *Acta Physiol. Scand.* **67**:253.

Eling, T. E., and Anderson, M. W., 1975, Studies on the biosynthesis, metabolism, and transport of prostaglandins by the lung, *Abst. Proc. IUPHAR Satellite Symp. (Turku)*, Finland, pp. 20–21.

Euler, U.S., von, 1934, Zur Kenntnis der pharmakologischen Wirkungen von Nativsekreten und Extrakten männlicher accessorischer Geschlechtsdrüsen, *Naunyn-Schmiedebergs Arch Exp. Pathol. Pharmakol.* **175**:78.

Euler, U.S., von, 1936, On the specific vaso-dilating and plain muscle stimulating substances from accessory genital glands in man and certain animals (prostaglandin and vesiglandin), *J. Physiol. (London)* **88**:213.

Euler, U.S., von, 1939. Weitere Untersuchungen über pg, die physiologisch aktive Substanz gewisser Genitaldrüsen, *Skand. Arch. Physiol.* **81**:65.

Even, P., Sors, H., Ruff, F., and Dray, F., 1975, Prostaglandin synthesis by the lung in the control of gas exchange, *Abst. 6th Int. Congr. Pharmacol. (Helsinki)*, p. 421.

Farmer, J. B., Farrar, D. G., and Wilson, J., 1972, The effect of indomethacin on the tracheal smooth muscle of the guinea pig, *Br. J. Pharmacol.* **46**:536.

Farmer, J. B., Farrar, D. G., and Wilson J., 1974, Antagonism of tone and prostaglandin-mediated responses in a tracheal preparation by indomethacin and SC-19220, *Br. J. Pharmacol.* **52**:559.

Ferreira, S. H., and Vane, J. R., 1967, Prostaglandins: Their disappearance from and release into the circulation, *Nature (London)* **216**:868.

Ferreira, S. H., and Vane, J. R., 1974, Aspirin and prostaglandins, in: *The Prostaglandins* (P. W. Ramwell, ed.), Vol. 2, p. 1, Plenum Press, New York.

Ferreira, S. H., Moncada, S., and Vane, J. R., 1973, Some effects of inhibiting endogenous prostaglandin formation on the responses of the cat spleen, *Br. J. Pharmacol.* **47**:48.

Ferreira, S. H., Herman, A. G., and Vane, J. R., 1976, Prostaglandin production by rabbit isolated jejunum and its relationship to the inherent tone of the preparation, *Br. J. Pharmacol.* **56**:467.

Fishbourne, J. I., Jr., Brenner, W. E., Braaksma, S. T., Staurovsky, L. G., Mueller, R. A., Hoffer, J. L., and Hendricks, C. H., 1972, Cardiovascular and respiratory responses to intravenous infusion of prostaglandin $F_{2\alpha}$ in the pregnant woman, *Am. J. Obstet. Gynecol.* **111**:765.

Finch, P. J. P., Douglas, J. S., and Bouhuys, A., 1974, Interaction of humoral vagal stimuli on airway smooth muscle of the guinea pig trachea *in vitro, Physiologist* **17**:222.

Fishman, A. P., and Pietra, G. G., 1974, Handling of bioactive materials by the lung, *N. Engl. J. Med.* **291**:884, 953.

Flower, R. J., 1974, Drugs which inhibit prostaglandin biosynthesis, *Pharmacol. Rev.* **26**:33.

Frey, H. H., and Schäfer, A., 1974, On the effect of prostaglandins E_2 and $F_{2\alpha}$ on bronchial tonus in cats, *Eur. J. Pharmacol.* **29**:267.

Goadby, P., and Smith, W. G., 1962, The effects of hydrocortisone on the changes in lipid metabolism induced in guinea pig lung tissue by anaphylaxis *in vivo, J. Pharm. Pharmacol.* **14**:739.

Goadby, P., 1975, Investigation of changes in the lipid content of guinea-pig lung after anaphylaxis, *J. Pharm. Pharmacol.* **27**:248.

Goldberg, N. D., Haddox, M. K., Nicol, S. E., Glass, D. B., Sanford, C. H., Kuehl, F. A. Jr., and Esyensen, R., 1975, Biological regulation through opposing influence of cyclic GMP and cyclic AMP: The yin yang hypothesis, in: *Advances in Cyclic Nucleotide Research,* Vol. 5 (G. I. Drummond, P. Greengard, and G. A. Robinson, eds.), pp. 307–330, Raven Press, New York.

Goldblatt, M. W., 1933, A depressor substance in seminal fluid, *J. Soc. Chem. Ind. (London)* **52**:1056.

Goldblatt, M. W., 1935, Properties of human seminal plasma, *J. Physiol. (London)* **84**:208.

Goldstein, A., 1949, Interactions of drugs and plasma protein, *Pharmacol. Rev.* **1**:102.

Golub, M., Zia, P., Matsuna, M., and Horton, R., 1974, Blood concentration, metabolic clearance rate (MCR), and sites of metabolism of prostaglandin A_1 (PGA_1) in man, *Clin. Res.* **22**:116A.

Gorman, R. R., 1975, Prostaglandin endoperoxides: possible new regulators of cyclic nucleotide metabolism, *J. Cyclic Nucleotide Res.* **1**:1.

Gréen, K., Hansson, E., and Samuelsson, B., 1967, Synthesis of tritium labeled prostaglandin $F_{2\alpha}$ and studies of its distribution by autoradiography, *Prog. Biochem. Pharmacol.* **3**:85.

Gréen, K., Hedqvist, P., and Svanborg, N., 1974, Increased plasma levels of 15-keto-13,14-dihydro-prostaglandin $F_{2\alpha}$ after allergen-provoked asthma in man, *Lancet* **2**:1419.

Greenberg, R., 1976, A comparison of the bronchodilator activity of an 11-deoxy prostaglandin (AY-23, 578) with its 15-methyl analogue (AY-24,559), in: *Advances in Prostaglandin and Thromboxane Research* (B. Samuelsson and R. Paoletti, eds.), p. 961, Raven Press, New York.

Greenberg, R., and Beaulieu, G., 1974, The bronchodilator activity of AY-22093, a prostanoic acid derivative, *Can. J. Physiol. Pharmacol.* **52**:1.

Greenberg S., Engelbrecht J. A., and Wilson W. R., 1973, Cardiovascular pharmacology of prostaglandin B_1 and B_2 in the intact dog, *Proc. Soc. Exp. Biol. Med.* **143**:1008.

Greenberg, S., Howard, L., and Wilson, W. R., 1974a, Comparative effects of prostaglandins A_2 and B_2 on vascular and airway resistances and adrenergic neurotransmission, *Can. J. Physiol. Pharmacol.* **52**:699.

Greenberg, S., Wilson, W. R., and Howard, L., 1974b, Mechanism of the vasoconstrictor action of prostaglandin B, *J. Pharmacol. Exp. Ther.* **190**:59.

Grodzinska, L., Panczenko, B., and Gryglewski, R. J., 1975, Generation of prostaglandin E-like material by the guinea pig trachea contracted by histamine, *J. Pharm. Pharmacol.* **27**:88.

Gross, K. B., and Gillis, C. N., 1975, Metabolism of prostaglandin A_1 by the perfused rabbit lung, *Biochem. Pharmacol.* **24**:1441.

Gryglewski, R. J., Panczenko, B., Korbut, R., Grodzinska, L., and Grodzinska, A., 1975, Corticosteroids inhibit prostaglandin release from perfused mesenteric blood vessels of rabbit and from perfused lungs of sensitized guinea pig, *Prostaglandins* **10**:343.

Gryglewski, R. J., Bunting, S., Moncada, S., Flower, R. J., and Vane, J. R., 1976, Arterial walls are protected against deposition of platelet thrombi by a substance (prostaglandin X) which they make from prostaglandin endoperoxides, *Prostaglandins* **12**:685.

Gutierrez-Cernosek, R. M., Morrill, L. M., and Levine, L., 1972, Prostaglandin $F_{2\alpha}$ levels in sera during human pregnancy, *Prostaglandins* **1**:331.

Hamberg, M., and Israelsson, M., 1970, Metabolism of prostaglandin E_2 in guinea pig liver, *J. Biol. Chem.* **245**:5107.

Hamberg, M., and Samuelsson, B., 1969, The structure of a urinary metabolite of prostaglandin E_2 in the guinea pig, *Biochem. Biophys. Res. Commun.* **34**:22.

Hamberg, M., and Samuelsson, B., 1973, Detection and isolation of an endoperoxide intermediate in prostaglandin biosynthesis, *Proc. Natl. Acad. Sci. U.S.A.* **70**:899.

Hamberg, M., and Samuelsson, B., 1974, Prostaglandin endoperoxides. VII. Novel transformations of arachidonic acid in guinea pig lung, *Biochem. Biophys. Res. Commun.* **61**:942.

Hamberg, M., Svensson, J., and Samuelsson, B., 1974, Prostaglandin endoperoxides. A new concept concerning the mode of action and release of prostaglandins, *Proc. Natl. Acad. Sci. U.S.A.* **71**:3824.

Hamberg, M., Svensson, J., and Samuelsson, B., 1975a, Thromboxanes: A new group of biologically active compounds derived from guinea pig endoperoxides, *Proc. Natl. Acad. Sci. U.S.A.* **72**:2994.

Hamberg, M., Hedqvist, P., Strandberg, K., Svensson, J., and Samuelsson, B., 1975*b*, Prostaglandin endoperoxides. IV. Effects on smooth muscle, *Life Sci.* **16**:451.

Hamberg, M., Svensson, J., Hedqvist, P., Strandberg, K., and Samuelsson, B., 1976, Involvement of the endoperoxides and thromboxanes in anaphylactic reactions, in: *Advances in Prostaglandin and Thromboxane Research* (B. Samuelsson and R. Paoletti, eds.), pp. 495–501, Raven Press, New York.

Hansen, H., 1974, Inhibition by indomethacin and aspirin of 15-hydroxy-prostaglandin dehydrogenase *in vitro, Prostaglandins* **8**:95.

Hansson, E., and Samuelsson, B., 1965, Autoradiographic distribution studies of ^{3}H-labeled prostaglandin E_1 in mice, *Biochim. Biophys. Acta* **106**:379.

Hauge, A., Lunde, P. K. M., and Waaler, B. A., 1967, Effects of prostaglandin E_1 and adrenaline on the pulmonary vascular resistance (PVR) in isolated rabbit lungs, *Life Sci.* **6**:673.

Hedqvist, P., 1970, Studies on the effect of prostaglandin E_1 and E_2 on the sympathetic neuromuscular transmission in some animal tissue, *Acta Physiol. Scand.* **79**:1 (Suppl. 345).

Hedqvist, P., Holmgren, A., and Mathé, A. A., 1971, Effect of prostaglandin $F_{2\alpha}$ on airway resistance in man, *Acta Physiol. Scand.* **82**:29A.

Heinemann, H. O., and Fishman, A. P., 1969, Nonrespiratory functions of mammalian lung, *Physiol. Rev.* **49**:1.

Herxheimer, R., and Roetscher, I., 1971, Effects of prostaglandin E_1 on lung function in bronchial asthma, *Eur. J. Clin. Pharmacol.* **3**:123.

Hitchcock, M., 1974, Regulation of immunologic histamine release from guinea pig lung by local production of prostaglandins, *Fed. Proc.* **33**:586.

Hitchcock, M., 1975, Stimulation of immunologic histamine(H) release and production of prostaglandin $F_{2\alpha}$ ($PGF_{2\alpha}$) by ascorbic acid, *Fed. Proc.* **34**:798.

Holmes, S. W., Horton, E. W., and Stewart, M. J., 1968, Observations on the extraction of prostaglandins from blood, *Life Sci.* **7**:349.

Hook, R., and Gillis, C. N., 1975, The removal and metabolism of prostaglandin B_1 by rabbit lung, *Prostaglandins* **9**:193.

Horton, E. W., 1972, The prostaglandins, *Proc. R. Soc. London B.* **182**:411.

Horton, E. W., and Jones, R. L., 1969, Prostaglandins A_1, A_2, and 19-hydroxy A_1; their actions on smooth muscle and their inactivation on passage through the pulmonary and hepatic portal vascular beds, *Br. J. Pharmacol.* **37**:705.

Hughes, J., Gillis, C. N., and Bloom, F. E., 1969. The uptake and disposition of d,l-norepinephrine in perfused rat lung, *J. Pharmacol. Exp. Ther.* **169**:237.

Hyman, A. L., 1969, The active responses of pulmonary veins in intact dogs to prostaglandin $F_{2\alpha}$ and E_1, *J. Pharmacol. Exp. Ther.* **165**:267.

Ishizaka, T., Ishizaka, K., Orange, R. P., and Austen, K. F., 1971, The pharmacologic inhibition of the antigen-induced release of histamine and slow reacting substance of anaphylaxis (SRS-A) from monkey lung tissues mediated by human IgE, *J. Immunol.* **106**:1267.

Jackson, H. R., Hall, R. C., Hodge, E., Gibson, F., Stevens, K., and Stevens, M., 1973, The effect of aspirin on the pulmonary extraction of $PGF_{2\alpha}$ and cardiovascular response to $PGF_{2\alpha}$, *Aust. J. Exp. Biol. Med. Sci.* **51**:837.

Jaffe, B. M., Parker, C. W., Marshall, G. R., and Needleman, P., 1972, Renal concentrations of prostaglandin E in acute and chronic renal ischemia, *Biochem. Biophys. Res. Commun.* **49**:799.

James, G. W. L., 1969, The use of the *in vitro* trachea preparation of the guinea pig to assess drug action on lung, *J. Pharm. Pharmacol.* **21**:379.

Junstad, M., and Wennmalm, Å., 1973, On the release of prostaglandin E_2 from the rabbit heart following infusion of noradrenaline, *Acta Physiol. Scand.* **87**:573.

Kadowitz, P. J., and Hyman, A. L., 1975, Differential effects of prostaglandins A_1 and A_2 on pulmonary vascular resistance in the dog, *Proc. Soc. Exp. Biol. Med.* **149**:282.

Kadowitz, P. J., Joiner, P. D., and Hyman, A. L., 1974*a*, Effects of prostaglandins E_1 and $F_{2\alpha}$ on the swine pulmonary circulation, *Proc. Soc. Exp. Biol. Med.* **145**:53.

Kadowitz, P. J., Joiner, P. D., and Hyman, A. L., 1974*b*, Influence of prostaglandins E_1 and $F_{2\alpha}$ on pulmonary vascular resistance in the sheep, *Proc. Soc. Exp. Biol. Med.* **145**:1258.

Kadowitz, P. J., Joiner, P. D., and Hyman, A. L., 1975*a*, Physiological and pharmacological roles of prostaglandins, in: *Annual Review of Pharmacology,* Vol. 15 (H. W. Elliott, R. George, and R. Okun, eds.), pp. 285–303, Annual Reviews Inc., Palo Alto, Calif.

Kadowitz, P. J., Joiner, P. D., and Hyman A. L., 1975*b*, Effect of prostaglandin E_2 on pulmonary vascular resistance in intact dog, swine, and lamb, *Eur. J. Pharmacol.* **31**:72.

Kadowitz, P. J., Joiner, P. D., Greenberg, S., and Hyman, A. L., 1975*c*, Effects of prostaglandins B_2 and B_1 on the pulmonary circulation in the intact dog, *J. Pharmacol. Exp. Ther.* **192**:157.

Kadowitz, P., Joiner, P. D., Matthews, C. S., and Hyman, A. L., 1975*d*, Effects of the 15-methyl analogs of prostaglandins E_2 and $F_{2\alpha}$ on the pulmonary circulation in the intact dog, *J. Clin. Invest.* **55**:937.

Kadowitz, P. J., Joiner, P. D., Greenberg, S., and Hyman, A. L., 1976, Comparison of the effects of prostaglandins A, E, F, and B on the canine pulmonary vascular bed, in: *Advances in Prostaglandin and Thromboxane Research* (B. Samuelsson and R. Paoletti, eds.), pp. 403–415, Raven Press, New York.

Kakiuchi, S., and Rall, T. W., 1968, The influence of chemical agents on the accumulation of adenosine 3′-5′-phosphate in slices of rabbit cerebellum, *Mol. Pharmacol.* **4**:367.

Kaliner, M., Orange, R. P., and Austen, K. F., 1972, Immunological release of histamine and slow-reacting substance of anaphylaxis from human lung. IV. Enhancement by cholinergic and alpha adrenergic stimulation, *J. Exp. Med.* **136**:556.

Kaliner, M., and Austen, K. F., 1974, Cyclic nucleotides and modulation of effector systems of inflammation, *Biochem. Pharmacol.* **23**:763.

Kaliner, M., and Austen, K. F., 1975, Immunologic release of chemical mediators from human tissues, in: *Annual Review of Pharmacology,* Vol. 15, (H. W. Elliott, R. George, and R. Okun, eds.), pp. 177–189, Annual Reviews Inc., Palo Alto, Calif.

Kantrowitz, F., Robinson, D. R., McGuire, M. B., and Levine, L., 1975, Corticosteroids inhibit prostaglandin production by rheumatoid synovia, *Nature (London)* **258**:737.

Karim, S. M. M., Sandler, M., and Williams, E. D., 1967, Distribution of prostaglandins in human tissues, *Br. J. Pharmacol.* **31**:340.

Karim, S. M. M., Hillier, K., and Devlin, J., 1968, Distribution of prostaglandins E_1, E_2, $F_{1\alpha}$, and $F_{2\alpha}$ in some animal tissues, *J. Pharm. Pharmacol.* **20**:749.

Kawakami, Y., Uchiyama, K., Irie, T., and Murao, M., 1973, Evaluation of aerosols of prostaglandins E_1 and E_2 as bronchodilators, *Eur. J. Clin. Pharmacol.* **6**:127.

Koopman, W. J., Orange, R. P., and Austen, K. F., 1971, Prostaglandin inhibition of the immunologic release of slow reacting substance of anaphylaxis in the rat, *Proc. Soc. Exp. Biol. Med.* **137**:64.

Kosterlitz, H. W., and Lees, G. M., 1972, Interrelationships between adrenergic and cholinergic mechanisms, in: *Catecholamines* (H. Blaschko and E. Muscholl, eds.), pp. 762–812, Springer-Verlag, Berlin.

Kuehl, F. A., Jr., 1974, Prostaglandins, cyclic nucleotides, and cell function, *Prostaglandins* **5**:325–340.

Kuehl, F. A., Jr., Oien, H. G., and Ham, E. A., 1974, Prostaglandins and prostaglandin synthetase inhibitors: Actions on cell function; in: *Prostaglandin Synthetase Inhibitors: Their Effects on Physiological Functions and Pathological States* (H. J. Robinson and J. R. Vane, eds.), pp. 53–65, Raven Press, New York.

Kunze, H., 1972, Stimulation and inhibition of phospholipase synaptosones of guinea pig brain, *Naunyn-Schmiedebergs Arch. Pharmacol.* **277**:R47.

Kunze, H., and Vogt, W., 1971, Significance of phospholipase A for prostaglandin formation, *Ann. N.Y. Acad. Sci.* **180**:123.

Kuo, W.-N., and Kuo, J.-F., 1973, Regulation of cyclic GMP and cyclic AMP levels in rat lung and other tissues by various agents as determined by double-prelabeling with radioactive guanine and adenine, *Fed. Proc.* **32**:773.

Lambley, J. E., and Smith, A. P., 1975, The effect of arachidonic acid, indomethacin and SC-19220 on guinea pig tracheal muscle tone, *Eur. J. Pharmacol.* **30**:148.

Large, B., Leswell, P. P., and Maxwell, D. R., 1969, Bronchodilator activity of an aerosol of prostaglandin E_1 in experimental animals, *Nature (London)* **4**:78.

Lavoisier, A.-L., 1778, Mémoire sur la nature du principe qui se combine avec les métaux pendant leur calcination et qui en augmente le poids, *Hist. Acad. R. Sci. Paris,* pp. 520–526.

Lee, J. B., 1976, The renal prostaglandins and blood pressure regulation, in: *Advances in Prostaglandin and Thromboxane Research* (B. Samuelsson and R. Paoletti, eds.), p. 573, Raven Press, New York.

Lee, S.-C., and Levine, L., 1974a, Cytoplasmic reduced nicotinamide adenine dinucleotide phosphate-dependent and microsomal reduced nicotinamide adenine dinucleotide-dependent prostaglandin E 9-ketoreductase activities in monkey and pigeon tissues, *J. Biol. Chem.* **249**:1369.

Lee, S.-C., and Levine, L., 1974b, Purification and regulatory properties of chicken heart prostaglandin E 9-ketoreductase, *J. Biol. Chem.* **250**:4549.

Lee, S.-C., and Levine, L., 1974c, Purification and properties of chicken heart prostaglandin Δ^{13}-reductase, *Biochem. Biophys. Res. Commun.* **61**:14.

Lee, S.-C., and Levine, L., 1975, Prostaglandin metabolism. II. Identification of two 15-hydroxy-prostaglandin dehydrogenase types, *J. Biol. Chem.* **250**:548.

Leslie, C. A., and Levine, L., 1973, Evidence for the presence of a prostaglandin E_2-9-keto reductase in rat organs, *Biochem. Biophys. Res. Commun.* **52**:717.

Levine, L., Gutierrez-Cernosek, R. M., and Van Vunakis, H., 1973, Specific antibodies: reagents for quantitative analysis of prostaglandins, *Adv. Biosci.* **9**:71.

Lewis, A. J., and Eyre, P., 1972, Some cardiovascular and respiratory effects of prostaglandins E_1, E_2, and $F_{2\alpha}$ in the calf, *Prostaglandins* **2**:55.

Lichtenstein, L. M., and DeBernardo, R., 1971, The immediate allergic response: *In vitro* action of cyclic AMP-active and other drugs on the two stages of histamine release, *J. Immunol.* **107**:1131.

Lichtenstein, L. M., and Margolis, S., 1968, Histamine release *in vitro:* Inhibition by catecholamines and methylxanthines, *Science* **161**:902.

Lichtenstein, L. M., Gillespie, E., Bourne, H. R., and Henney, C. S., 1972, The effects of a series of prostaglandins on *in vitro* models of the allergic response and cellular immunity, *Prostaglandins* **2**:519.

Liebig, R., Bernauer, W., and Peskar, B. A., 1974, Release of prostaglandins, a prostaglandin metabolite, slow-reacting substance and histamine from anaphylactic lungs, and its modification by catecholamines, *Naunyn-Schmiedebergs Arch. Pharmacol.* **284**:279.

Lindsey, J. H., and Wyllie, J. H., 1970, Release of prostaglandins from embolized lungs, *Br. J. Surg.* **57**:738.

MacManus, J. P., Whitfield, J. F., and Rixon, A., 1974, Control of normal cell proliferation *in vivo* and *in vitro* by agents that use cyclic AMP as their mediator, in: *Cyclic AMP, Cell Growth, and the Immune Response* (W. Braun, L. M. Lichtenstein, and C. Parker, eds.), pp. 302–317, Springer-Verlag, New York.

Main, I. H. M., 1964, The inhibitory actions of prostaglandins on respiratory smooth muscle, *Br. J. Pharmacol.* **22**:511.

Makman, M. H., 1971, Conditions leading to enhanced response to glucagon, epinephrine, or prostaglandins by adenylate cyclase of normal and malignant cultured cells, *Proc. Natl. Acad. Sci. U.S.A.* **68**:2127.

Malpighius, M., 1661, *De pulmonibus observationes anatomicae,* Bologna. English translation by J. Young, 1929–1930, *Proc. R. Soc. Med.* **23**:1.

Marrazzi, M. A., and Matschinsky, F. M., 1972, Properties of 15-hydroxy prostaglandin dehydrogenase: Structural requirements for substrate binding, *Prostaglandins* **1**:373.

Marrazzi, M. A., and Andersen, F., 1974, Prostaglandin dehydrogenase, in: *The Prostaglandins,* Vol. 2 (P. W. Ramwell, ed.), pp. 99–155, Plenum Press, New York.

Mathé, A. A., 1971, Decreased circulating epinephrine, possibly secondary to decreased hypothalamic adrenalmedullary discharge; a supplementary hypothesis of bronchial asthma pathogenesis, *J. Psychosom. Res.* **15**:349.

Mathé, A. A., and Hedqvist, P., 1975, Effect of prostaglandins $F_{2\alpha}$ and E_2 on airway conductance in healthy subjects and asthmatic patients, *Am. Rev. Resp. Dis.* **111**:313.

Mathé, A. A., and Levine, L., 1973, Release of prostaglandins and metabolites from guinea pig lung: Inhibition by catecholamines, *Prostaglandins* **4**:877.

Mathé, A. A., and Volicer, L., 1976a, Effects of anaphylaxis, phenoxybenzamine (PBA) and cocaine (COC) on uptake of norepinephrine (NE) in lung, *Fed. Proc.* **35**:600.

Mathé, A. A., and Volicer, L., 1976b, Norepinephrine uptake in guinea pig lung: Effects of anaphylaxis, phenoxybenzamine and cocaine, *Int. Arch. Allergy Immunol.* (in press).

Mathé, A. A., Strandberg, K., and Åström, A., 1971, Blockade by polyphloretin phosphate of the prostaglandin $F_{2\alpha}$ action on isolated human bronchi, *Nature (London) New Biol.* **230**:215.

Mathé, A. A., Hedqvist, P., Holmgren, A., and Svanborg, N., 1973a, Prostaglandin $F_{2\alpha}$: Effect on airway conductance in healthy subjects and patients with bronchial asthma, *Adv. Biosci.* **9**:241.

Mathé, A. A., Hedqvist, P., Holmgren, A., and Svanborg, N., 1973b, Bronchial hyperreactivity to prostaglandin $F_{2\alpha}$ and histamine in patients with asthma, *Br. Med. J.* **1**:193.

Mathé, A. A., Puri, S. K., and Volicer, L., 1974a, Sensitized guinea-pig lung: Altered adenylate cyclase stimulation by epinephrine, *Life Sci.* **15**:1917.

Mathé, A. A., Volicer, L., and Puri, S. K., 1974b, Effect of anaphylaxis and histamine, pyrilamine and burimamide on levels of cyclic AMP and cyclic GMP in guinea-pig lung. *Res. Commun. Chem. Pathol. Pharmacol.* **8**:635.

Mathé, A. A., Levine, B. I., and Antonucci, M. J., 1975, Uptake of catecholamines in guinea pig lung: Influence of cortisol and anaphylaxis, *J. Allergy Clin. Immunol.* **55**:170.

Mathé, A. A., Sohn, R. J., and Volicer L., 1976a, Effect of histamine on cyclic AMP levels in control and antigen sensitized guinea pig lung, *Pharmacology* (in press).

Mathé, A. A., Strandberg, K., and Yen, S.-S., 1976b, Prostaglandin release by slow-reacting substance from guinea pig and human lung tissue, *Prostaglandins* (submitted).

Mathé, A. A., Tong, E. Y., and Fisher, P. W., 1976c, Sympathetic-parasympathetic balance in the lung: Release of nonepinephrine by nerve stimulation, *Fed. Proc.* **35**:600.

Mathé, A. A., Tong, E., and Tisher, P. W., 1976d, Norepinephrine release from the lung by sympathetic nerve stimulation. Inhibition by vagus and methacholine, *Life Sci.* (in press).

Mathé, A. A., Puri, S. K., Volicer, L., and Sohn, R. J., 1976e, Effect of epinephrine on cyclic AMP levels and adenylate cyclase and prosphodiesterase activities in control and antigen sensitized guinea pig lungs, *Pharmacology* **14**:511.

Mathé, A. A., Yen, S.-S., Sohn, R. J., and Hedqvist, P., 1976f, Release of prostaglandins and histamine from sensitized and anaphylactic guinea pig lungs; Changes in cyclic AMP levels, *Biochem. Pharmacol.* (in press).

Matsumura, S., Mizuno, K., Miyamoto, T., Suzuki, T., and Oshima, Y., 1972, The effects of ozone, nitrogen dioxide, and sulfur dioxide on experimentally induced allergic respiratory disorder in guinea pigs, *Am. Rev. Resp. Dis.* **105**:262.

McGiff, J. C., Terragno, N. A., Strand, J. C., Lee, J. B., Lonigro, A. J., and Ng, K. K. F., 1969, Selective passage of prostaglandins across the lung, *Nature (London)* **223**:742.

McGiff, J. C., Crowshaw, K., Terragno, N. A., and Lonigro, A. J., 1970, Release of prostaglandin-like substance into renal venous blood in response to angiotensin II, *Circ. Res. Suppl.* **26**:121.

Moncada, S., Gryglewski, R. J., Bunting, S., and Vane, J. R., 1976, A lipid peroxide inhibits the enzyme in blood vessel microsomes that generates from prostaglandin endoperoxides the substance (prostaglandin X) which prevents platelet aggregation, *Prostaglandins* **12**:715.

Murad, F., and Kimura, H., 1974, Cyclic nucleotide levels in incubations of guinea pig trachea, *Biochem. Biophys. Acta* **343**:275.

Muscholl, E., 1973, Muscarinic inhibition of the norepinephrine release from peripheral sympathetic fibers, in: *Pharmacology and the Future of Man* (Proc. 5th Int. Congr. Pharmacol., San Francisco), Vol. 4 (F. E. Bloom and G. H. Acheson, eds.) pp. 440–457, Karger, Basel.

Nakano, J., 1973a, Cardiovascular actions, in: *The Prostaglandins*, Vol. 1 (P. W. Ramwell, ed.). pp. 239–316, Plenum Press, New York.

Nakano, J., 1973b, General pharmacology of prostaglandins, in: *The Prostaglandins* (M. F. Cuthbert, ed.), pp. 23–124, Lippincott, Philadelphia.

Nakano, J., and Cole, B., 1969, Effects of prostaglandins E_1 and $F_{2\alpha}$ on systemic, pulmonary and splanchnic circulation in dogs, *Am. J. Physiol.* **217**:222.

Nakano, J., and Prancan, A. V., 1973, Metabolic degradation of prostaglandin E_1 in the lung and kidney of rats in endotoxin shock, *Proc. Soc, Exp. Biol. Med.* **144**:506.

Newball, H. H., Keiser, H. R., and Lenfant, C. J., 1974, Influence of atropine on prostaglandin $F_{2\alpha}$ induced airway constriction in normal and asthmatics, *Fed. Proc.* **33**:366.

Northover, B. J., 1971, Mechanism of the inhibitory action of indomethacin on smooth muscle, *Br. J. Pharmacol.* **41**:540.

Northover, B. J., 1973, Effect of anti-inflammatory drugs on the binding of calcium to cellular membranes in various human and guinea pig tissues, *Br. J. Pharmacol.* **48**:496.

Nugteren, D. H., and Hazelhof, E., 1973, Isolation and properties of intermediates in prostaglandin biosynthesis, *Biochim. Biophys. Acta* **326**:448.

Oesterling, T. O., Morozowich, W., and Roseman, T. J., 1972, Prostaglandins, *J. Pharm. Sci.* **61**:1861.

Okpako, D. T., 1972, The actions of histamine and prostaglandins $F_{2\alpha}$ and E_2 on pulmonary vascular resistance of the lung of the guinea pig, *J. Pharm. Pharmacol.* **24**:40.

Orange, R. P., Kaliner, M. A., Laraia, P. J., and Austen, K. F., 1971, Immunological release of histamine and slow reacting substance of anaphylaxis from human lung. II. Influence of cellular levels of cyclic AMP, *Fed. Proc.* **30**:1725.

Orehek, J., Douglas, J. S., Lewis, J., and Bouhuys, A., 1973, Prostaglandin regulation of airway smooth muscle tone, *Nature (London) New Biol.* **245**:84.

Orehek, J., Douglas, J. S., and Bouhuys, A., 1975, Contractile responses to the guinea pig trachea *in vitro:* Modification by prostaglandin synthesis-inhibiting drugs, *J. Pharmacol. Exp. Ther.* **194**:554.

Pace-Asciak, C., and Miller, D., 1973, Prostaglandins during development. I. Age-dependent activity profiles of prostaglandin 15-hydroxydehydrogenase and 13,14-reductase in lung tissue from late prenatal, early postnatal and adult rats, *Prostaglandins* **4**:351.

Palmer, G. C., 1972, Cyclic 3'-5'-adenosine monophosphate response in the rabbit lung—Adult properties and development, *Biochem. Pharmacol.* **21**:2907.

Palmer, M. A., Piper, P. J., and Vane, J. R., 1973, Release of rabbit aorta contracting substance (RCS) and prostaglandins induced by chemical or mechanical stimulation of guinea pig lungs, *Br. J. Pharmacol.* **49**:226.

Parker, C. W., Sullivan, T. J., and Wédner, H. J., 1974, Cyclic AMP and the immune response, in: *Advances in Cyclic Nucleotide Research,* Vol. 4. (P. Greengard and G. A. Robinson, eds.), pp. 1–80, Raven Press, New York.

Parkes, D. G., and Eling, T. E., 1974, Characterization of prostaglandin synthesis in guinea pig lung. Isolation of a new prostaglandin derivative from arachidonic acid, *Biochemistry* **13**:2598.

Parkes, D. G., and Eling, T. E., 1975, The influence of environmental agents on prostaglandin biosynthesis and metabolism in the lung, *Biochem. J.* **146**:549.

Pasargiklian, M., Bianco, S., and Allegra, L., 1976, Clinical, functional and pathogenetic aspects of bronchial reactivity to prostaglandins $F_{2\alpha}$, E_1 and E_2, in: *Advances in Prostaglandin and Thromboxane Research* (B. Samuelsson and R. Paoletti, eds.), pp. 461–475, Raven Press, New York.

Patel, K. B., 1975, Atropine, sodium chromoglycate, and thymoxamine in $PGF_{2\alpha}$ induced bronchocontraction in extrinsic asthma, *Br. Med. J.* **2**:360.

Piper, P. J., and Vane, J. R., 1969, Release of additional factors in anaphylaxis and its antagonism by anti-inflammatory drugs, *Nature (London)* **223**:29.

Piper, P. J., and Vane, R. J., 1971, The release of prostaglandins from lung and other tissues, *Ann. N.Y. Acad. Sci.* **180**:363.

Piper, P. J., and Walker, J. L., 1973, The release of spasmogenic substances from human chopped lung tissue and its inhibition, *Br. J. Pharmacol.* **47**:291.

Piper, P. J., Vane, J. R., and Wyllie, J. H., 1970, Inactivation of prostaglandins by the lungs, *Nature (London)* **225**:600.

Pong, S.-S., and Levine, L., 1977, Prostaglandin biosynthesis and metabolism as measured by radioimmunoassay, *The Prostaglandins* (P. W. Ramwell, ed.), Vol. 3, p. 41, Plenum Press, New York.

Popa, V., Douglas, J. S., and Bouhuys, A., 1973, Airway response to histamine, acetylcholine and propranolol in anaphylactic hypersensitivity in guinea pigs, *J. Allergy Clin. Immunol.* **51**:344.

Puglisi, L., 1973, Opposite effects of prostaglandins E and F on tracheal smooth muscle and their interaction with calcium ions, *Adv. Biosci.* **9**:219.

Puglisi, L., Berti, F., Bosisio, E., Longiave, D., and Nicosia, S., 1976, Ascorbic acid and $PGF_{2\alpha}$

antagonism on tracheal smooth muscle, in: *Advances in Prostaglandin and Thromboxane Research* (B. Samuelsson and R. Paoletti, eds.), pp. 503–506, Raven Press, New York.

Ramwell, P. W., and Shaw, J. E., 1967, Prostaglandin release from tissues by drug, nerve, and hormone stimulation, in: *Prostaglandins* (Proc. 2nd Nobel Symp.) (S. Bergström and B. Samuelsson, eds.), pp. 283–292, Almqvist and Wiksell, Stockholm.

Robinson, J. H., and Vane, J. R., 1974, *Prostaglandin Synthetase Inhibitors: Their Effect on Physiological Functions and Pathological States,* Raven Press, New York.

Rosenthale, M. E., Dervinis, A., Begany, A. J., Lapidus, M., and Gluckman, M. J., 1968, Bronchodilator activity of prostaglandin PGE_2, *Pharmacologist* **10**:175.

Rosenthale, M. E., Dervinis, A., Begany, A. J., Lapidus, M., and Gluckman, M. J., 1970, Bronchodilator activity of prostaglandin E_2 when administered by aerosol to three species, *Experientia* **26**:1119.

Rosenthale, M. E., Dervinis, A., and Kassarich, J., 1971, Bronchodilator activity of the prostaglandins E_1 and E_2, *J. Pharmacol. Exp. Ther.* **178**:541.

Rosenthale, M. E., Dervinis, A., Kassarich, J., Singer, S., and Gluckman, M. J., 1973, Comparative studies on the bronchodilating properties of the prostaglandin $F_{2\beta}$, *Adv. Biosci.* **9**:229.

Rosenthale, M. E., Dervinis, A., and Strike, D., 1976, Actions of prostaglandins on the respiratory tract of animals, in: *Advances in Prostaglandin and Thromboxane Research* (B. Samuelsson and R. Paoletti, eds.), pp. 477–493, Raven Press, New York.

Ruff, F., Gray, R., Santais, M., Allouche, G., Foussard, C., Florence, A., and Parrott, J., 1975, Measurements of pulmonary prostaglandins, histamine and cyclic AMP during induced anaphylactic shock in guinea pig, *Proc. 6th Int. Congr. Pharmacol., (Helsinki),* p. 379.

Ryan, J. W., Niemeyer, R. S., and Ryan, U., 1975, Metabolism of prostaglandin $F_{1\alpha}$ in the pulmonary circulation, *Prostaglandins* **10**:101.

Said, S. I., 1969, Some respiratory effects of prostaglandins E_2 and $F_{2\alpha}$, in: *Prostaglandin Symposium of the Worcester Foundation for Experimental Biology* (P. W. Ramwell and J. E. Shaw, eds.), pp. 267–277, Interscience, New York.

Said, S. I., 1973, The lung in relation to vasoactive hormones, *Fed. Proc.* **32**:1972.

Said, S. I., and Yoshida, T., 1974, Release of prostaglandins and other humoral mediators during hypoxic breathing and pulmonary edema, *Chest (Suppl.)* **66**:12S.

Said, S. I., Yoshida, T., Kitamura, S., and Vreim, C., 1974, Pulmonary alveolar hypoxia: Release of prostaglandins and other humoral mediators, *Science* **185**:1181.

Samuelsson, B., 1964, Prostaglandins and related factors. 27. Synthesis of tritium-labeled prostaglandin E_1 and studies on its distribution and excretion in the rat, *J. Biol. Chem.* **239**:4091.

Samuelsson, B., 1965, On the incorporation of oxygen in the conversion of 8,11,14-eicosatrienoic acid to prostaglandin E_1, *J. Am. Chem. Soc.* **87**:3011.

Samuelsson, B., 1972, Biosynthesis of prostaglandins, *Fed. Proc.* **31**:1442.

Samuelsson, B., and Paoletti, R., 1976, *Advances in Prostaglandin and Thromboxane Research,* Raven Press, New York.

Samuelsson, B., Granström, E., Gréen, K., and Hamberg, M., 1971, Metabolism of prostaglandins, *Ann. N.Y. Acad. Sci.* **180**:138.

Sandler, M., 1972, Migraine: A pulmonary disease?, *Lancet* **1**:618.

Schlegel, W., Demers, L. M., Hilderbrandt-Stark, H. E., Behrman, H. R., and Greep, R. O., 1974, Partial purification of human placental 15-hydroxyprostaglandin dehydrogenase: Kinetic properties, *Prostaglandins* **5**:417.

Sheppard, H., 1974, The role of cyclic AMP in the control of cell division, in: *Cyclic AMP, Cell Growth and the Immune Response* (W. Braun, L. M. Lichtenstein, and C. W. Parker, eds.), pp. 290–301, Springer-Verlag, New York.

Sih, C. J., and Takeguchi, C. A., 1973, Biosynthesis, in: *The Prostaglandins,* Vol. 1 (P. W. Ramwell, ed.), Plenum Press, New York.

Sjöstrand, N. O., and Swedin, G., 1974, On the mechanism of the enhancement by smooth muscle stimulants of the motor responses of the guinea pig vas deferens to nerve stimulation, *Acta Physiol. Scand.* **90**:513.

Smith, W. G., 1962, The effect of ethanolamine on changes in lung lipids induced by anaphylaxis, *Biochem. Pharmacol.* **11**:183.

Smith, A. P., 1973a, Lungs, in: *The Prostaglandins,* Vol. 1 (P. W. Ramwell, ed.), pp. 203–218, Plenum Press, New York.

Smith, A. P., 1973*b*, The effects of intravenous infusion of graded doses of prostaglandins $F_{2\alpha}$ and E_2 on lung resistance in patients undergoing termination of pregnancy, *Clin. Sci.* **44**:17.

Smith, A. P., 1974*a*, A comparison of the effects of prostaglandin E_2 and salbutamol by intravenous infusion on the airways obstruction of patients with asthma, *Br. J. Clin. Pharmacol.* **1**:399.

Smith, A. P., 1974*b*, Effects of three prostaglandin analogues on airway tone in asthmatics and normal subjects, *IRCS Libr. Compend.* **2**:1457.

Smith, A. P., and Cuthbert, M. F., 1972, The antagonistic action of prostaglandin $F_{2\alpha}$ and E_2 aerosols on bronchial muscle tone in man, *Br. Med. J.* **3**:212.

Smith, A. P., Cuthbert, M. F., and Dunlop, L. S., 1975, Effects of inhaled prostaglandin E_1, E_2 and $F_{2\alpha}$ on the airway resistance of healthy and asthmatic man, *Clin. Sci. Mol. Med.* **48**:421.

Sohn, R. J., Mathé, A. A., and Volicer, L., 1976, Role of the cyclic nucleotides in the normal lung and in bronchial asthma, in: *Clinical Aspects of Cyclic Nucleotides* (L. Volicer, ed.), p. 193, Spectrum Publishing Co., Hollister, N.Y.

Söndergaard J., and Greaves, M. W., 1971, Prostaglandin E_1: Effect on human cutaneous vasculature and skin histamine, *Br. J. Dermatol.* **84**:424.

Starling, E. H., and Verney, E. B., 1925, The secretion of urine as studied on the isolated kidney, *Proc. R. Soc. London Ser. B* **97**:321.

Stechschulte, D. J., and Austen, K. F., 1974, Phosphatidylserine enhancement of antigen-induced mediator release from rat mast cells, *J. Immunol.* **112**:970.

Stoner, J., Manganiello, V. C., and Vaughan, M., 1973, Effects of bradykinin and indomethacin on cyclic GMP and cyclic AMP in lung slices, *Proc. Natl. Acad. Sci. U.S.A.* **70**:3830.

Stoner, J., Manganiello, V. C., and Vaughan, M., 1974, Guanosine cyclic 3'-5'-monophosphate and guanylate cyclase activity in guinea pig lung: Effect of acetylcholine and cholinesterase inhibitors, *Mol. Pharmacol.* **10**:155.

Stjärne, L., Kaijser, L., Mathé, A. A., and Birke, G., 1975, Specific and unspecific removal of circulating noradrenaline in pulmonary and systemic vascular beds in man, *Acta Physiol. Scand.* **95**:46.

Strandberg, K., and Hamberg, K., 1974, Increased excretion of $5\alpha,7\alpha$-dihydroxy-11-ketotetranor-prostanoic acid on anaphylaxis in the guinea pig, *Prostaglandins* **6**:159.

Strandberg, K., and Hedqvist, P., 1975, Airway effects of slow reacting substance, $PGF_{2\alpha}$ and histamine in guinea pig, *Acta Physiol. Scand.* **94**:105.

Strandberg, K., Mathé, A. A., and Yen, S.-S., 1976, Release of histamine and formation of prostaglandins in human lung tissue and rat mast cells, *Int. Arch. Allergy Immunol.* (in press).

Sun, F. F., and Armour, S. B., 1974, Prostaglandin 15-hydroxy dehydrogenase and Δ-13-reductase levels in the lungs of maternal, fetal, and neonatal rabbits, *Prostaglandins* **7**:327.

Sun, F. F., Armour, S. B., Bockstanz, V. R., and McGuire, J. C., 1976, Studies on 15-hydroxyprostaglandin dehydrogenase from monkey lung, in: *Advances in Prostaglandin and Thromboxane Research* (B. Samuelsson and R. Paoletti, eds.), pp. 163–169, Raven Press, New York.

Svensson, J., Hamberg, M., and Samuelsson, B., 1975, Prostaglandin endoperoxides. IX. Characterization of rabbit aorta contracting substance (RCS) from guinea pig lung and human platelets, *Acta Physiol. Scand.* **94**:222.

Sweatman, W. J. F., and Collier, H. O. J., 1968, Effects of prostaglandins on human bronchial muscle, *Nature (London)* **217**:69.

Szczeklik, A., Gryglewski, R. J., and Czerniawska-Mysik, G., 1975, Relationship of inhibition of prostaglandin biosynthesis by analgesics to asthma attacks in aspirin-sensitive patients, *Br. Med. J.* **1**:67.

Tai, H.-H., Tai, C. O., and Hollander, C. S., 1974, Regulation of prostaglandin metabolism: Inhibition of 15-hydroxyprostaglandin dehydrogenase by thyroid hörmones, *Biochem. Biophys. Res. Commun.* **57**:457.

Tashjian, A. H., Jr., Voelkel, E. F., McDonough, J., and Levine, Ł., 1975, Hydrocortisone inhibits prostaglandin production in mouse fibrosarcoma cells, *Nature (London)* **258**:739.

Tauber, A. I., Kaliner, M., Stechschulte, D. J., and Austen, K. F., 1973, Immunologic release of histamine and slow reacting substance of anaphylaxis from human lung. V. Effects of prostaglandins on release of histamine, *J. Immunol.* **111**:27.

Thomas, G., and West, G. B., 1973, Prostaglandins as regulators of bradykinin responses, *J. Pharm. Pharmacol.* **25**:747.

Tiffeneau, R., 1958, Hypersensibilité cholinergo-histaminique pulmonaire de l'asthmatique: Relation avec l'hypersensibilité allergénique pulmonaire, *Acta Allergol.* **12**(Suppl.):187.

Trendelenburg, U., 1972, Factors influencing the concentrations of catecholamines at the receptor, in: *Catecholamines* (H. Blaschko and E. Muscholl, eds.), pp. 726–761, Springer-Verlag, Berlin.

Vane, J. R., 1971, Inhibition of prostaglandin synthesis as a mechanism of action for aspirin-like drugs, *Nature (London) New Biol.* **231**:232.

Vargaftig, B. B., 1974, Search for common mechanisms underlying the various effects of putative inflammatory mediators, in: *The Prostaglandins,* Vol. 2 (P. W. Ramwell, ed.), pp. 205–276, Plenum Press, New York.

Vargaftig, B. B., and Dao-Hai, N., 1971, Release of vasoactive substances from guinea pig lungs by slow reacting substance C and arachidonic acid, *Pharmacology* **6**:99.

Vargaftig, B. B., and Dao-Hai, N., 1972, Selective inhibition by mepacrine of the release of "rabbit aorta contracting substance" evoked by the administration of bradykinin, *J. Pharm. Pharmacol.* **24**:159.

Walker, J. L., 1973, The regulatory function of prostaglandins in the release of histamine and SRS-A from passively sensitized human lung tissue, *Adv. Biosci.* **9**:235.

Wasserman, W. A., 1975, Bronchopulmonary effects of prostaglandin $F_{2\alpha}$ and three of its metabolites in the dog, *Prostaglandins* **9**:959.

Weinryb, I., Michel, I. M., and Hess, S. M., 1973, Adenylate cyclase from guinea pig lung: Further characterization and inhibitory effects of substrate analogs and cyclic nucleotides, *Arch. Biochem. Biophys.* **154**:240.

Weir, E. K., Reeves, J. T., Droegemueller, W., and Grover, R. F., 1975, 15-Methylation augments the cardiovascular effects of prostaglandin $F_{2\alpha}$, *Prostaglandins* **9**:369.

Wennmalm, Å., 1975, Prostaglandin release and mechanical performance in the isolated rabbit heart during induced changes in the internal environment, *Acta Physiol. Scand.* **93**:15.

Wennmalm, Å., Phan-Hou-Chank, and Junstad, M., 1974, Hypoxia causes prostaglandin release from perfused rabbit hearts, *Acta Physiol. Scand.* **91**:133.

White, G., 1974, Stimulation of rat and monkey lung adenyl cyclase by various prostaglandins, *Fed. Proc.* **33**:590.

Widdicombe, J. G., 1963, Regulation of tracheobronchial smooth muscle, *Physiol. Rev.* **43**:1.

Willis, A. L., Davison, P., and Ramwell, P. W., 1974, Inhibition of intestinal tone, motility and prostaglandin biosynthesis by 5,8,11,14-eicosatetraynoic acid (TYA), *Prostaglandins* **5**:355.

Yen, S.-S., Mathé, A. A., and Dugan, J. J., 1976, Release of prostaglandins from healthy and sensitized guinea pig lung and trachea by histamine, *Prostaglandins* **11**:227.

Zurier, R. B., 1974, Prostaglandins, inflammation, and asthma, *Arch. Intern. Med.* **133**:101.

NOTE ADDED IN PROOF

Since this chapter was written many papers on the material reviewed have been published. The effects of PGEs, PGFs, and their metabolites, as well as of PGAs and PGBs on the tone of the airways and the pulmonary vasculature have been further investigated. Additional evidence has been presented for the existence of an active uptake process for prostaglandins. In the field of neurotransmission, preliminary experiments indicate that the presynaptic sympathetic–parasympathetic balance also exists in the lung, and that PGEs inhibit release of norepinephrine elicited by nerve stimulation in that organ.

The complicated role of prostaglandins, beneficial and/or deleterious, in the inflammatory and allergic processes is being further elucidated. Whether there are any changes, such as increased or decreased bioavailability of PGE_2, $PGF_{2\alpha}$, or other prostaglandins and their metabolites in any type of bronchial asthma, has not been fully clarified. It is now well documented that in addition to the classical prostaglandins, endoperoxides and thromboxanes play a significant role in respiratory and nonrespiratory lung function.

A new, unstable metabolic product of endoperoxides, named prostacyclin (PGX), which is catabolized into the stable 6-keto-$PGF_{1\alpha}$, has just been discovered (Gryglewski *et al.*, 1976; Moncada *et al.*, 1976). Compared to the aorta and to tissues of other organs, only small quantities of PGX appear to be produced by the lung. The actions of PGX are apparently opposite to those of endoperoxides and TxA_2. Since PGX probably stimulates adenylate cyclase and antagonizes the effects of PGG_2, PGH_2, TxA_2, and $PGF_{2\alpha}$ in the lung, it can be speculated that relatively less (compared to the other endoperoxide products) is synthesized in allergic conditions. Further, it seems plausible to suggest that PGX analogs may be useful in the treatment of obstructive lung diseases. Another important consequence of the discovery of PGX (and of the establishment of the TxA_2 and HHT pathways) is that the relative ratios of, i.e., balance between the many compounds deriving from the arachidonic acid metabolism, and not the actual, local concentration of any single substance, might determine the physiological and pathophysiological state of the lung and, by extension, of the rest of the organism.

Prostaglandins and the Digestive System

André Robert

Department of Experimental Biology
The Upjohn Company
Kalamazoo, Michigan 49001

I. INTRODUCTION

The effects of prostaglandins (PGs) on the digestive system have been extensively studied, in part because they were found to be present in the gastrointestinal tract in large amounts [the human gastric mucosa contains 1 μg per gram of wet tissue of PGE_2 (Bennett *et al.*, 1968*c*)] and in part because they tend to stimulate intestinal smooth muscle *in vitro*. As is generally the case for other organs, PGs are synthesized within the gastrointestinal tract (Bartels *et al.*, 1970; Bennett *et al.*, 1967, 1968*c*; Coceani *et al.*, 1967; Miyazaki, 1968*a,b*, 1969; Pace-Asciak *et al.*, 1968; Wolfe *et al.*, 1967) and seem to exert their activity locally, at the site where they are produced. This chapter summarizes the main findings on the effects of PGs on the digestive system. Several reviews have already appeared on this subject (Bennett, 1972; Main, 1973; Matuchansky and Bernier, 1973*c*; Waller, 1973; Wilson, 1974).

II. SALIVARY GLANDS

$PGF_{2\alpha}$ stimulated salivation in the anesthetized dog, after either intravenous administration (0.1–16.0 μg/kg) or direct intraarterial infusion (0.05–0.5 μg) into the lingual artery supplying the submaxillary gland (Hahn, 1972; Hahn and Patil, 1972, 1974; Taira *et al.*, 1975). $PGF_{1\alpha}$ and $PGF_{2\beta}$ were much less active. PGE_2 was very weak, whereas PGE_1 and 15-epi-$PGF_{2\alpha}$ were inactive. The salivary response was similar to that produced by stimulation of the chorda tympani nerve (parasympathetic). Anticholinergic agents such as atropine or L-hyoscyamine abolished the salivary responses of the PGs, of chorda tympani stimulation, and of administration of acetylcholine. These studies suggest that salivation produced by $PGF_{2\alpha}$ results from liberation of acetylcholine from cholinergic nerve terminals.

III. ESOPHAGUS

The lower esophageal sphincter (also called cardiac sphincter) plays an important role in deglutition and is also believed to prevent reflux of gastric contents into the esophagus. Reflux esophagitis can therefore result from lower esophageal sphincter incompetence. The frog esophagus releases large amounts of PGs of the E type *in vitro* after electrical stimulation (Rashid, 1971). The opossum has been used for the study of lower esophageal sphincter motility since, as in man, its lower third contains smooth muscle (Christensen and Lund, 1969). In animals lightly anesthetized with sodium pentobarbital, PGE_1 and PGE_2 injected intravenously caused a dose-dependent fall in the sphincter pressure. PGA_2 was much less potent (Goyal and Rattan, 1973; Goyal *et al.*, 1972, 1973). The threshold doses for PGE_2 and PGA_2 were 0.15 μg/kg and 1 μg/kg, respectively. The blood pressure dropped, but only transiently, whereas the effect on the lower esophageal sphincter was much more prolonged. The duration of action was also dose dependent. On the other hand, $PGF_{2\alpha}$ increased the lower esophageal sphincter pressure in a dose-dependent manner; the threshold dose was 1–2 μg/kg (Goyal *et al.*, 1972; Rattan *et al.*, 1972). Further studies in the opossum suggested that PGE_1 may relax the sphincter by stimulating intracellular cyclic AMP (cAMP) since the effect of PGE_1 was mimicked by theophylline (an inhibitor of phosphodiesterase) and by dibutyryl cAMP itself (Goyal and Rattan, 1973). Moreover, both isoproterenol, an adenyl cyclase stimulator, and theophylline increased the action of PGE_1 on the sphincter.

On the basis of these results in opossums, PGE_2 and $PGF_{2\alpha}$ were administered to humans. $PGF_{2\alpha}$, given intravenously as a single bolus (2.5–40 μg/kg) to healthy volunteers, did not affect the lower esophageal sphincter (Dilawari *et al.*, 1973, 1975). When administered by constant intravenous infusion (0.05–0.8 μg/kg-min), however, marked, sustained, and dose-related increases in the cardiac sphincter pressure resulted, approximately 10 min after the start of the infusion, and persisted for 15–20 min after the end of $PGF_{2\alpha}$ infusion. Esophageal peristalsis and gastric fundal motility were not affected (Dilawari *et al.*, 1975). Plasma gastrin levels were unchanged (Dilawari *et al.*, 1975); a finding of significance since gastrin has been shown to increase cardiac sphincter pressure (Cohen and Lipshutz, 1971). This result suggests that the effect of $PGF_{2\alpha}$ is not mediated through release of gastrin. Interestingly, rectal administration of indomethacin (200 mg) resulted in a rise of cardiac sphincter pressure, presumably because of inhibition of endogenous synthesis of E-type PGs (Dilawari *et al.*, 1975). Intravenous infusion of PGE_2 (0.01–0.08 μg/kg-min) failed to affect the resting sphincter pressure (Dilawari *et al.*, 1973, 1975), although other investigators, using doses of 0.01–0.4 μg/kg-min of PGE_2, observed a marked fall in lower esophageal pressure in normal subjects (Goyal *et al.*, 1974). A reduction in lower esophageal sphincter pressure was recorded in three patients with achalasia after intravenous infusion of PGE_2 (0.4 μg/kg-min) (Goyal *et al.*, 1974). Finally, intravenous infusion of PGE_2 (0.08 μg/kg-min) inhibited sphincter contractions to serial bolus intravenous injections of pentagastrin (Dilawari *et al.*, 1975). 16,16-Dimethyl-PGE_2, administered orally to normal subjects (125 μg/subject), did not alter basal lower esophageal sphincter pressure, but inhibited the rise in pressure produced by pentagastrin (McCallum *et al.*, 1975).

These various results suggest that certain PGs of the E and F types may play a role in the regulation of the lower esophageal sphincter tone in humans, that PGE_2 may be useful in the treatment of achalasia, whereas $PGF_{2\alpha}$ may be useful in the treatment of gastroesophageal reflux.

IV. STOMACH

A. Motility

In this section, studies on the motility of the stomach, the small intestine, and the large intestine are combined since the effect is generally the same for the three segments.

1. *In Vitro* Studies

PGE_1 and PGE_2 were found to contract the longitudinal smooth muscle but to inhibit contraction of the circular muscle elicited by either acetylcholine or potassium (Bennett and Posner, 1971; Bennett *et al.*, 1967, 1968a; Fleshler and Bennett, 1969; Horton and Jones,

1969). On the other hand, PGs of the F series, such as $PGF_{2\alpha}$, contract both longitudinal and circular muscle (Bennett and Posner, 1971; Bergström *et al.*, 1968; Fleshler and Bennett, 1969; Vanasin *et al.*, 1970). PGs of the A series behave like PGF (Bergström *et al.*, 1968), although they are much less potent than PGs of the E and F series (Horton and Jones, 1969; Weeks *et al.*, 1969). Arachidonic acid, a PG precursor, was also found to contract the rat stomach, presumably by conversion into PGE_2 (Splawinski *et al.*, 1971). These opposing effects of PG on the contraction of longitudinal and circular muscles may play a role in gastric emptying and intestinal peristalsis.

2. *In Vivo* Studies

a. Natural Prostaglandins. The effect of PG on gastrointestinal motility *in vivo* is not as clear-cut as *in vitro*. PGE_1 and PGE_2 were found to contract the intestinal longitudinal smooth muscle of rats; the effect was more marked if the substances were injected intraarterially than intravenously (Bennett *et al.*, 1968*b*). When given intraluminally, no contraction was observed. In guinea pigs, however, the results were not as constant. Only one-half of the animals studied did respond with increased intestinal motility, although all showed a reduction in blood pressure following PG administration. As for the rat, intraarterial was more effective than intravenous injection. Spontaneous intestinal motility was abolished in dogs after intramesenteric arterial infusion of PGE_1 (Shehadeh *et al.*, 1969). When $PGF_{2\alpha}$ was infused in a similar manner, intestinal motility was increased. In anesthetized dogs, intravenous infusion of PGE_1 (1 μg/kg-min) inhibited antral motility stimulated by the intravenous injection of 2-deoxy-*d*-glucose, a substance acting through vagal stimulation (Chawla and Eisenberg, 1969*a,b*).

Human studies on gastrointestinal motility are very limited. When PGs of the E type were first administered to humans, abdominal cramps were sometimes reported after intravenous infusion (Bergström *et al.*, 1959*b*). Whether such effect was due to gastrointestinal contractions was not investigated. Also, in several studies in which PGs of the E and F types were administered intravenously for labor induction, vomiting often occurred; it was assumed that this reflected increased gastric motility. Abdominal colics were experienced from 1 to 3 hr after oral administration of PGE_1 (0.2–2.8 mg/subject) (Bergström *et al.*, 1959*b*; Horton *et al.*, 1968). In almost every case, watery stools were passed after 2–4 hr. The colics might have been caused by increased intestinal motility. The diarrhea, as discussed below, was probably due to formation of large amounts of fluid in the lumen of the small intestine ("enteropooling") and its eventual expulsion from the large intestine as diarrhea. No cardiovascular effects were noted. In these same studies, bile appeared in gastric juice after PGE_1 administration. We have repeatedly observed that after administration of either PGE_1 (1 μg/kg-min intravenously) or certain methyl analogues of PGE_2 (orally or intravenously) to gastric fistula dogs similar duodenal reflux occurred. These observations suggest that PGs of the E type may stimulate duodenal contractions and cause reflux of duodenal contents into the stomach.

   b. *Prostaglandin Analogues.* 16,16-Dimethyl-PGE_2 administered subcutaneously to rats retarded gastric emptying, as measured by the amount of food remaining in the stomach at autopsy (Robert *et al.*, 1974*a*). The effect was seen after a single injection, but was more pronounced after a 4-day treatment (2 injections/day), and was dose-dependent. Food intake was not affected, but water intake increased five- to sixfold. The increased water intake may be compensatory, due to the enteropooling effect of the PG (see p. 249). The increase in the amount of gastric contents was not due to the presence of water (as a result of increased drinking), since it was as marked when the contents were dried to constant weight. On the other hand, 16,16-dimethyl-PGE_2 given orally did not retard gastric emptying, regardless of the dose, and food intake was not appreciably affected, although again water intake was increased. Although the mechanism for this retardation of gastric emptying after subcutaneous treatment is unexplained, such retardation is unlikely to develop in humans when this antisecretory PG is given orally. Actually, 16,16-dimethyl-PGE_2 (140 μg/subject) given orally to human volunteers accelerated the gastric emptying of a barium meal (Nylander and Mattsson, 1975). When given orally at the antisecretory dose of 80 μg/subject, 16,16-dimethyl-PGE_2 did not affect gastric and duodenal motility as measured by pressure recordings from rubber balloons (Nylander and Andersson, 1975). A dose of 140 μg/subject, however, completely inhibited motor activity of the stomach and the duodenum. When 15(S)-15-methyl-PGE_2 methylester was introduced directly into the duodenum, duodenal motility was inhibited for more than 2½ hr (Nylander and Andersson, 1975).

B. Gastric Secretion

1. Animal Studies

   a. In Vivo Studies

   i. Natural Prostaglandins. Several natural PGs of the E and A types were found to inhibit gastric secretion in dogs stimulated with histamine, food, pentagastrin, and 2-deoxy-d-glucose (Robert, 1968; Robert *et al.*, 1967). The inhibition affected volume of secretion, acid concentration and output, and pepsin output. Pepsin concentration was not significantly changed. Similar results were obtained in rats (Nezamis *et al.,* 1971; Robert *et al.,* 1968*a*, 1976*c*). In rats and dogs, the PGs were administered parenterally, either by continuous intravenous infusion (dogs) or by subcutaneous injection (rats). In rats, the gastric output of fucose (a hydrolytic by-product of mucus) was markedly reduced by PGE_1, whereas the secretion of hexosamine and sialic acid was not changed (Robert *et al.,* 1968*a*). The body temperature of rats was not altered by PGE_1 infusion. In dogs, PGE_1, and PGE_2 inhibited gastric acid secretion in a dose-dependent manner, and the inhibition was maximal after 30–45 min following the start of the infusion (Nezamis *et al.,* 1971; Robert, 1968). Within 1–2 hr following discontinuation of infusion, the secretion had returned to the pretreatment value. The ED_{50} (dose inhibiting acid output by 50%) was around 0.5–0.75 μg/kg-min after intravenous infusion, and 10–15 μg/kg after intravenous injection (Nezamis *et al.,* 1971; Robert, 1973; Robert *et al.,* 1976*c*). PGE_1 administered intravenously also inhibited gastric secretion stimulated by carbachol and reserpine (Robert *et al.,* 1968*b*). PGA_1 and PGE_1, infused intravenously at 2.5 μg/kg-min, inhibited gastric secretion elicited by CCK-octapeptide from gastric fistula dogs (Kaminski *et al.,* 1975). PGE_1, perfused directly into the gastric lumen of rats, inhibited gastric secretion induced with either pentagastrin (Banerjee *et al.,* 1972; Shaw and Ramwell, 1968; Shaw and Urquhart, 1972; Whittle, 1972), histamine (Shaw and Ramwell, 1968; Shaw and Urquhart, 1972; Whittle, 1972), or vagal stimulation (Shaw and Ramwell, 1968; Shaw and Urquhart, 1972). Perfusion of cAMP through the rat stomach stimulated gastric secretion, and this stimulation was also inhibited by concomitant perfusion of PGE_1 (1 μg/min) (Shaw and Ramwell, 1968). Others, however, found that the slight increase in acid secretion produced in rats by intravenously administered cAMP was partially inhibited by PGE_2 given intravenously or perfused through the stomach (Whittle, 1972). In the pylorus-ligated rat, PGE_2 given orally at high doses also inhibited gastric secretion (acid concentration and output, but not volume or pepsin output) (Lee *et al.,* 1973; Robert *et al.,* 1976*c*). The inhibition of gastric secretion in dogs and rats was repeatedly confirmed (Banerjee *et al.,* 1972; Hakanson *et al.,* 1973; Hohnke, 1974; Jacobson, 1970; Kowalewski and Kolode, 1974; Levine, 1971; Main, 1969, 1973; Main and Whittle, 1974*a*; Murai *et al.,* 1970; Whittle, 1976; Wilson and Levine, 1969, 1972; Yamagata *et al.,* 1970).

    ii. Prostaglandin Analogues. For effective antisecretory activity, the natural PGs had to be administered parenterally. If PGs were ever to be considered as therapeutic agents for reducing gastric secretion, they would need to be active orally and also to have a longer duration of action than that shown for the natural prostaglandins. Consequently, several analogues of natural PGs were prepared. Among these were analogues of PGE_2 with methyl groups at either C-15 (Yankee and Bundy, 1972; Yankee *et al.*, 1974) or C-16 (Magerlein *et al.*, 1973).

iia. Methyl Analogues. Systemic Effect in Dogs, Cats, and Rats from Intravenous, Oral, or Intrajejunal Administration: Using PGE_2 analogues with methyl groups at either C-15 or C-16, the following observations were made:

1. When administered intravenously to Heidenhain pouch dogs, $15(S)$-15-methyl-PGE_2 methylester and 16,16-dimethyl-PGE_2 at a dose of 0.5 μg/kg were 30 and 50 times, respectively, more potent than the parent compound PGE_2 (Fig. 1) (Robert and Magerlein, 1973; Robert *et al.*, 1976c).
2. Whereas the effect of PGE_2 (10 μg/kg) lasted for 1 hr, 15-methyl-PGE_2 (0.5 μg/kg) inhibited gastric secretion for 2 hr, and 16,16-dimethyl-PGE_2 (0.5 μg/kg) for still longer, the inhibi-

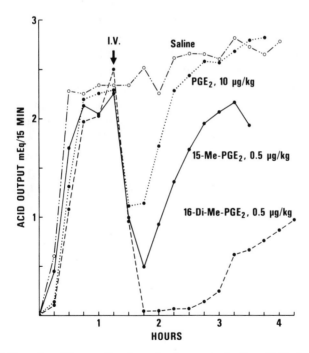

Fig. 1. Inhibition of gastric secretion in a dog by prostaglandins. Heidenhain pouch dog stimulated by histamine dihydrochloride, 1 mg/hr, infused intravenously. Acid secretion was inhibited by the three prostaglandins given intravenously; the methyl analogues were much more potent than PGE_2, and their effect lasted longer.

tion being 43% at 3 hr and 15 min after injection (Fig. 1) (Robert and Magerlein, 1973; Robert *et al.*, 1976c).

3. When these same analogues were given orally to Heidenhain pouch dogs, gastric secretion was markedly inhibited (Robert and Magerlein, 1973; Robert *et al.*, 1976c). The inhibition started 30 min after treatment and lasted for several hours. The degree of activity and the duration of action were dose dependent. At an oral dose of 20 μg/kg, gastric secretion was still reduced by 50% at 2 hr after treatment. PGE_2, on the other hand, was inactive orally even at a dose as high as 400 μg/kg (10 mg/dog).

4. These methyl analogues sometimes elicited vomiting in the dogs, after oral administration, as did the natural prostaglandins (PGE_1, PGE_2), occasionally, when given intravenously. Since earlier studies had shown that vomiting by itself could reduce gastric secretion (Grossman *et al.*, 1945), 16,16-dimethyl-PGE_2 was then administered directly into the jejunum through a jejunostomy which had been prepared at the time the gastric pouch was made. When given intrajejunally, both methyl analogues inhibited histamine-induced gastric secretion, 16,16-dimethyl-PGE_2 being 2.8 times more potent than 15(S)-15-methyl-PGE_2 (Robert and Magerlein, 1973; Robert *et al.*, 1976c). Moreover, the dogs did not vomit, nor did they seem to experience nausea. It was concluded that, first, unlike PGE_2, the two methyl analogues are absorbed through the intestinal mucosa and inhibit gastric secretion via systemic action, and, second, the antisecretory effect of these analogues after oral administration is not mediated through a vomiting reflex. Further evidence that the antisecretory property of PGs is not related to a vomiting reflex was provided by two additional observations: an antiemetic agent, such as atropine, prevented vomiting without interfering with the effect of PGE_1; and profuse vomiting elicited by apomorphine did not reduce gastric secretion stimulated by either histamine or pentagastrin (Nezamis *et al.*, 1971).

5. As for PGE_2, 16,16-dimethyl-PGE_2 infused intravenously to gastric fistula dogs inhibited gastric secretion stimulated by histamine, pentagastrin, urecholine, and 2-deoxy-d-glucose (Mihas *et al.*, 1975, 1976).

6. The gastric antisecretory ED_{50}'s (i.e., dose reducing acid output by 50%) of 16,16-dimethyl-PGE_2 were as follows: continuous intravenous infusion, 0.005 μg/kg-min (Mihas *et al.*, 1975); single intravenous injection, 0.5 μg/kg; intrajejunal administra-

tion, 15 μg/kg; oral administration, 5 μg/kg (Robert and Magerlein, 1973; Robert *et al.*, 1976c). Therefore, this analogue is 50–100 times more potent, than the parent compound, PGE_2.

7. 16(S)-Methyl-13-dehydro-PGE_2, given intravenously, also inhibited gastric secretion in dogs and cats stimulated with either pentagastrin, histamine, betanechol, bombesin, or caerulein (Impicciatore *et al.*, 1976).

8. In conscious cats with chronic gastric fistula, 15(S)-15-methyl-PGE_2 methylester (0.12–1.0 μg/kg, intravenously) caused a dose-related inhibition of maximal acid and pepsin response to pentagastrin, histamine, and peptone meal (Konturek *et al.*, 1974). Inhibition against either of the three stimuli occurred also after intraduodenal administration.

9. In rats, 15(R)-15-methyl-PGE_2 (Carter *et al.*, 1974; Whittle, 1976), 15(S)-15-methyl-PGE_2, and 16,16-dimethyl-PGE_2 inhibited gastric secretion (Robert *et al.*, 1976c; Whittle, 1976). 16,16-Dimethyl-PGE_2, perfused into the gastric lumen of anesthetized rats, inhibited gastric secretion stimulated by histamine and pentagastrin, whereas burimamide and metiamide, H_2 receptor antagonists, did not consistently inhibit gastric secretion (Banerjee *et al.*, 1975).

Systemic Effect in Dogs, Cats, and Rats from Epimerization after Oral Administration: As mentioned below, 15(S)-15-methyl-PGE_2 methylester (Karim *et al.*, 1973b; Nylander and Andersson, 1974; Robert *et al.*, 1974b, 1976a) and 15(R)-15-methyl-PGE_2 methylester (Carter *et al.*, 1973; Karim *et al.*, 1973a,b; Karim and Fung, 1975), when given orally, were found to inhibit gastric secretion in humans, the 15(S) compound being about 5 times more potent than the 15(R) (Karim *et al.*, 1973a). The 15(R) compound was inactive when given intravenously. The mechanism by which the 15(R) compound was active orally was investigated in Heidenhain pouch dogs (Robert and Yankee, 1975). The following results were obtained:

1. As in humans, the 15(R) inhibited gastric secretion (histamine induced) after oral administration but was almost inactive when given intravenously, whereas the 15(S) was active by both routes.

2. When given directly into the small intestine via a jejunostomy, the 15(S) was active but not the 15(R).

3. When the 15(R) was diluted in acid and then administered intrajejunally, it became active in inhibiting gastric secretion, whereas the 15(S) diluted in acid and administered intrajejunally lost half of its activity.

4. When either analogue was incubated *in vitro* in an acid medium (pH 1), each was found to epimerize to give approximately a 1:1 mixture of both 15(*R*) and 15(*S*), whereas incubation of the 15(*R*) in pH 3.0 buffer resulted in formation of only a trace of the 15(*S*) (Robert and Yankee, 1975). These results explain why the 15(*R*) is active orally but not intrajejunally or intravenously. When given orally, the 15(*R*) comes in contact with gastric juice, and at the low pH of gastric juice the 15(*R*) is epimerized into the 15(*S*), which is active by any route. No such epimerization can take place after intravenous or intrajejunal administration since the pH of blood and intestinal contents is around 7. These results also suggest that the degree of acidity of gastric contents may determine whether the 15(*R*) will exert an antisecretory effect, and to what extent.

Systemic Effect in Dogs, Cats, and Rats from Intravaginal Administration: In Heidenhain pouch dogs, intravaginal administration of PGE_2 and of 16,16-dimethyl-PGE_2 inhibited histamine-induced gastric secretion (Robert *et al.*, 1976*a*). In the case of the methyl analogue, the effect was obtained at doses much lower than those required after oral administration, and approached efficacy obtained after intramuscular administration. Moreover, the methylester of either PGE_2 or 16,16-dimethyl-PGE_2 was 4–5 times more potent than the free acid. These results show that these PGs are absorbed through the vaginal mucosa, and exert gastric antisecretory activity after circulating through the blood and reaching the stomach. The enhanced biological activity of the methylesters of each of the two PGs may be due mainly to their increased penetration through the vaginal mucosa rather than increased intrinsic activity, because these esters are more lipophilic than their corresponding free acids.

Local Effect: 16,16-Dimethyl-PGE_2 was also found to exert a local antisecretory effect when administered directly into a gastric pouch. In dogs with two Heidenhain pouches, administration of 100 μg of 16,16-dimethyl-PGE_2 in pouch No. 1 totally inhibited gastric secretion from that pouch within 15 min, and for the duration of the study (2½ hr), and also inhibited secretion from the distal pouch (pouch No. 2) (Robert, 1973). In the distal pouch, the inhibition started 15 min after treatment of pouch No. 1 was maximal after 30 min and remained total for 45 min, after which the values rapidly returned to the baseline. When lower doses (5–50 μg) were administered into pouch No. 1, only that pouch was inhibited; the ED_{50} was 5–10 μg (Robert, unpublished). These studies show that (1) 16,16-dimethyl-PGE_2, at high doses, is absorbed from the gastric mucosa and enters the bloodstream, as shown by the antisecretory effect produced in the second pouch; (2) 16,16-dimethyl-PGE$_2$ also exerts a local antisecretory effect which is much stronger and more prolonged than the inhibition produced by systemic absorption. The local inhibitory effect of 16,16-dimethyl-PGE$_2$ was confirmed (Andersson and Nylander, 1976).

iib. Other Analogues. A group of four isomers, under the code name of AY-22093, were reported to inhibit gastric secretion in pylorus-ligated rats after subcutaneous administration (Lippmann, 1971). The isomers have the general formula 11-deoxy-13,14-dihydro-PGE$_1$. This mixture also prevented the stimulation of gastric secretion induced by subcutaneous injections of pentagastrin in rats with an esophageal and pyloric ligation. AY-22469, another mixture of four analogues with the general formula 11-deoxy-13,14-dihydro-15-methyl-PGE$_1$, inhibited gastric secretion in the same rat preparations, even after oral administration (Lippmann and Seethaler, 1973). AY-22093 and AY-22469, however, were less potent than PGE$_1$ (Lippmann, 1970, 1971). $\Delta^{8(12)13}$-PGE$_1$ (or SC-24665), given orally, inhibited gastric secretion in pylorus-ligated rats and in chronic fistula rats stimulated with pentagastrin (Lee and Bianchi, 1972).

b. In Vitro Studies

i. Frog. PGE$_1$ inhibited acid secretion from the bullfrog (*Rana catesbeiana*) gastric mucosa incubated *in vitro*. The inhibition was demonstrated against gastrin and to a lesser degree against histamine, when PGE$_1$ (10^{-7}M) was added to the incubation medium (Way and Durbin, 1969). Both the stimuli (gastrin, histamine) and PGE$_1$ were added to the nutrient solution (i.e., serosal side). On the other hand, secretion stimulated by addition of cAMP was not inhibited by PGE$_1$. The authors concluded that, at least in the case of amphibian gastric secretion, PGE$_1$ may inhibit acid secretion by decreasing the mucosal level of cAMP. The finding that PGE$_1$ does not block stimulation by exogenous cAMP supports the hypothesis that prostaglandins may inhibit the formation of cAMP by adenylate cyclase rather than accelerate its breakdown. The inhibition of pentagastrin-stimulated acid secretion by PGE$_1$ was confirmed (Charters *et al.*, 1975). Metiamide also inhibited pentagastrin-induced secretion. The concentration of cAMP in the mucosa was nearly doubled by pentagastrin, and the addition of PGE$_1$ prevented this increase. On the other hand, metiamide, although inhibiting pentagastrin-induced acid secretion, did not prevent the rise in tissue (Charters *et al.*, 1975). These results suggest that PGE$_1$ inhibits acid secretion by blocking the formation of cAMP whereas metiamide acts by another mechanism. Contrary to these results, others found that superfusion of isolated frog gastric mucosa with PGE$_1$ produced a two- to threefold increase in acid secretion, whether the PGE_1 was added to the nutrient or the secretory surface (Shaw and Ramwell, 1969). Pentagastrin produced a similar degree of stimulation, whereas simultaneous perfusion of PGE_1 (on either nutrient or mucosal side) and pentagastrin (nutrient side) slightly reduced the response to pentagastrin alone.

ii. Dog. Using an isolated canine stomach perfused with blood, PGE_2 ($0.5-10 \times 10^{-8} \mu$g/ml added to the perfusion blood) inhibited acid secretion elicited by vagus stimulation, histamine, or pentagastrin (Shaw and Urquhart, 1972). When added to the mucosal side, however, PGE_2 ($1-2 \times 10^{-6} \mu$g/ml) was ineffective.

2. Human Studies

a. Natural Prostaglandins. PGA_1, PGE_1, and PGE_2, administered to humans by continuous intravenous infusion, inhibited gastric secretion stimulated with either histamine (PGA_1, Wilson *et al.*, 1970, 1971), pentagastrin (PGE_1, Classen *et al.*, 1971; PGE_2, Newman *et al.*, 1975), or tetragastrin (PGE_1, PGE_2, Wada and Ishizawa, 1970), as well as basal secretion (Classen *et al.*, 1970, Wada and Ishizawa, 1970). When given orally, PGE_1, (10–40 μg/kg) did not inhibit pentagastrin-induced gastric secretion (Horton *et al.*, 1968). At the higher doses, gastric juice contained large amounts of bile, and diarrhea occurred 2–4 hr after treatment. PGE_2 given orally (2.5–4.0 mg/subject) did not inhibit basal secretion (Karim *et al.*, 1973*b*). In unstimulated subjects, PGA_1 given orally did not consistently affect gastric acid secretion, whereas PGA_2 produced a transient inhibition (Bhana *et al.*, 1973). $PGF_{2\alpha}$ infused intravenously (0.08 μg/kg-min) did not affect basal or pentagastrin-induced gastric secretion, but increased the frequency of antral contractions (Classen *et al.*, 1971). At a dose of 0.8 μg/kg-min, $PGF_{2\alpha}$ inhibited transiently the volume of secretion elicited by a submaximal dose of pentagastrin, but not that of a maximal dose. Pentagastrin increased the frequency of antral contractions, but $PGF_{2\alpha}$ had no effect on the stimulated antral motility (Classen *et al.*, 1971).

b. Prostaglandin Analogues. $15(R)$-15-Methyl-PGE_2 (free acid or methylester), given orally, reduced gastric acid secretion in healthy volunteers, either in the basal state or after maximal stimulation with pentagastrin (Carter *et al.*, 1973; Karim *et al.*, 1973*a,b*; Karim and Fung, 1975). When administered orally to pregnant women at term, $15(R)$-15-methyl-PGE_2 methylester markedly inhibited basal gastric secretion without producing excessive uterine stimulation (Amy *et al.*, 1973). $15(S)$-15-Methyl-PGE_2 methylester was at least 5 times more potent than $15(R)$-15-methyl-PGE_2 methylester (Karim *et al.*, 1973*b*; Nylander and Andersson, 1974; Nylander *et al.*, 1974; Robert *et al.*,

1974). Mild side effects (nausea, borborygmi) were noted at 50 μg/ subject of the 15(S), whereas there were no side effects at doses up to 200 μg/subject of the 15(R) (Karim *et al.*, 1973*b*). Similarly, 16,16-dimethyl-PGE$_2$ (or its methylester) administered orally to healthy volunteers inhibited basal (Karim *et al.*, 1973*c*,*d*; Wilson *et al.*, 1976) secretion and secretion stimulated with either pentagastrin (Karim *et al.*, 1973*c*,*d*; Nylander and Andersson, 1974; Nylander *et al.*, 1974; Robert *et al.*, 1974) or histamine (Wilson *et al.*, 1975*b*). Administration of gelatin capsules containing 16,16-dimethyl-PGE$_2$ to human volunteers markedly inhibited gastric secretion (volume, acid concentrations, and output) stimulated with histamine (Wilson *et al.*, 1976).

In these various studies, the oral dose of 16,16-dimethyl-PGE$_2$ reducing acid output by 50% was 50–100 μg/subject (around 1 μg/kg). When given intravenously, the ED$_{50}$ was 12.5 μg/subject (Karim *et al.*, 1973*c*,*d*). 15(S)-15-Methyl-PGE$_2$ methylester and 16,16-dimethyl-PGE$_2$ methylester were equally potent in inhibiting pentagastrin-induced gastric secretion (Nylander and Andersson, 1974; Robert *et al.*, 1974). Interestingly, when these two methyl analogues were administered directly into the intestine, their effect was much reduced; actually, 16,16-dimethyl-PGE$_2$ was almost ineffective when given at 200 μg/subject directly into the upper jejunum (Nylander *et al.*, 1974; Robert *et al.*, 1974*b*). This observation is at variance with results obtained in dogs, where both 15(S)-15-methyl-PGE$_2$ methylester and 16,16-dimethyl-PGE$_2$ given intrajenunally inhibit gastric secretion from a Heidenhain pouch (Robert and Magerlein, 1973; Robert *et al.*, 1976*c*). The reasons for this decreased activity after intestinal administration in humans is unknown, but the following possibilities can be considered: (1) The greater inhibition after intragastric administration may be due to a local action on the parietal cells. In support for this explanation, topical application of 16,16-dimethyl-PGE$_2$ into a Heidenhain pouch, in dogs, was found to inhibit gastric secretion from that pouch (Robert, 1973). Such local effect may take place in humans after intragastric administration. (2) The methyl analogues used may be more inactivated (e.g., metabolized, bound) when given directly into the intestine than when administered orally. (3) These analogues may be absorbed better by the stomach than by the intestine. The reason why the analogues retain their antisecretory activity when given intrajejunally to dogs is unknown. Even in dogs, however, 16,16-dimethyl-PGE$_2$ is only half as potent after intrajejunal than after oral administration in inhibiting secretion from a Heidenhain pouch (Robert *et al.*, 1976*c*).

Meal-stimulated acid secretion in duodenal ulcer patients was strongly inhibited by oral administration of 15(R)-15-methyl-PGE$_2$

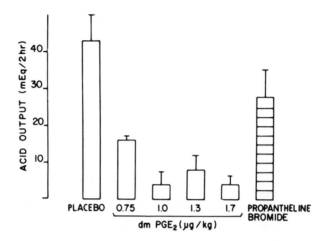

Fig. 2. Inhibition of gastric secretion in humans by 16,16-dimethyl-PGE₂. Mean 2-hr meal-stimulated acid output. Vertical bars: standard error. 16,16-Dimethyl-PGE₂ given orally inhibits secretion. The effect is more marked than that obtained by the optimally effective dose of propantheline bromide. From Ippoliti *et al.* (1976).

methylester or 15(*S*)-15-methyl-PGE₂ methylester (Konturek *et al.*, 1976*b*), or 16,16-dimethyl-PGE₂ (Ippoliti *et al.*, 1976; Konturek *et al.*, 1976*b*). 16,16-Dimethyl-PGE₂ was more potent than the highest toler-ated dose of an anticholinergic agent, probanthine, which produced only 35% inhibition (Ippoliti *et al.*, 1976) (Fig. 2). No side effects were noted at a dose of 1 μg/kg of 16,16-dimethyl-PGE₂. This prostaglandin analogue also prevented the rise in serum gastrin levels normally produced by a meal (Ippoliti *et al.*, 1976; Konturek *et al.*, 1976*b*). 15(*R*)-15-Methyl-PGE₂ methylester, administered to unstimulated sub-jects, slightly increased the serum gastrin level (Adaikan *et al.*, 1974). In histamine-stimulated subjects, 16,16-dimethyl-PGE₂ did not change significantly the serum gastrin level (Wilson *et al.*, 1976). 15(*R*)-15-Methyl-PGE₂ methylester (Konturek *et al.*, 1976*b*; Salmon *et al.*, 1975), 15(*S*)-15-methyl-PGE₂ methylester, or 16,16-dimethyl-PGE₂ (Konturek *et al.*, 1976*b*), administered orally to pentagastrin-stimu-lated subjects, inhibited pepsin output, and the effect was dose dependent. In meal-stimulated subjects, these same analogues mark-edly inhibited pepsin concentration (Konturek *et al.*, 1976*b*).

C. Ulcer Formation

The gastric antisecretory effect of certain PGs in animals and humans prompted their study as possible antiulcer agents. In this section, we are including results obtained with both gastric and duo-denal ulcers.

1. Animal Studies

a. Gastric Ulcers. In the rat, Shay ulcers, produced by pylorus ligation, were prevented by administration of PGE_1 subcutaneously (Robert, 1968; Robert *et al.*, 1968*a*), PGE_2 orally or subcutaneously (Lee *et al.*, 1973; Robert *et al.*, 1976*e*), $\Delta^{8(12)13}$-PGE_1 orally (Lee and Bianchi, 1972), 15(*R*)-15-methyl-PGE_2 methylester orally (Carter *et al.*, 1974), and 16,16-dimethyl-PGE_2 orally or subcutaneously (Robert *et al.*, 1973, 1976*c*). Similarly, PGE_1, PGE_2, or 16,16-dimethyl-PGE_2 inhibited gastric ulcers produced in rats by a variety of methods, such as administration of glucocorticoids ("steroid-induced ulcers") (Robert *et al.*, 1968*a*, 1976*c*) (Fig. 3), of reserpine (Lee *et al.*, 1973), of serotonin (Ferguson *et al.*, 1973), of indomethacin (Daturi *et al.*, 1974; Robert, 1975*a*; Whittle, 1976), of bile (Mann, 1975), and by restraint (Karawada *et al.*, 1975) and exertion (Lee and Bianchi, 1972). PGE_2 given orally (Banerjee *et al.*, 1975), 15(*S*)-15-methyl PGE_2 given subcutaneously (Whittle, 1976), and 16,16-dimethyl-PGE_2 given either subcutaneously (Banerjee *et al.*, 1975) or orally (Whittle, 1976) inhibited indomethacin-induced gastric ulcers, whereas metiamide, an H_2 receptor antagonist, was a weak inhibitor (Banerjee *et al.*, 1975). Gastric ulcers produced in rats by nonsteroidal anti-inflammatory compounds (NOSAC) such as indomethacin and flurbiprofen were prevented by oral or subcutaneous administration of several PGs of the A, B, D, E, Fα, and Fβ types (Robert, 1975*a*).

b. Duodenal Ulcers. Duodenal ulcers produced in rats by gastric secretogogues such as histamine, carbachol, and pentagastrin were inhibited by either PGE_2 (Robert, 1971, 1974*a*; Robert *et al.*, 1971, 1973; Robert and Standish, 1973) or 16,16-dimethyl-PGE_2 (Robert, 1973; Robert *et al.*, 1973, 1976*c*) (Fig. 4). Duodenal ulcers produced by constant infusion of pentagastrin in guinea pigs and cats were prevented by either PGE_2 (Lee *et al.*, 1973), 15(*S*)-15-methyl-PGE_2 methylester (Konturek *et al.*, 1974), or SC-24665 (Lee and Bianchi, 1972). Other types of duodenal ulcers produced in rats by cysteamine and propionitrile were prevented by either PGE_2 or 16,16-dimethyl-PGE_2 (Robert, 1974*a*; Robert *et al.*, 1975*a*,*b*). Histamine ulcers produced in guinea pigs were inhibited by PGE_2 (Lee *et al.*, 1973) and by $\Delta^{8(12)13}$-PGE_1 (SC-24665) (Lee and Bianchi, 1972).

2. Human Studies

15(*R*)-15-Methyl-PGE_2 methylester was administered orally at a single dose of 150 μg/subject to ten patients with proven peptic ulcer who were experiencing typical epigastric pain. In most patients, the severity of the pain and the epigastric tenderness were reduced, and the reduction usually correlated with a marked rise in gastric pH

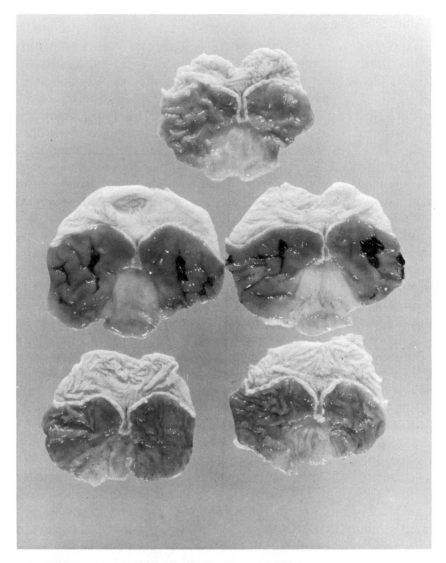

Fig. 3.   Inhibition of steroid-induced ulcers in rats by 16,16-dimethyl-PGE$_2$. Rats' stomachs shown after opening along the greater curvature. Top: Normal stomach. Middle: Prednisolone, 5 mg/rat, subcutaneously, for 3 days. Multiple, severe ulcers in the corpus (glandular portion of the stomach). Bottom: Same doses of prednisolone plus 16,16-dimethyl-PGE$_2$, 0.4 mg/kg orally, 3 times a day. Complete prevention of ulcers.

(Fung and Karim, 1974). The same authors later performed an identical study, but double blind. The mean duration of pain relief following a single oral dose of 150 μg of 15(R)-15-methyl-PGE$_2$ methylester was 81 min and for epigastric tenderness 93 min. By comparison, following a single dose of an antacid the times were 70 and 82 min, respectively

(Fung *et al.*, 1974*a*). Endoscopic and histological study of the gastric mucosa in gastric ulcer patients was carried out after a single administration of 150 μg of 15(*R*)-15-methyl-PGE$_2$ methylester. One hour after administration of the PG, a marked increase in viscid mucoid secretion was noted in six out of eight cases. Histologically, in all eight cases increased amount of mucus was found in epithelial cells (Fung *et al.*, 1974*c*). 15(*R*)-15-Methyl-PGE$_2$ methylester was administered orally to ten gastric ulcer patients at doses of 150 μg 4 times/day for 2 weeks. A control group consisted of nine subjects with proven gastric ulcer but not receiving PG. As assessed by endoscopic examination, performed just before and 2 weeks after treatment, in the PG group complete

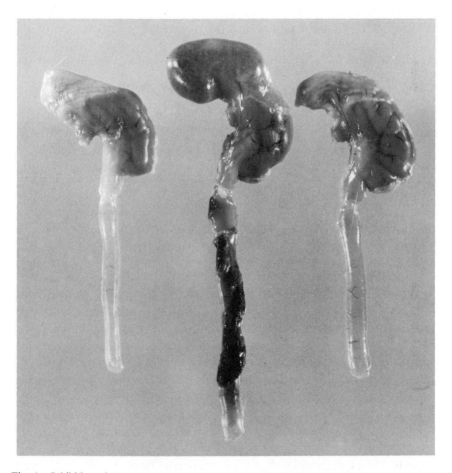

Fig. 4. Inhibition of duodenal ulcers in rats by 16,16-dimethyl-PGE$_2$. Left: Stomach and duodenum of control animal. Middle: Continuous subcutaneous infusion of histamine dihydrochloride (0.1 mg/kg-min) plus carbachol (0.35 μg/kg-min) for 24 hr. Severe duodenal ulcers. Right: same treatment with histamine and carbachol, plus 16,16-dimethyl-PGE$_2$, 0.75 mg/kg orally, only once, at the start of infusion. Complete protection.

healing was seen in three cases, considerable healing in six cases, and slight healing in one case. In the control group, complete healing was seen in none, considerable healing in two cases, slight healing in four cases, and no healing in three cases. The difference was highly significant. No side effects were noted except in two patients who had transient diarrhea during the first 4 days of treatment (Fung *et al.*, 1974*b*). It was later pointed out that this study was not double blind and that an improper statistical test had been used (Parkin *et al.*, 1974).

In a preliminary study in five patients with medium to large gastric ulcers, oral administration of 150 μg of 15(*R*)-15-methyl PGE_2, 4 times a day for 14 days, resulted in complete healing of ulcers in two, considerable healing in two, and slight healing in one. The ulcer size was assessed endoscopically before and 2 weeks after treatment. The authors stated that a double-blind study was being undertaken (Karim and Fung, 1975). This same analogue also inhibited gastric acid secretion both in the basal state and after maximal stimulation with pentagastrin.

In a double-blind study, PGE_2 was given orally to gastric ulcer patients for 2 weeks (1 mg/subject, 4 times a day). As judged by endoscopy, the rate of healing was 42% in the treated group and 14% in the controls (Karim and Fung, 1976). This is of particular significance since PGE_2, given orally, does not inhibit gastric secretion (Karim *et al.*, 1973*b*). The acceleration of healing could be due to cytoprotection by PGE_2, as discussed later.

D. Mechanism of the Antisecretory Action

Unlike most other inhibitors of gastric secretion, the PGs block secretion induced by most known stimulants: histamine, gastrin, vagal stimulation (electrical current or 2-deoxy-*d*-glucose), cholinergic agents, food, reserpine, cholecystokinin, bombesin, and caerulein. This lack of selectivity suggests that PGs act as a protective shield for the parietal cells against various kinds of stimulation. The effect is limited to parietal cells; the chief cells seem unaffected since pepsin concentration is not appreciably reduced. In retrospect, the unique properties of the 15-methyl and 16,16-dimethyl analogues of PGE_2 may be due, in part, to the fact that the methyl radical(s) protect the hydroxyl at C-15 from being degraded by the 15-hydroxydehydrogenase into 15-ketoprostaglandins which are usually of diminished biological activity. Although the mechanism by which PGs inhibit gastric secretion is not understood, the following possibilities can be considered.

1. Gastric Mucosal Blood Flow

PGE_1 administered intravenously was found to decrease gastric mucosal blood flow in dogs, as shown by the aminopyrine technique (Jacobson, 1970; Wilson and Levine, 1969, 1972). Since this decrase coincided with inhibition of gastric secretion, it was first thought that the antisecretory effect of PGE_1 might be due to the decrease in mucosal blood flow. Further studies, however, showed that the ratio R of blood flow to secretory rate increased with PGE_1 (Jacobson, 1970). This increase indicated that PGE_1 acted primarily on secretory cells, and that the decrease in blood flow was secondary to the declining secretory activity. This was not the case for norepinephrine, another agent that can inhibit gastric acid secretion; norepinephrine depressed both acid secretion and the R value (Jacobson, 1970). Thus the primary effect of norepinephrine is on the circulation; i.e., the fall in blood flow exceeds the decrease in secretory rate, so that blood flow becomes a limiting factor for secretion. PGE_1 therefore is a potent vasodilator of the gastric circulation in the dog. These studies were confirmed for PGE_1 (Nakano and Prancan, 1972; Wilson and Levine, 1972), and PGA_1 was found to have the same effect (Nakano and Prancan, 1972).

Similarly, several PGs of the E type reduced the mucosal blood flow in rats without lowering the R value (Banerjee *et al.*, 1975; Main and Whittle, 1972*a*,*b*, 1973*a*,*b*). In fact, PGA_1, PGA_2, PGE_1, and PGE_2 increased the R value at a time when acid secretion was inhibited (Main and Whittle, 1972*a*, *b*, 1973*a*,*b*). In these studies, gastric secretion was stimulated with either pentagastrin or histamine. PGE_1, PGE_2, PGA_1, and PGA_2, administered during basal secretion, actually increased gastric mucosal blood flow, an effect showing a direct vasodilator action on the gastric mucosa (Main and Whittle, 1973*a*,*b*). 15(S)-15-Methyl-PGE_2, administered intravenously at doses inhibiting completely acid secretion in histamine-stimulated rats, did not prevent the increase in mucosal blood flow elicited by histamine (Main and Whittle, 1974*b*). The results of these various studies show that in the rat, as in the dog, PGs do not inhibit gastric secretion by reducing mucosal blood flow, and appear to exert a primary effect on parietal cells.

2. Absence of Nervous Mediation

PGs do not have anticholinergic activity. This was shown by the following studies: (1) the hypotensive effect of PGE_1 in rabbits was not prevented by atropine given at doses that abolished the action of acetylcholine (Bergström *et al.*, 1959*b*); (2) PGE_1 applied into the anterior chamber of the rabbit eye produced miosis, and even reversed the mydriatic effect of atropine (Waitzman and King, 1967); (3) PGE_1 was equally effective in inhibiting gastric secretion in denervated gastric pouches (Nezamis *et al.*, 1971; Robert, 1968) as in gastric fistula preparations (in which the stomach is innervated) (Jacobson, 1970; Robert, 1968); (4) finally, PGE_1 was found to inhibit gastric secretion *in vitro,* in the isolated (denervated) frog gastric mucosa. Therefore, the antisecretory property of PGs is not mediated through the nervous system.

3. Local vs. Systemic Action

Natural PGs of the E type can inhibit gastric secretion when applied locally on the gastric mucosal surface, under certain conditions [isolated frog mucosa (Way and Durbin, 1969), gastric perfusion at high concentrations in anesthetized rats (Barrowman *et al.*, 1975; Shaw and Ramwell, 1968; Shaw and Urquhart, 1972; Whittle, 1972), oral administration at high doses to conscious rats (Lee *et al.*, 1973; Robert *et al.*, 1976c)]. In humans and dogs, however, intragastric administration of PGE_1 (Horton *et al.*, 1968) and PGE_2 (Karim *et al.*, 1973b; Robert *et al.*, 1976c) is not antisecretory. In the case of 16,16-dimethyl-PGE_2, a local antisecretory effect was demonstrated in dogs (Andersson and Nylander, 1976; Robert, 1973). The fact that this same PG analogue is more active when given orally than intrajejunally, in both dogs (Robert and Magerlein, 1973; Robert *et al.*, 1976c) and humans (Ippoliti *et al.*, 1976; Nylander and Andersson, 1974; Nylander *et al.*, 1974; Robert *et al.*, 1974), suggests a local inhibitory effect on parietal cells, or possibly greater absorption of this PG from the stomach than from the small intestine. Such local effect would be desirable in the treatment of peptic ulcer since it would minimize the possibility of systemic side effects.

4. Role of Cyclic AMP

The role of cAMP in the regulation of gastric secretion is much debated. Several PGs were found to increase the tissue level of cAMP in many organs by stimulating adenylate cyclase activity (Butcher, 1970). PGE_1 was reported to increase adenylate cyclase activity in the guinea pig gastric mucosa (Perrier and Laster, 1970; Wollin *et al.*, 1974). More recently, PGE_2, incubated with a broken-cell preparation of canine gastric fundic mucosa, was found to increase adenylate cyclase activity by 150% (Dozois and Wollin, 1975). Cyclic AMP can initiate gastric secretion when applied *in vitro* to the gastric mucosa of certain strains of frogs (Harris and Alonso, 1965; Harris *et al.*, 1969). It was suggested by some that in the dog cAMP can also initiate acid secretion (Bieck *et al.*, 1973), whereas others found that it inhibits such secretion (Levine and Wilson, 1971; Levine *et al.*, 1967); still others concluded that cAMP is not a mediator of gastric secretion at all (Mao *et al.*, 1972). In rats, cAMP was found to elicit a small acid response when perfused intraluminally (Shaw and Ramwell, 1968) and after intravenous injection (Main and Whittle, 1975; Whittle, 1972) and to potentiate the secretogogue effects of theophylline (Shaw and Ramwell, 1968) and pentagastrin and histamine (Whittle, 1972).

PGE$_2$ failed to block acid secretion stimulated in rats by a combination of pentagastrin plus cAMP, although it inhibited secretion produced by pentagastrin alone (Hohnke, 1974). In other rat studies, PGE$_2$ inhibited the secretion induced by cAMP plus theophylline (Shaw and Ramwell, 1968), cAMP plus pentagastrin, and cAMP plus histamine (Whittle, 1972), but only partially reduced the response to cAMP alone (Whittle, 1972). The findings that PGE$_1$ in frogs inhibits acid secretion stimulated by histamine and gastrin but not by locally applied cAMP (Way and Durbin, 1969), and that PGE$_2$ does not block completely the effect of cAMP in rats (Hohnke, 1974; Whittle, 1972), suggest that in these species these PGs may act by inhibiting the formation of cAMP in the gastric mucosa, presumably by blocking adenylate cyclase activity. In view, however, of the many contradictory observations, further studies, especially *in vivo* and in several species, are needed to clarify the interaction of PGs with the cAMP system.

The effect of PGs on cGMP has not been studied yet. Since, however, cGMP is increased in the canine gastric fundus after vagal stimulation (Eichhorn *et al.*, 1974), it would be of interest to find out whether antisecretory PGs would decrease the basal tissue level of cGMP or would prevent its rise upon vagal stimulation. In this connection, a role for cGMP as a mediator for stimulation, and for cAMP as a mediator for inhibition of acid secretion, has been proposed (Amer, 1972, 1974).

5. Gastric Mucosal Barrier

Local application of a high concentration of 16,16-dimethyl-PGE$_2$ into a Heidenhain pouch was reported to break the gastric mucosal barrier (O'Brien and Carter, 1975). However, the gastric antisecretory property of PGs does not appear to be related to hydrogen ion back-diffusion, for the following reasons: (1) The ED$_{50}$, dose reducing gastric acid secretion by 50%, after local application into a Heidenhain pouch is 60 times lower than that shown to break the mucosal barrier (5 μg vs. 300 μg) (Robert, unpublished). (2) The mucosal barrier was not affected by 16,16-dimethyl-PGE$_2$ given intravenously, even at a dose (2

$\mu g/kg$) that is 20 times the intravenous ED_{50} (O'Brien and Carter, 1975). Therefore, although 16,16-dimethyl-PGE_2 can affect the mucosal barrier when very high concentrations are placed in prolonged contact with the gastric mucosa (probably because of irritating properties of the compound), this effect is distinct from the antisecretory activity shown by this (and probably other) PG given at therapeutic doses.

6. Gastrin Release

In Pavlov pouch dogs, the rise in serum gastrin following a meal was further increased by PGE_1 given intravenously at a dose inhibiting acid output by 73% (Reeder *et al.*, 1972). Similarly, 15(R)-15-methyl-PGE_2 methylester, given orally to human volunteers at a dose that raised basal pH from 2 to over 7, also increased the serum gastrin level (Adaikan *et al.*, 1974). Such rises in serum gastrin are probably due to lowered acidity of the fluid bathing the antral mucosa as a result of PG treatment. In histamine-stimulated subjects, 16,16-dimethyl-PGE_2 did not significantly affect serum gastrin levels (Wilson *et al.*, 1975*b*). Serum gastrin was decreased in meal-stimulated subjects given 15(R)-15-methyl-PGE_2 methylester, 15(S)-15-methyl-PGE_2 methylester, or 16,16-dimethyl-PGE_2, orally (Ippoliti *et al.*, 1976; Konturek *et al.*, 1976*b*). Although this lowering in gastrin level does not explain the effect of PGs, it may further increase their antisecretory activity.

E. Physiopathological Role of Prostaglandins

The fact that PGs are normally present in gastric mucosa (Bennett *et al.*, 1968*c*) and gastric juice (Bennett *et al.*, 1967; Shaw and Ramwell, 1968) raises the question of whether these substances play a physiological role in the regulation of gastric acid secretion and in the pathogenesis of peptic ulcer. The following observations favor such a role:

1. Cholinergic stimulation of the rat stomach *in vitro* releases PG into the lumen (Bennett *et al.*, 1967). In humans, vagal stimulation (insulin hypoglycemia) of gastric secretion also increased the output of PGE in gastric juice (Cheung *et al.*, 1975). Such releases of inhibitors of gastric secretion can be viewed as a compensatory mechanism aimed at restricting the secretogogue effect of cholinergic stimulation.
2. In rats, administration of indomethacin, an inhibitor of PG synthesis, increased acid secretion stimulated by pentagastrin

and by cAMP, although by itself it did not affect acid secretion (Main and Whittle, 1975). This effect was observed at a time when the PG content of the mucosa was markedly reduced. The increase in gastric secretion may have been due to a depletion of endogenous PG produced by indomethacin, which blocks PG synthetase activity. PGs could therefore be considered natural inhibitors. On the other hand, the repeated observation that PG synthetase inhibitors, such as aspirin and indomethacin, do not by themselves stimulate gastric acid secretion is evidence against a physiological role for PG. In fact, in most studies in rats (Brodie and Chase, 1969; Main and Whittle, 1975; Vanasin *et al.*, 1970), cats (Lynch *et al.*, 1964), dogs (Nicoloff, 1968; Stephens *et al.*, 1966), and humans (Bennett *et al.*, 1972) these agents decreased acid production. If PG restricted the rate of acid secretion, one would expect these drugs to stimulate secretion by removing endogenous inhibitors.

3. In duodenal ulcer patients, the concentration of PGE in plasma and gastric juice was found to be lower than in healthy subjects (Hinsdale *et al.*, 1974). This reduction was correlated with gastric hyperacidity in such patients. Similar changes in PGE concentration were found by others (Tonnesen *et al.*, 1974). These findings suggest an inverse relationship between the amount of PGE (inhibitors of gastric secretion) and acid secretion.

4. Conversely, in gastritis and gastric ulcer, the concentration of PGA, PGE, and PGF in the antrum (obtained by endoscopic biopsy) was increased three- to sixfold (Schlegel *et al.*, 1976). Since hypoacidity is the rule in such cases, an inverse correlation exists between acid secretion and PG content, as in the case of duodenal ulcer.

These findings are suggestive that PGs may act as local regulators of gastric secretion. If so, an imbalance in their rate of formation may predispose to ulcer formation. Whether this proves to be the case or not, it remains that some of the methyl analogues mentioned in this chapter may be of particular value in the treatment of peptic ulcer. Their broad-spectrum antiulcer activity in many animal models and their oral and long-acting properties would make them suitable for chronic treatment aimed at reducing gastric acid secretion. The possible role of PGs in other digestive diseases was previously discussed (Bennett, 1976).

V. INTESTINE

A. Intestinal Secretion and Absorption

Several PGs, administered either into the superior mesenteric artery (PGA_1, PGE_1, $PGF_{2\alpha}$ in dogs) (Greenough *et al.*, 1969; Pierce *et al.*, 1971), intravenously ($PGF_{2\alpha}$ in humans) (Cummings *et al.*, 1973), or directly into the lumen of the small intestine [PGE_1 in humans (Matuchansky and Bernier, 1971, 1973*a*) and dogs (Pierce *et al.*, 1971)], were found to evoke the accumulation of fluid (water and electrolytes) within the lumen. The effect was similar to that produced by cholera exotoxin (Pierce *et al.*, 1971), although cholera toxin does not seem to act through local release of PGs (Hudson *et al.*, 1975; Wilson *et al.*, 1975*a*). In dogs, the response obtained when PGs were introduced into the jejunal lumen was not as pronounced as after injection into the mesenteric artery (Pierce *et al.*, 1971). In humans, the accumulation of fluid after PGE_1 was not accompanied by a change in intestinal transit time (Matuchansky *et al.*, 1972). As for PGE_1, intrajejunal administration of PGE_2 to healthy volunteers produced net excretion of water, sodium, potassium, and chloride. The mean transit time, however, was prolonged by 36%. Within 2 hr after perfusion of 5 μM of PGE_2 (325 μg), all subjects experienced one to three episodes of watery diarrhea (Hinsdale *et al.*, 1974). In subjects undergoing both jejunal and colonic perfusion, intravenous administration of $PGF_{2\alpha}$ increased the ileal flow (i.e., excretion of water and electrolytes into the lumen) whereas colonic absorption was not changed (Cummings *et al.*, 1974). $PGF_{2\alpha}$ perfused through the colon did not increase the rate of colonic absorption, whereas perfusion of the ileum markedly increased the flow rate (Newman *et al.*, 1974). As for cholera toxin, PGE_1 also reduced jejunal absorption of glucose and sodium in humans (Matuchansky and Bernier, 1973*b*). Using the distal rabbit ileum incubated *in vitro*, PGE_1 and $PGF_{2\alpha}$, added to the serosal side, inhibited the absorption of sodium from the mucosal side (Al Awqati and Greenough, 1972). Similarly, PGE_2 inhibited sodium and chloride absorption (fluxes from mucosa to serosa) in rat jejunum *in vitro* (Declusin *et al.*, 1974).

The mechanism by which PGs increase the unidirectional fluxes of water and electrolytes from blood to lumen, and inhibit absorption in the other direction, appears to involve local stimulation of cAMP (Kimberg *et al.*, 1971). However, although PGE_1, PGE_2, and $PGF_{1\alpha}$ perfused through the rat jejunum inhibited glucose absorption by 25%, similar perfusion with dibutyryl cAMP increased glucose absorption by 13% (Coupar and McColl, 1972). This suggests that the effect of these PGs on glucose absorption may not be mediated through stimulation of adenylate cyclase.

B. Enteropooling Assay

As a result of these various studies showing that PGs stimulated water and electrolyte secretion into the small intestine, an assay was developed to test the diarrheogenic property of PGs. The accumulation of fluid into the small intestine was called "enteropooling" (Robert, 1975a,b; Robert et al., 1976b,e). It is the sum of (1) the fluid being excreted from the blood into the lumen and (2) to a lesser extent the portion of fluid already present in the lumen of the small intestine but whose absorption is inhibited by the PG given. The enteropooling assay uses rats fasted for 24 hr. A PG is administered either orally or subcutaneously, and the animals are sacrificed 30 min later. The entire small intestine is removed and its contents are collected into a graduated test tube. The greater the volume of this intestinal fluid, the more diarrheogenic is the PG. The assay is quantitative, and predictive of diarrhea since by waiting longer than 30 min actual diarrhea can be observed if enough fluid has accumulated into the small intestine. It can be used to grade the relative diarrheogenic activity of PGs as well as to test agents that may block this effect.

Table 1 shows the enteropooling effect of several PGs. The ED_{50} is the dose eliciting accumulation of 1.75 ml of intestinal fluid 30 min after oral administration. This amount of 1.75 ml was chosen because the maximum volume for PGs is around 3.5 ml. Although all the PGs shown in Table 1 were enteropooling, their potency varied greatly. Among the natural PG, $PGF_{2\beta}$ and PGD_2 were the least active.

Agents other than PG were also tested for enteropooling activity. Laxatives such as castor oil, hypertonic solutions and bile salts were enteropooling, the degree of which was dose related. Carbachol was also enteropooling, but the maximum volume formed into the intestine was less than that produced by laxatives or PGs (Robert et al., 1976e).

C. Diarrhea

PGs may play a role in the development of diarrhea. Administration of high doses of certain PGs was shown to produce diarrhea, both in humans and in animals. The studies discussed earlier indicated that the origin of PG-induced diarrhea was the small, not the large intestine. If the degree of enteropooling is large enough, the fluid formed in the

Table I. Comparative Enteropooling Activity of Various Prostaglandins[a]

	Dose (mg/kg)	Times as active as PGE_2
A. Prostaglandins other than analogues		
PGE_2	0.1	1
PGC_2	0.25	½
PGB_2	1	⅙
PGA_1	5	1/33
PGA_2	5	1/33
$PGF_{2\alpha}$	5	1/33
$PGF_{2\beta}$	12.5	1/83
PGD_2	25	1/166
B. Prostaglandin analogues		
16,16-DiMe-E_2	0.0005	200
15(S)-15-Me-E_2	0.0005	200
15(R)-15-Me-E_2	0.005	30
15(S)-15-Me-$F_{2\beta}$	0.5	⅓
15(R)-15-Me-$F_{2\beta}$	1.1	1/7

[a] All prostaglandins were given orally in 1 ml ethanol–water. Intestinal fluid collected 30 min later. Doses indicate amount of a prostaglandin stimulating the accumulation of 1.75 ml of intestinal fluid (i.e., ED_{50}, see text).

small intestine is transported to the large intestine and then expelled as diarrhea. Contractions of the intestinal smooth muscle, elicited either because of the large amount of fluid or because of a direct effect of the PG, probably carry the watery contents through the intestine, and favor their expulsion. However, hypermotility by itself cannot liquefy the contents, a necessary change for diarrhea to occur; it can only contribute to the transport of the contents.

In favor of a role for PGs in the development of diarrhea is the observation that certain tumors, found to contain high levels of PGs, are associated with the presence of diarrhea. This is the case for the medullary carcinoma of the thyroid and certain peptide-secreting tumors (Barrowman *et al.*, 1975; Bernier *et al.*, 1969; Bigazzi *et al.*, 1973; Grahame-Smith, 1973; Sandler *et al.*, 1968, 1969; Williams *et al.*, 1968). A case with watery diarrhea and other symptoms attributable to PGs was reported in which hyperprostaglandinemia was demonstrated (500 pg/ml of PGE_1 vs. normal values of 7.5 pg/ml) (Labrum *et al.*, 1976). Nutmeg has been used with some success to treat the diarrhea associated with some of these tumors (Barrowman *et al.*, 1975; Bennett *et al.*, 1974). Since nutmeg was found to inhibit PG synthesis in human colon (Bennett *et al.*, 1974), it was hypothesized that its antidiarrheal effect might be due to a reduction in the production of

PGs. However, an anticholinergic effect of nutmeg may have contributed to the antidiarrheal activity, since some toxic effects of nutmeg do resemble those of atropine (Fawell and Thompson, 1973). Still, other workers attributed this antidiarrheal effect of nutmeg to its content in sympathomimetic agents (Schlemmer *et al.*, 1973). Although diarrhea is generally considered as an undesirable side effect, it was suggested that PGs be used as purgatives in certain clinical situations such as prior to intestinal surgery or radiography (Bennett, 1973).

VI. BILE AND PANCREATIC SECRETION

A. Bile Secretion

PGA$_1$ (2.5 μg/kg per min intravenously) increased bile flow and bile content in bicarbonate and sodium in dogs stimulated with taurocholate (Kaminski and Jellinek, 1974; Kaminski *et al.*, 1975, 1976). In secretin-stimulated dogs, PGA$_1$ reduced bile flow and biliary sodium, whereas the only effect of PGA$_1$ in animals stimulated with taurocholate plus CCK-octapeptide was an increase in the output of sodium. PGE$_1$ (2.5 μg/kg-min) also increased bile flow and bicarbonate and sodium content in taurocholate-stimulated dogs, but did not affect bile stimulated by either secretin or bile salt plus CCK-octapeptide. The choleretic effect of PGA$_1$, PGE$_1$, and PGE$_2$ was confirmed after intraportal [and sometimes intraarterial (hepatic)] administration, in cats (Krarup *et al.*, 1975) and rats (Lauterberg *et al.*, 1975).

B. Pancreatic Secretion

PGE$_1$ given intravenously inhibited pancreatic secretion (volume, bicarbonate concentration, and output) in dogs stimulated with either secretin or secretin plus CCK (Rudick *et al.*, 1970, 1971). The ED$_{50}$ was 23 μg/kg by bolus injection and 1.8 μg/kg-min by continuous infusion; these doses are about twice the gastric antisecretory ED$_{50}$ for the dog (Hirshowitz *et al.*, 1976; Mihas *et al.*, 1975; Robert *et al.*, 1976c). By contrast, enzyme output was increased by PGE$_1$ in secretin-stimulated animals, although the maximum response was less than half the response produced by CCK (Rudick *et al.*, 1970, 1971). 16,16-Dimethyl-PGE$_2$, given intravenously or intraduodenally to dogs, affected pancreatic secretion similarly, except that it was about 30 times more potent than PGE$_1$ (Rosenberg *et al.*, 1974). On the other hand, 15(S)-15-methyl-PGE$_2$ methylester given intravenously to cats did not affect pancreatic secretion stimulated with secretin (Konturek *et al.*, 1974). PGE$_1$ and PGE$_2$ injected into the aorta (above the celiac axis) of cats reduced pancreatic flow and bicarbonate stimulated by secretin (Case and Scratcherd, 1972). PGF$_{1\alpha}$ and PGF$_{2\alpha}$ were inactive. Neither of the PGs affected enzyme secretion. In saline-perfused cat pancreas, addition of either of these four PGs stimulated electrolyte secretion (Case and Sratcherd, 1972). In humans, 15(R)-15-methyl-PGE$_2$ methylester and 16,16-dimethyl-PGE$_2$ given orally (1.5 μg/kg) had no effect on bicarbonate secretion stimulated by secretin, but slightly increased secretion of pancreatic enzymes (protein, amylase, trypsin) (Konturek *et al.*, 1976c). The discrepancy between animal and human studies is unexplained.

VII. CYTOPROTECTION BY PROSTAGLANDINS

A. Intestine

1. Lesions Produced by Nonsteroidal Antiinflammatory Compounds (NOSACs)

Several PGs were found (Robert, 1974b,c, 1975a,c) to prevent the development of lesions of the small intestine produced in rats by administration of NOSACs such as indomethacin, flufenamic acid, ibuprofen, flurbiprofen, naproxen, carbazole, and phenylbutazone. These lesions, characterized by multiple ulcerations of the jejunum and the ileum, appear within 12 hr after treatment. After 3–4 days, the entire small intestine is transformed into a solid mass of adhesions due to the presence of hundreds of ulcerations that have perforated and caused peritonitis (Kent *et al.*, 1969). Abundant exudate is present in the abdominal cavity, and most animals die within 4 days. Several PG such as PGA$_1$, PGA$_2$, PGB$_2$, PGC$_2$, PGD$_2$, PGF$_{2\alpha}$, PGF$_{2\beta}$, and certain methyl analogues of PGA$_2$, PGE$_2$, PGF$_{2\alpha}$, and PGF$_{2\beta}$, prevented this syndrome (Robert, 1974b,c, 1975a,c). The term "cytoprotection"* was used to refer to this property of certain PGs to protect the cells of the intestinal epithelium against agents that would otherwise produce damage and eventual necrosis of these cells. Although these PGs were all cytoprotective, their potency varied greatly. In general, PGs of the E and A types were the most potent (Robert, 1974b,c, 1975a,c). The findings that (1) several NOSACs with unrelated chemical structures produce an identical syndrome, (2) these NOSACs are all inhibitors of PG synthetase (Takeguchi and Sih, 1972; Vane, 1971) and therefore

---

*Dr. Eugene D. Jacobson, University of Texas Medical School, coined the term "cytoprotection" while discussing these data with the author.

prevent the formation of PG in the body, and (3) administration of several PGs prevents NOSAC-induced intestinal lesions strongly support the hypotesis that this intestinal disease is due to a PG deficiency. By giving PGs to NOSAC-treated animals, an optimal tissue level of PGs was presumably maintained, with the result that no PG deficiency, and hence no lesions, developed.

It was found that a single oral treatment with 16,16-dimethyl-PGE_2, 30 min after oral treatment with indomethacin, was sufficient to prevent completely the development of the intestinal lesions (Robert, 1975c). This suggests that (1) a transient PG deficiency can be detrimental and even fatal in the rat and (2) replacement therapy during the few hours of PG deficiency (caused by a single administration of indomethacin) protects because it prevents such a PG depletion.

The intestinal microflora seems to play a necessary role in the development of NOSAC-induced intestinal lesions, as shown by the following observations: (1) Bacteremia develops within 3 days after treatment with indomethacin, at a time when severe intestinal lesions are present (Kent *et al.,* 1969; Robert *et al.,* 1976d). (2) Certain antibiotics prevent indomethacin-induced intestinal lesions (Kent *et al.,* 1969). (3) Indomethacin does not produce intestinal lesions in germfree rats (Robert and Asano, 1976).

2. Lesions Produced by Glucocorticoids

The agents against which PGs may protect are not limited to NOSACs; they may include microorganisms and their toxins, and other noxious chemicals present in the intestine. In this connection, cytoprotection by PGs was demonstrated in the case of intestinal lesions produced by prednisolone, an agent that does not block PG synthetase activity. Prednisolone administered to rats daily, subcutaneously at high doses (10 mg/rat) for 8 days, produced in 75% of animals severe necrotic lesions of the small intestine (Lancaster and Robert, 1977). Histologically, a localized area of the intestinal mucosa first became necrotic and the process progressed toward the serosa. The full thickness of the intestinal wall was then involved, with necrotic cells and extensive polymorphonuclear infiltration. Bacterial colonies were often prominent on the surface of and within the lesion. These necrotic foci were located away from the mesenteric attachment, either opposite to (the most frequent location) or midway between the mesenteric and antimesenteric portion of the gut; they seldom involved the full circle of the lumen. Conversely, indomethacin-induced lesions are localized at the mesenteric attachment only.

The prednisolone-induced lesions appeared to be infectious. They always started in the terminal ileum, usually for a distance of 3–6 cm, and in most cases were limited to this segment of the intestine, thus reminiscent of regional ileitis. In some cases, lesions were found throughout the ileum and in the distal jejunum, but this was uncommon. Pulmonary abscesses were also found in many animals, as was reported before (Selye, 1951).

Oral administration of 16,16-dimethyl-PGE$_2$ concomitantly with prednisolone treatment reduced the incidence of intestinal lesions, and the degree of protection was dose dependent (Figs. 5 and 6) (Lancaster and Robert, 1977). A dose of 300 μg/kg twice a day afforded complete protection. The incidence of lung infection was not reduced.

B. Stomach and Duodenum

The antiulcer property of certain PGs may be due in part to their antisecretory effect, but also to cytoprotection. PGs may increase the cellular resistance of the gastric and duodenal mucosa to the damaging

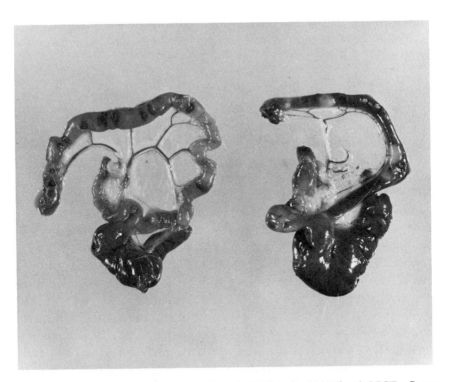

Fig. 5. Prevention of prednisolone-induced intestinal lesions by 16,16-dimethyl-PGE$_2$. Cecum and terminal ileum are shown. Left: Prednisolone, 10 mg/rat subcutaneously, once a day for 8 days. Multiple ileal lesions consisting of necrosis, ulcerations, and bacterial infiltration. Right: Same doses of prednisolone plus 16,16-dimethyl-PGE$_2$ 200 μg/kg orally twice a day. Complete prevention of the lesions.

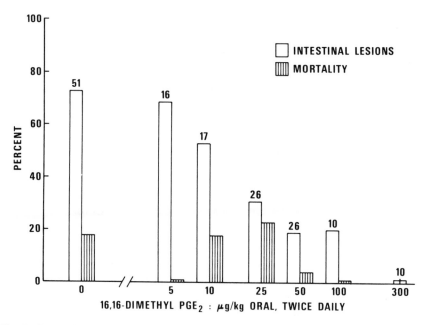

Fig. 6. Prevention of prednisolone-induced intestinal lesions in rats by 16,16-dimethyl-PGE$_2$: dose response. All groups received prednisolone, 10 mg/rat subcutaneously, once a day, for 8 days. The incidence of intestinal lesions and of mortality decreased with the dose of prostaglandin, given orally twice a day. Numbers on tops of columns indicate number of animals.

effect of gastric acid and pepsin and of other components of gastric juice. The prevention of indomethacin-induced gastric lesions by PGF$_{2\beta}$ (Robert, 1975a) favors this hypothesis, since PGF$_{2\beta}$ was not shown to inhibit gastric secretion. It is likely that this PG protected the stomach by preventing the breaking of the gastric mucosal barrier by indomethacin. Similar results were reported for two other PGs. First, 15(S)-15-methyl-PGE$_2$ methylester applied into a dog Heidenhain pouch protected the gastric mucosa from damage (increased permeability to hydrogen ions) produced by either aspirin or indomethacin (Cohen, 1975, 1976). Second, gastric erosive gastritis produced in rats by oral administration of bile was prevented by PGE$_2$ (1 mg/rat) given orally (Mann, 1975). While an antacid (Maalox) and a bile acid sequestrant (cholestyramine) reduced the numer of erosions, PGE$_2$ protected completely.

In other studies, gastric mucosal damage produced by aspirin was correlated with a reduction in PG formation by the gastric mucosa. In a dog preparation, instillation of aspirin on the gastric mucosa reduced by 40% the concentration of PGE in the gastric venous blood (Cheung *et al.*, 1974).

C. Mechanism of Cytoprotection

The mechanism by which certain PG are cytoprotective is unknown. Among the possibilities are the following:

1. PGs may act as trophic hormones for the small intestine.
2. The intestinal lesions may be related to water and electrolyte fluxes known to be increased after PG administration. In a PG-depleted animal, such as produced by treatment with a large dose of NOSAC, excessive water and electrolyte absorption from the intestine may develop and be manifested morphologically by necrotic areas throughout the small intestinal tract. If increased absorption does exist in these conditions, it may involve chemicals that are normally either not absorbed or absorbed in only small amounts. Their passage through the intestinal epithelium could produce cellular injury.
3. PGs may prevent the spreading of noxious agents from the intestinal lumen, such as microorganisms and/or their toxins, or irritating chemicals.

The presence of PGs inside the cells of the intestinal epithelium appears to be necessary to maintain cellular integrity. A depletion in PGs seems to render these cells sensitive to the unfavorable milieu of the intestinal lumen. It is to be noted, however, that prednisolone is not an inhibitor of PG synthetase. Therefore, the cytoprotection produced by 16,16-dimethyl-PGE_2 is not due to prevention of a PG deficiency, as may be the case for NOSACs such as indomethacin.

It appears, therefore, that certain PGs have the property of protecting the cellular epithelium of the small intestine against injury produced by either PG synthetase inhibitors, such as indomethacin, or noxious agents that act through a different mechanism, such as prednisolone. Both types of agents, however, produce intestinal mucosal necrosis and ulcerations that involve penetration and proliferation of microorganisms. By increasing the cellular resistance of the intestinal epithelium, a PG such as 16,16-dimethyl-PGE_2 prevents the development of a fatal infection. Not only did 16,16-dimethyl-PGE_2 prevent formation of intestinal lesions by indomethacin, it also prevented completely the development of bacteremia (Robert *et al.*, 1976*d*).

Whether cells other than those of the gastrointestinal epithelium require PGs for their resistance to noxious agents is unknown.

As practical implications of these findings, one can envisage the combination of certain PGs with a NOSAC such as aspirin or indomethacin in order to prevent the development of gastrointestinal side effects so often encountered during treatment with the latter. On the basis of a PG deficiency involved in the pathogenesis of certain inflammatory bowel diseases, certain PGs could be administered to patients with enteritis, Crohn's disease, or colitis to maintain cellular integrity of the intestinal mucosa, thus providing cytoprotection. Even if the PG tissue levels are normal, an increase in PG content, by exogenous administration, can still be cytoprotective. As an analogy, glucocorticoids are useful in the treatment of arthritis and asthma although there is no corticoid deficiency in these diseases.

VIII. REFERENCES

Adaikan, G. P., Salmon, J. A., Ng, B. K., and Karim, S. M. M, 1974, The effect of 15(R)-15-methyl PGE$_2$ methyl ester on serum gastrin levels in man, *Int. Res. Comm. Syst.* **2**:1587.

Al Awqati, Q., and Greenough, W. B., III, 1972, Prostaglandins inhibit intestinal sodium transport, *Nature (London) New Biol.* **238**:26.

Amer, M. S., 1972, Cyclic AMP and gastric secretion, *Am. J. Digest. Dis.* **17**:945.

Amer, M. S., 1974, Cyclic GMP and gastric secretion, *Am. J. Digest. Dis.* **19**:71.

Amy, J. J., Jackson, D. M., Ganesan, P. A., and Karim, S. M. M., 1973, Prostaglandin 15(R)-15-methyl-E$_2$ methyl ester for suppression of gastric activity in gravida at term, *Br. Med. J.* **4**:208.

Andersson, S., and Nylander, B., 1976, Local inhibitory action of 16,16-dimethyl PGE$_2$ on gastric acid secretion in the dog, in: *Advances in Prostaglandin and Thromoboxane Research*, p. 943, Raven Press, New York.

Banerjee, A. K., Phillips, J., and Winning, W. W., 1972, E-type prostaglandins and gastric acid secretion in the rat, *Nature (London) New Biol.* **238**:177.

Banerjee, A. K., Christmas, A. J., and Hall, C. E., 1975, Effects of H2-receptor antagonists, prostaglandins E$_2$ and 16,16-dimethyl ester on gastric acid secretion, mucosal blood flow and ulceration, *Abst. 6th Int. Congr. Pharmacol.*, p. 120.

Barrowman, J. A., Bennett, A., Hillenbrand, P., Rolles, K., Pollock, D. J., and Wright, J. T., 1975, Diarrhoea in thyroid medullary carcinoma: Role of prostaglandins and therapeutic effect of nutmeg, *Br. Med. J.* **3**:11.

Bartels, J., Kunze, H., Vogt, W., and Wille, G., 1970, Prostaglandin: Liberation from and formation in perfused frog intestine, *Naunyn-Schmiedebergs Arch. Pharmakol.* **266**:199.

Bennett, A., 1972, Effects of prostaglandins on the gastrointestinal tract, in: *The Prostaglandins: Progress in Research* (S. M. M. Karim, ed.), pp. 205–221, MTP Med. and Tech. Publ. Co., Oxford.

Bennett, A., 1973, Gastrointestinal disorders involving prostaglandins, *Med. Today* **7**:21.

Bennett, A., 1976, Prostaglandins as factors in diseases of the alimentary tract, in: *Advances in Prostaglandin and Thromoboxane Research*, p. 547, Raven Press, New York.

Bennett, A., and Posner, J., 1971, Studies on prostaglandin antagonists, *Br. J. Pharmacol.* **42**:584.

Bennett, A., Friedmann, C. A., and Vane, J. R., 1967, Release of prostaglandin E$_1$ from the rat stomach, *Nature (London)* **216**:873.

Bennett, A., Eley, K. G., and Scholes, G. B., 1968*a*, Effects of prostaglandins E$_1$ and E$_2$ on human, guinea-pig and rat isolated small intestine, *Br. J. Pharmacol.* **34**:630.

Bennett, A., Eley, K. G., and Scholes, G. B., 1968*b*, Effects of prostaglandins E$_1$ and E$_2$ on intestinal motility in the guinea-pig and rat, *Br. J. Pharmacol.* **34**:639.

Bennett, A., Murrary, J. G., and Wyllie, J. H., 1968*c*, Occurrence of prostaglandin E$_2$ in the human stomach, and a study of its effects on human isolated gastric muscle, *Br. J. Pharmacol. Chemother.* **32**:339.

Bennett, A., Stamford, I. F., and Unger, W. G., 1972, Prostaglandins and gastric secretion in man, *Adv. Biosci. (Suppl.)* **9**:43.

Bennett, A., Gradidge, C. F., and Stamford, I. F., 1974, Prostaglandins, nutmeg and diarrhea, *N. Engl. J. Med.* **290**:110.

Bergström, S., Duner, H., Von Euler, U. S., Pernow, B., and Sjövall, J., 1959*a*, Observations on the effects of infusion of prostaglandin E in man, *Acta Physiol. Scand.* **45**:145.

Bergström, S., Eliasson, R., Von Euler, U. S., and Sjövall, J., 1959*b*, Some biological effects of two crystalline prostaglandin factors, *Acta Physiol. Scand.* **45**:133.

Bergström, S., Carlson, L. A., and Weeks, J. R., 1968, The prostaglandins: A family of biologically active lipids, *Pharmacol. Rev.* **20**:1.

Bernier, J. J., Rambaud, J. C., Cattan, D., and Prost, A., 1969, Diarrhoea associated with medullary carcinoma of the thyroid, *Gut* **10**:980.

Bhana, D., Karim, S. M. M., Carter, D. C., and Ganesan, P. A., 1973, The effect of orally administered prostaglandins A_1, A_2 and 15 EPI-A_2 on human gastric acid secretion, *Prostaglandins* **3**:307.

Bieck, P. R., Oates, J. A., Robison, G. A., and Adkins, R. B., 1973, Cyclic AMP in the regulation of gastric secretion in dogs and humans, *Am. J. Physiol.* **224**:158.

Bigazzi, M., Casciano, S., Carini, L., Marzocca, U., and Zurli, A., 1973, Medullary carcinoma: Considerations about a case of familial pathogenesis, *J. Gerontol.* **21**:1049.

Brodie, D. A., and Chase, B. J., 1967, Role of gastric acid in aspirin-induced gastric irritation in the rat, *Gastroenterology* **53**:604.

Brodie, D. A., and Chase, B. J., 1969, Evaluation of gastric acid as a factor in drug-induced gastric hemorrhage in the rat, *Gastroenterology* **56**:206.

Butcher, R. W., 1970, Prostaglandins and cyclic AMP, *Adv. Biochem. Psychopharmacol.* **3**:173.

Carter, D. C., Karim, S. M. M., Bhana, D., and Ganesan, P. A., 1973, Inhibition of human gastric secretion by prostaglandin, *Br. J. Surg.* **60**:828.

Carter, D. C., Ganesan, P. A., Bhana, D., and Karim, S. M. M., 1974, The effect of locally administered prostaglandin 15(*R*) 15-methyl-E_2 methyl ester on gastric ulcer formation in the shay rat preparation, *Prostaglandins* **5**:455.

Case, R. M., and Scratcherd, T., 1972, Prostaglandin action on pancreatic blood flow and on electrolyte and enzyme secretion by exocrine pancreas *in vivo* and *in vitro*, *J. Physiol. (London)* **226**:393.

Charters, A. C., Brown, B. N., and Orloff, M. J., 1975, Metiamide and prostaglandin E_1 inhibition of pentagastrin stimulated acid secretion, *Gastroenterology* **68**:872.

Chawla, R. C., and Eisenberg, M. M., 1969*a*, Effect of prostaglandin E_1 on the motility of innervated antral pouches in dogs, *Clin. Res.* **17**:299.

Chawla, R. C., and Eisenberg, M. M., 1969*b*, Prostaglandin inhibition of innervated antral motility in dogs, *Proc. Soc. Exp. Biol. Med.* **132**:1081.

Cheung, L. Y., Jubiz, W., Torma, M. J., and Frailey, J., 1974, Effects of aspirin on canine gastric prostaglandin output and mucosal permeability, *Surg. Forum* **25**:407.

Cheung, L. Y., Jubiz, W., Moore, J. G., and Frailey, J., 1975, Gastric prostaglandin E (PGE) output during basal and stimulated acid secretion in normal subjects and patients with peptic ulcer, *Gastroenterology* **68**:873.

Christensen, J., and Lund, G. F., 1969, Esophageal responses to distension and electrical stimulation, *J. Clin. Invest.* **48**:408.

Classen, M., Koch, H., Deyhle, P., Weidenhiller, S., and Demling, L., 1970, The effect of prostaglandin E_1 on basal gastric secretion in humans, *Klin. Wochenschr.* **48**:876.

Classen, M., Koch, H., Bickhardt, J., Topf, G., and Demling, L., 1971, The effect of prostaglandin E_1 on the pentagastrin-stimulated gastric secretion in man, *Digestion* **4**:333

Coceani, F., Pace-Asciak, C., Volta, F., and Wolfe, L. S., 1967, Effect of nerve stimulation on prostaglandin formation and release from the rat stomach. *Am. J. Physiol.* **213**:1056.

Cohen, M. M., 1975, Prostaglandin E_2 prevents gastric mucosal barrier damage, *Gastroenterology* **68**:876.

Cohen, M. M., 1976, 15-Methyl prostaglandin E_2 protects gastric mucosal barrier, in: *Advances in Prostaglandin and Thromboxane Research,* p. 937, Raven Press, New York.

Cohen, S., and Lipshutz, W., 1971, Hormonal regulation of human lower esophageal sphincter competence: Interaction of gastrin and secretin, *J. Clin. Invest.* **50**:449.

Coupar, I. M., and McColl, I., 1972, Inhibition of glucose absorption by prostaglandins E_1, E_2 and $F_{2\alpha}$, *J. Pharm. Pharmacol.* **24**:254.

Cummings, J. H., Newman, A., Misiewicz, J. J., Milton-Thompson, G. J., and Billings, J. A., 1973, Effect of intravenous prostaglandin $F_{2\alpha}$ on small intestinal function in man, *Nature (London)* **243**:169.

Cummings, J. H., Milton-Thompson, G. J., Billings, J., Newman, A., and Misiewicz, J. J., 1974, Studies on the site of production of diarrhoea induced by prostaglandins, *Clin. Sci. Mol. Med.* **46**:15P.

Daturi, S., Franceschini, J., Mandelli, V., Mizzotti, B., and Usardi, M. M., 1974, A proposed role for PGE_2 in the genesis of stress-induced gastric ulcers, *Br. J. Pharmacol*, **52**:464P.

Declusin, R., Wall, M., and Whalen, G., 1974, Effect of prostaglandin E_2 (PGE_2) on sodium and chloride transport in rat jejunum, *Clin. Res.* **22**:635A.

Dilawari, J. B., Newman, A., Poleo, J., and Misiewicz, J. J., 1973, The effect of prostaglandins and of anti-inflammatory drugs on the oesophagus and the cardiac sphincter in man, *Gut* **14**:822.

Dilawari, J. B., Newman, A., Poleo, J., and Misiewicz, J. J., 1975, Response of the human cardiac sphincter to circulating prostaglandins F_{2A} and E_2 and to anti-inflammatory drugs, *Gut* **16**:137.

Dozois, R. R., and Wollin, A., 1975, Prostaglandin E_2, adenylate cyclase et secretion gastrique chez le chien, *Biol. Gastro-enterol.* **8**:122.

Eichhorn, J. H., Salzman, E. W., and Silen, W., 1974, Cyclic GMP response in vivo to cholinergic stimulation of gastric mucosa, *Nature* **248**:238.

Fawell, W. N., and Thompson, G., 1973, Nutmeg for diarrhea of medullary carcinoma of thyroid, *New. Engl. J. Med.* **289**:108.

Ferguson, W. W., Edmonds, A. W., Starling, J. R., and Wagensteen, S. L., 1973, Protective effect of prostaglandin E_1 (PGE_1) on lysosomal enzyme release in serotonin-induced gastric ulceration, *Ann. Surg* **177**:648.

Fleshler, B., and Bennett, A., 1969, Responses of human, guinea pig, and rat colonic circular muscle to prostaglandins, *J. Lab. Clin. Med.* **74**:872.

Fung, W. P., and Karim, S. M. M., 1974, Treatment of peptic ulcer pain with prostaglandin 15(R), 15-methyl-E_2 methyl ester, *J. Int. Res. Commun.* **2**:1001.

Fung, W. P., Karim, S. M. M., and Tye, C. Y., 1974*a*, Double-blind trial of 15(R)-15-methyl prostaglandin E_2 methyl ester in the relief of peptic ulcer pain. *Ann. Acad. Med.* **3**:375.

Fung, W. P., Karim, S. M. M., and Tye, C. Y., 1974*b*, Effect of 15(R)-15 methyl prostaglandin in E_2 methyl ester on healing of gastric ulcers: Controlled endoscopic study, *Lancet* **2**:10.

Fung, W. P., Lee, S. K., and Karim, S. M. M., 1974*c*, Effect of prostaglandin 15(R)-15-methyl-E_2-methyl ester on the gastric mucosa in patients with peptic ulceration—an endoscopic and histological study, *Prostaglandins* **5**:465.

Gibinski, K., Rybicka, J., Mikos, E., and Nowak, A., 1976, Anti-ulcer effects of methyl analogs of prostaglandin E_2 in man, *10th Int. Cong. Gastroenterol. Budapest,* June 23–27.

Goyal, R. K., and Rattan, S., 1973, Mechanism of the lower esophageal sphincter relaxation action of prostaglandin E_1 and theophylline, *J. Clin. Invest.* **52**:337.

Goyal, R. K., Rattan, S., and Hersh, T., 1973, Comparison of the effects of prostaglandins E_1, E_2, prostaglandins on the lower esophageal sphincter, *Clin. Res.* **20**:454.

Goyal, R. K., Rattan, S., and Hersh, T., 1973, Comparison of the effects of prostaglandins E_1, E_2, and A_2, and of hypovolumic hypotension on the lower esophageal sphincter, *Gastroenterology* **65**:608.

Goyal, R. K., Mukhopadhyay, A., and Rattan, S., 1974, Effect of prostaglandin E_2 on the lower esophageal sphincter in normal subjects and patients with achalasia, *Clin. Res.* **22**:358A.

Grahame-Smith, D. G., 1973, Endocrine tumors producing gastrointestinal symptoms, *Int. Encycl. Pharmacol. Ther.* **2**:639.

Greenough, W. B., Pierce, N. F., A1 Awqati, Q., and Carpenter, C. C. J., 1969, Stimulation of gut electrolyte secretion by prostaglandins, theophylline and cholera exotoxin, *J. Clin. Invest.* **48**:32A.

Grossman, M. I., Wooley, J. R., Dutton, D. F., and Ivy, A. C., 1945, The effect of nausea on gastric secretion and a study of the mechanism concerned, *Gastroenterology* **4**:347.

Hahn. R. A., 1972, An investigation into the interaction of prostaglandin $F_{2\alpha}$ with cholinergic mechanisms in canine salivary glands, dissertation, Ohio State University.

Hahn, R. A., and Patil, P. N., 1972, Salivation induced by prostaglandin $F_{2\alpha}$ and modification of the response by atropine and physotigmine, *Br. J. Pharmacol.* **44**:527.

Hahn, R. A., and Patil, P. N., 1974, Further observations on the interaction of prostaglandin $F_{2\alpha}$ with cholinergic mechanisms in canine salivary glands, *Eur. J. Pharmacol.* **25**:279.

Hakanson, R., Liedberg, G., and Oscarson, J., 1973, Effects of prostaglandin E_1 on acid secretion, mucosal histamine content and histidine decarboxylase activity in rat stomach, *Br, J. Pharmacol.* **47**:498.

Harris, J. B., and Alonso, D., 1965, Stimulation of the gastric mucosa by adenosine-3,5′-monophosphate, *Fed. Proc.* **24**:1368.

Harris, J. B., Nigon, K., and Alonso, D., 1969, Adenosine-3′,5′-monophosphate: Intracellular mediator for methylxanthine stimulation of gastric secretion, *Gastroenterology* **57**:377.

Hinsdale, J. G., Engel, J. J., and Wilson, D. E., 1974, Prostaglandin E in peptic ulcer disease, *Prostaglandins* **6**:495.

Hohnke, L. A., 1974, Interaction of dibutyryl cyclic AMP and PGE_2 on stimulated gastric acid secretion in rats, *Fed. Proc.* **33**:329.

Horton, E. W., and Jones, R. L., 1969, The biological assay of prostaglandins A_1 and A_2, *J. Physiol. (London)* **200**:56P.

Horton, E. W., Main, I. H. M., Thompson, C. J., and Wright, P. M., 1968, Effect of orally administered prostaglandin E_1 on gastric secretion and gastrointestinal motility in man, *Gut* **9**:655.

Hudson, N., Hindi, E. L. S., Wilson, D. E., and Poppe, L., 1975, Prostaglandin E in cholera toxin-induced intestinal secretion—Lack of an intermediary role, *Am. J. Digest. Dis.* **20**:1035.

Impicciatore, M., Bertaccini, G., and Usardi, M. M., 1976, Effect of a new synthetic prostaglandin on acid gastric secretion in different laboratory animals, in: *Advances in Prostaglandin and Thromboxane Research.*, p.945, Raven Press, New York.

Ippoliti, A. F., Isenberg, J. I., Maxwell, V. J., and Walsh, J. H., 1976, The effect of 16,16-dimethyl PGE_2 on meal-stimulated gastric acid secretion and serum gastrin in duodenal ulcer patients, *Gastroenterology* **70**:488.

Jacobson, E. D., 1970, Comparison of prostaglandin E_1 and norepinephrine on the gastric mucosal circulation, *Proc. Soc. Exp. Biol. Med.* **133**:516.

Kaminski, D., and Jellinek, M., 1974, The effect of prostaglandins A_1 (PGA_1) and E_1 (PGE_1) on canine hepatic bile flow, *Clin. Res.* **22**:361A.

Kaminski, D. L., Ruwart, M., and Willman, V. L., 1975, The effect of prostaglandin A_1 and E_1 on canine hepatic bile flow, *J. Surg. Res.* **18**:391.

Kaminski, D. L., Ruwart, M. J., and Willman, V. L., 1976, The effect of prostaglandin A_1 on canine hepatic bile flow, in: *Advances in Prostaglandin and Thromboxane Research,* p. 940, Raven Press, New York.

Kaplan, E. L., Saxena, N., and Peskin, G. W., 1970, Prostaglandins: A possible mediator of diarrhea in endocrine syndromes, *Surg. Forum* **21**:94.

Karim, S. M. M., and Fung, W. P., 1975, Effect of 15(R) 15-methyl prostaglandin E_2 on gastric secretion and a preliminary study on the healing of gastric ulcers in man, *Int. Res. Commun. Syst. Med. Sci.* **3**:348.

Karim, S. M. M., and Fung, W. P., 1976, Effects of some naturally occurring prostaglandins and synthetic analogs on gastric secretion and ulcer healing in man, in: *Advances in Prostaglandin and Thromboxane Research,* p. 529, Raven Press, New York.

Karim, S. M. M., Carter, D. C., Bhana, D., and Ganesan, P. A., 1973*a*, Effect of orally and intravenously administered prostaglandin 15(R)-15-methyl E_2 on gastric secretion in man, *Adv. Biosci.* **9**:255.

Karim, S. M. M., Carter, D. C., Bhana, D., and Ganesan, P. A., 1973*b*, Effect of orally administered prostaglandin E_2 and its 15-methyl analogues on gastric secretion, *Br. Med. J.* **1**:143.

Karim, S. M. M., Carter, D. C., Bhana, D., and Ganesan, P. A., 1973*c*, Inhibition of basal and pentagastrin induced gastric acid secretion in man with prostaglandin 16:16 dimethyl E_2 methyl ester, *Int. Res. Commun. Syst.,* (73-3) 8-3-2.

Karim, S. M. M., Carter, D. C., Bhana, D., and Ganesan, P. A., 1973*d*, The effect of orally and intravenously administered prostaglandin 16:16 dimethyl E_2 methyl ester on human gastric acid secretion, *Prostaglandins* **4**:71.

Kawarada, Y., Lambek, J., and Matsumoto, T., 1975, Pathophysilogy of stress ulcer and its prevention. II. Prostaglandin E_1 and microcirulatory responses in stress ulcer, *Am. J. Surg.* **129**:217.

Kent, T. H., Cardelli, R. M., and Stamler, F. W., 1969, Small intestinal ulcers and intestinal flora in rats given indomethacin, *Am. J. Pathol.* **54**:237.

Kimberg, D. V., Field, M., Johnson, J., Henderson, A., and Gershon, E., 1971, Stimulation of intestinal mucosal adenyl cyclase by cholera enterotoxin and prostaglandins, *J. Clin. Invest.* **50**:1218.

Konturek, S. J., Radecki, T., Demitrescu, T., Kwiecien, N., Pucher, A., and Robert, A., 1974, Effect of synthetic 15-methyl analog of prostaglandin E₂ on gastric secretion and peptic ulcer formation, *J. Lab. Clin. Med.* **84**:716.

Konturek, S. J., Kwiecen, N., Swierczek, J., Sito, E., Oleksy, J., and Robert, A., 1976*a*, Inhibition of pentagastrin-induced gastric secretion by orally administered 15-methyl analogs of PGE₂ in man, in: *Advances in Prostaglandin and Thromboxane Research,* p. 944, Raven Press, New York.

Konturek, S. J., Kwiecien, N., Swierczek, J., Oleksy, J., Sito, E., and Robert, A., 1976*b*, Comparison of methylated prostaglandin E₂ analogs given orally in the inhibition of gastric responses to pentagastrin and peptone meal in man, *Gastroenterology* **70**:683.

Konturek, S. J., Kwiecien, N., Swierczek, J., and Oleksy, J., 1976*c*, Effect of methylated PGE₂ analogs given orally on pancreatic response to secretion in man, *Am. J. Digest. Dis.* (in press).

Kowalewski, K., and Kolode, J. A., 1974, Effect of prostaglandin-E₂ on gastric secretion and on gastric circulation of totally isolated ex vivo canine stomach, *Pharmacology* **11**:85.

Krarup, N., Larsen, J. A., and Munck, A., 1975, Choleretic effect of prostaglandin PGE₁ and PGE₂ in cats, *Digestion* **12**:272.

Labrum, A. H., Lipkin, M., and Dray, F., 1976, Hyperprostaglandinemia: A previously unrecognized syndrome, in: *Advances in Prostaglandins and Thromboxane Research* (B. Samuelsson and R. Paoletti, eds.) p. 88, Raven Press, New York.

Lancaster, C., and Robert, A., 1977, Intestinal lesions produced by prednisolone: prevention by 16,16-dimethyl PGE₂, *Fed. Proc.* (in press).

Lauterburg, B., Paumgartner, G., and Preisig, R., 1975, Prostaglandin-induced choleresis in the rat, *Experientia* **31**:1191.

Lee, Y. H., and Bianchi, R. G., 1972, The anti-secretory and anti-ulcer activity of a prostaglandin analog, SC-24665, in experimental animals, *Abst. 5th Int. Congr. Pharmacol. (San Francisco),* p. 136.

Lee, Y. H., Cheng, W. D., Bianchi, R. G., Mollison, K., and Hansen, J., 1973, Effects of oral administration of PGE₂ on gastric secretion and experimental peptic ulcerations, *Prostaglandins* **3**:29.

Levine, R., 1971, Effect of prostaglandins and cyclic AMP on gastric secretion, *Ann. N.Y. Acad. Sci.* **180**:336.

Levine, R. A., and Wilson, D. E., 1971, The role of cyclic AMP in gastric secretion, *Ann. N.Y. Acad. Sci.* **185**:363.

Levine, R. A., Cafferata, E. P., and McNally, E. F., 1967, Inhibitory effect of adenosine-3′,5′-monophosphate on gastric secretion and gastrointestinal motility *in vivo., Third World Cong. Gastroenterol.* **1**:408.

Lippmann, W., 1970, Inhibition of gastric acid secretion by a potent synthetic prostaglandin, *J. Pharm. Pharmacol.* **22**:65.

Lippmann, W., 1971, Inhibition of gastric acid secretion in the rat by synthetic prostaglandin analogues, *Ann. N.Y. Acad. Sci.* **180**:332.

Lippmann, W., and Seethaler, K., 1973, Oral anti-ulcer activity of a synthetic prostaglandin analogue (9-oxoprostanoic acid: AY-22, 469), *Experientia* **29**:993.

Lynch, A., Shaw, H., and Milton, G. W., 1964, Effect of aspirin on gastric secretion, *Gut* **5**:230.

Magerlein, B. J., Ducharme, D. W., Magge, W. E., Miller, W. L., Robert, A., and Weeks, J. R., 1973, Synthesis and biological properties of 16-alkylprostaglandins, *Prostaglandins* **4**:143.

Main, I. H. M., 1969, Effects of prostaglandin E₂ (PGE₂) on the output of histamine and acid in rat gastric secretion induced by pentagastrin or histamine, *Br. J. Pharmacol.* **36**:214P.

Main, I. H. M., 1973, Prostaglandins and the gastro intestinal tract, in: *The Prostaglandins: Pharmacological and Therapeutic Advances,* (M. F. Cuthbert, ed.), pp. 287–323, Heinemann, London.

Main, I. H. M., and Whittle, B. J. R., 1972*a*, Effects of prostaglandins of the E and A series on rat gastric mucosal blood flow as determined by ¹⁴C-aniline clearance, *Abst. 5th Int. Congr. Pharmacol. (San Francisco),* p. 145.

Main, I. H. M., and Whittle, B. J. R., 1972*b*, Effects of prostaglandin E₂ on rat gastric muscosal blood flow, as determined by ¹⁴C-aniline clearance, *Br. J. Pharmacol.* **44**:331P.

Main, I. H. M., and Whittle, B. J. R., 1973*a*, The relationship between rat gastric mucosal blood flow and acid secretion during oral or intravenous administration of prostaglandins and dibutyryl cyclic AMP, *Adv. Biosci.* **9**:271.

Main, I. H. M., and Whittle, B. J. R., 1973*b*, The effects of E and A prostaglandins on gastric mucosal blood flow and acid secretion in the rat, *Br. J. Pharmacol.* **49**:428.

Main, I. H. M., and Whittle, B. J. R., 1974*a*, Prostaglandin E₂ and the stimulation of rat gastric acid secretion by dibutyryl cyclic 3′,5′-AMP, *Eur. J. Pharmacol.* **26**:204.

Main, I. H. M., and Whittle, B. J. R., 1974*b*, Failure of burimamide and (15*S*)-15-methyl prostaglandin E₂ to prevent histamine-induced gastric mucosal hyperaemia in the rat, *J. Physiol. (London)*, **239**:118P.

Main, I. H. M., and Whittle, B. J. R., 1975, Investigation of the vasodilator and anti-secretory role of prostaglandins in the rat gastric mucosa by use of non-steroidal anti-inflammatory drugs, *Br. J. Pharmacol.* **53**:217.

Mann, N. S., 1975, Prevention of bile induced acute erosive gastritis by prostaglandin E₂, maalox and cholestryamine, *Gastroenterology* **68**:946.

Mao, C. C., Shanbour, L. L., Hodgins, D. S., and Jacobson, E. D., 1972, Cyclic adenosine-3′,5′-monophosphate (cyclic AMP) and secretion in the canine stomach, *Gastroenterology* **63**:427.

Matuchansky, C., and Bernier, J. J., 1971, Effects of prostaglandin E₁ on net and unidirectional movements of water and electrolytes across jejunal mucosa in man, *Gut* **12**:854.

Matuchansky, C., and Bernier, J. J., 1973*a*, Effect of prostaglandin E₁ on glucose, water, and electrolyte absorption in the human jejunum, *Gastroenterology* **64**:1111.

Matuchansky, C., and Bernier, J. J., 1973*b*, Effects of prostaglandin E₁ on jejunal absorption in man, *Digestion* **9**:86.

Matuchansky, C., and Bernier, J. J., 1973*c*, General review: Prostaglandins and the digestive tract, *Biol. Gastroenterol. (Paris)* **6**:251.

Matuchansky, C., Mary, J. Y., and Bernier, J. J., 1972, Effect of prostaglandin E₁ on glucose absorption and on transintestinal movements of water and electrolytes in human jejunum, *Biol. Gastroenterol. (Paris)* **5**:636C.

McCallum, R. W., Ippoliti, A. F., and Sturdevant, R. A. L., 1975, Effect of oral 16,16-dimethyl prostaglandin E₂ (PGE₂) on the lower esophageal sphincter in man, *Gastroenterology* **68**:949.

Mihas, A. A., Gibson, R., and Hirschowitz, B. I., 1975, Inhibition of gastric secretion in dogs by a synthetic prostaglandin (PGE₂), *Fed. Proc.* **34**:442.

Mihas, A. A., Gibson, R., and Hirschowitz, B. I., 1975, Inhibition of gastric secretion in dogs by a dog by 16,16-dimethyl prostaglandin E₂, *Am. J. Physiol.* **230**:351.

Misiewicz, J. J., Waller, S. L., Kiley, N., and Horton, E. W., 1969, Effect of oral prostaglandin E₁ on intestinal transit in man, *Lancet* **1**:648.

Miyazaki, Y., 1968*a*, Isolation of prostaglandin E like substances from the mucous membrane layer of large intestine of pig, *Sapporo Med. J.* **34**:141.

Miyazaki, Y., 1968*b*, Occurrence of prostaglandin E₁ in the mucous membrane layer of swine large intestine, *Sapporo Med. J.* **34**:321.

Miyazaki, Y., 1969, On the extraction of a prostaglandin E-like substance from the large intestine of the pig, in: *The Prostaglandins* (Symp., Kyoto), pp. 22–24, Ono Pharmaceutical Co., Osaka.

Mize, B. F., Wu, W. C., and Whalen, G. E., 1974, The effect of prostaglandin E₂ (PGE₂) on net jejunal transport and mean transit time, *Gastroenterology* **66**:A-93/747.

Murai, S., Mizushima, N., Nukaga, A., Komaya, S., and Sano, T., 1970, Effect of prostaglandin E₁ on the acidity of gastric juices, *Gastroentrol. Japon.* **5**:274.

Nakano, J., and Prancan, A. V., 1972, Effect of prostaglandins E₁ and A₁ on the gastric circulation in dogs, *Proc. Soc. Exp. Biol. Med.* **139**:1151.

Newman, A., Milton-Thompson, G., Cummings, J. H., Billings, J. A., and Misiewicz, J. J., 1974, Differential response of the human small and large intestine to prostaglandin F₂α, *Gastroenterology* **66**:A-100/754.

Newman, A., De Moraes-Fil, J. P. P., Philippakos, D., and Misiewicz, J. J., 1975, The effect of intravenous infusions of prostaglandins E₂ and F₂α on human gastric function, *Gut* **16**:272.

Nezamis, J. E., Robert, A., and Stowe, D. F., 1971, Inhibition by prostaglandin E₁ of gastric secretion in the dog, *J. Physiol. (London)* **218**:369.

Nicoloff, D. M., 1968, Indomethacin effect on gastric secretion: Parietal cell population and ulcer provocation in the dog, *Arch. Surg.* **97**:809.

Nylander, B., and Andersson, S., 1974, Gastric secretory inhibition induced by three methyl analogs of prostaglandin E₂ administered intragastrically to man, *Scand. J. Gastroenterol.* **9**:751.

Nylander, B., and Andersson, S., 1975, Effect of two methylated prostaglandin E₂ analogs on gastroduodenal pressure in man, *Scand. J. Gastroenterol.* **10**:91.

Nylander, B., and Mattsson, O., 1975, Effect of 16,16-dimethyl PGE₂ on gastric emptying and intestinal transit of a barium-food test meal in man, *Scand. J. Gastroenterol.* **10**:289.

Nylander, B., Robert, A., and Andersson, S., 1974, Gastric secretory inhibition by certain methyl analogs of prostaglandin E₂ following intestinal administration in man, *Scand. J. Gastroenterol.* **9**:759.

O'Brien, P. E., and Carter, D. C., 1975, Effect of gastric secretory inhibitors on the gastric mucosal barrier, *Gut* **16**:437.

Pace-Asciak, C., Morawska, K., Coceani, F., and Wolfe, L. S., 1968, The biosynthesis of prostaglandins E₂ and F₂α in homogenates of the rat stomach, in: *Prostaglandin Symposium of Worcester Foundation for Experimental Biology,* (P. W. Ramwell and J. E. Shaw, eds.), pp. 371–378, Interscience, New York.

Parkin, J. V., Faber, R. G., and Hobsley, M., 1974, Synthetic prostaglandin in gastric ulcer, *Lancet* **2**:161,

Perrier, C. V., and Laster, L., 1970, Adenyl cyclase activity of guinea pig gastric mucosa: Stimulation by histamine and prostaglandins, *J. Clin. Invest.* **49**:73A.

Pierce, N. F., Carpenter, C. C. J., Jr., Elliot, H. L., and Greenough, W. B., III, 1971, Effects of prostaglandins, theopylline, and cholera exotoxin upon transmucosal water and electrolyte movement in the canine jejunum, *Gastroenterology* **60**:22.

Rashid, S., 1971, The release of prostaglandin from the oesophagus and the stomach of the frog *(Rana temporaria)*, *J. Pharm. Pharmacol.* **23**:456.

Rattan, S., Hersh, T., and Goyal, R. K., 1972, Effect of prostaglandin F₂α and gastrin pentapeptide on the lower esophageal sphincter, *Proc. Soc. Exp. Biol. Med.* **141**:573.

Reeder, D. D., Becker, H. D., and Thompson, J. C., 1972, Effect of prostaglandin E₁ on food stimulated gastrin and gastric secretion in dogs, *Physiologist* **15**:246.

Robert, A., 1968, Anti-secretory property of prostaglandins, in: *Prostaglandin Symposium of Worcester Foundation for Experimental Biology* (P. W. Ramwell and J. E. Shaw, eds.), pp. 47–54, Interscience, New York.

Robert, A., 1971, Duodenal ulcers in the rat: Production and prevention, in: *Peptic Ulcer* (C. J. Pfeiffer, ed.), pp. 21–33, Munksgaard, Copenhagen.

Robert, A., 1973, Prostaglandins and the digestive system, in: *Prostaglandins 1973,* pp. 297–315, Inserm, Paris.

Robert, A., 1974*a*, Prevention by prostaglandins of duodenal ulcers produced experimentally in rats, *Fifth World Congr. Gastroenterol. (Mexico City),* p. 123, October 13–18.

Robert, A., 1974*b*, An intestinal disease in the rat probably caused by a prostaglandin deficiency, *Gastroenterology* **66**:A-111/765.

Robert, A., 1974*c*, Effects of prostaglandins on the stomach and the intestine, *Prostaglandins* **6**:523.

Robert, A., 1975*c*, An intestinal disease produced experimentally by a prostaglandin deficiency, *Gastroenterology* **69**:1045.

Robert, A., 1975*a*, Anti-secretory, anti-ulcer, cytoprotective and diarrheogenic properties of prostaglandins, in: *Advances in Prostaglandin and Thromboxane Research,* p. 507, Raven Press, New York.

Robert, A., 1975*b*, A quantitative assay for diarrhea produced by prostaglandins, in: *Advances in Prostaglandin and Thromboxane Research,* p. 947, Raven Press, New York.

Robert, A., and Asano, T., 1976, Resistance of germ-free rats to intestinal lesions produced by indomethacin, *10th Int. Congr. Gastroenterol. Budapest,* June 23–29.

Robert, A., and Magerlein, B. J., 1973, 15-Methyl PGE₂ and 16,16-dimethyl PGE₂: Potent inhibitors of gastric secretion, *Adv. Biosci.* **9**:247.

Robert, A., and Standish, W. L., 1973, Production of duodenal ulcers, in rats, with one injection of histamine, *Fed. Proc.* **32**:322.

Robert, A., and Yankee, E. W., 1975, Gastric anti-secretory effect of 15(*R*)-15-methyl PGE₂, methyl ester and of 15(*S*)-15-methyl PGE₂, methyl ester, *Proc. Soc. Exp. Biol. Med.* **148**:1155.

Robert, A., Nezamis, J. E., and Phillips, J. P., 1967, Inhibition of gastric secretion by prostaglandins, *Am. J. Digest. Dis.* **12**:1073.

Robert, A., Nezamis, J. E., and Phillips, J. P., 1968*a*, Effect of prostaglandin E₁ on gastric secretion and ulcer formation in the rat, *Gastroenterology* **55**:481.

Robert, A., Phillips, J. P., and Nezamis, J. E., 1968*b*, Inhibition by prostaglandin E₁ on gastric secretion in the dog, *Gastroenterology* **54**:1263.

Robert, A., Stowe, D. F., and Nezamis, J. E., 1971, Prevention of duodenal ulcers by administration of prostaglandins E₂ (PGE₂), *Scand. J. Gastroenterol.* **6**:303.

Robert, A., Lancaster, C., Nezamis, J. E., and Badalamenti, J. N., 1973, A gastric anti-secretory and anti-ulcer prostaglandin with oral and long-acting activity, *Gastroenterology* **64**:790.

Robert, A., Nylander, B., and Andersson, S., 1974, Marked inhibition of gastric secretion by two prostaglandin analogs given orally to man, *Life Sci.* **14**:533.

Robert, A., Nezamis, J. E., Lancaster, C., and Badalamenti, J. N., 1975*a*, Cysteamine-induced duodenal ulcers: A new model to test anti-ulcer agents, *Digestion* **11**:199.

Robert, A., Nezamis, J. E., and Lancaster, C., 1975*b*, Duodenal ulcers produced in rats by propionitrile: Factors inhibiting and aggravating such ulcers, *Toxicol. Appl. Pharmacol.* **31**:201.

Robert, A., Lancaster, C., and Nezamis, J. E., 1976*a*, Inhibition of gastric secretion after intravaginal administration of prostaglandins, in: *Advances in Prostaglandin and Thromboxane Research,* p. 946, Raven Press, New York.

Robert, A., Nezamis, J. E., Hanchar, A. J., Lancaster, C., and Klepper, M. S., 1976*b*, The enteropooling assay to test for diarrhea due to prostaglandins, *Fed. Proc.* **35**:457.

Robert, A., Schultz, J. R., Nezamis, J. E., and Lancaster, C., 1976*c*, Gastric anti-secretory and anti-ulcer properties of PGE₂, 15-methyl PGE₂, and 16, 16-dimethyl PGE₂: Intravenous, oral and intrajejunal administration, *Gastroenterology* **70**:359.

Robert, A., Sokolski, W. T., and Nezamis, J. E., 1976*d*, Prevention of bacteremia and anemia by 16,16-dimethyl PGE₂, unpublished.

Robert, A., Nezamis, J. E., Lancaster, C., Hanchar, A. J., and Klepper, M. S., 1976*e*, Enteropooling assay: A test for diarrhea produced by prostaglandins, *Prostaglandins* **11**:809.

Robert, A., Nezamis, J. E., and Lancaster, C., 1976*f*. Effect of 16,16-dimethyl PGE₂ on gastric emptying, in: *Advances in Prostaglandin and Thromboxane Research,* p. 946, Raven Press, New York.

Rosenberg, V., Robert, A., Gonda, M., Dreiling, D. A., and Rudick, J., 1974, Synthetic prostaglandin analog and pancreatic secretion, *Gastroenterology* **66**:A-113/767.

Rudick, J., Gonda, M., and Janowitz, H. D., 1970, Prostaglandin E₁: An inhibitor of electrolyte and stimulant of enzyme secretion in the pancreas, *Fed. Proc.* **29**:445.

Rudick, J., Gonda, M., Dreiling, D. A., and Janowitz, H. D., 1971, Effects of prostaglandin E₁ on pancreatic exocrine function, *Gastroenterology* **60**:272.

Salmon, J. A., Karim, S. M. M., Carter, D. C., Ganesan, P. A., and Bhana, D., 1975, Effect of 15(*R*) 15-methyl prostaglandin E₂ methyl ester on basal and pentagastrin stimulated pepsin secretion in man, *Int. Res. Commun. Syst. Med. Sci.* **3**:83.

Sandler, M., Karim, S. M. M., and Williams, E. D., 1968, Prostaglandins in amine-peptide-secreting tumours, *Lancet* **2**:1053.

Sandler, M., Williams, E. D., and Karim, S. M. M., 1969, The occurrence of prostaglandins in amine-peptide-secreting tumours, in: *Prostaglandins, Peptides and Amines* (P. Mantegazza and E. W. Horton, eds.) pp. 3–7, Academic Press, London.

Schlegel, W., Wenk, K., and Dollinger, H. C., 1976, Mucosal concentrations of prostaglandin PGA, PGE, PGF in gastric diseases, *V Int. Congr. Endocrinol. (Hamburg),* July 18–24.

Schlemmer, R. F., Farnsworth, N. R., Cordell, G. A., and Bederka, J. P., 1973, Nutmeg pharmacognosy, *N. Engl. J. Med.* **289**:922.

Selye, H., 1951, In: *First Annual Report on Stress,* p. 374, Acta Inc. Medical Publishers, Montreal.

Shaw, J. E., and Ramwell, P. W., 1968, Inhibition of gastric secretion in rats by prostaglandin E₁, in: *Prostaglandin Symp. of Worcester Found. for Exp. Biology,* (P. W. Ramwell and J. E. Shaw, eds.), pp. 55–66, Interscience, New York.

Shaw, J. E., and Ramwell, P. W., 1969, Direct effect of prostaglandin E$_1$ on the frog gastric mucosa, *Abst. 4th Int. Congr. Pharmacol. (Basel),* pp. 109–110.

Shaw, J. E., and Urquhart, J., 1972, Parameters of the control of acid secretion in the isolated blood-perfused stomach, *J. Physiol. (London)* **226**:107P.

Shehadeh, Z., Price, W. E., and Jacobson, E. D., 1969, Effects of vasoactive agents on intestinal blood flow and motility in the frog, *Am. J. Physiol.* **216**:386.

Splawinski, J. A., Nies, A. S., Bieck, P. R., and Oates, J. A., 1971, Mechanism of the contraction induced by arachidonic acid (AA) on the rat stomach longitudinal muscle strip, *Pharmacologist* **13**:291.

Stephens, F. O., Milton, G. W., and Loewenthal, J., 1966, Effect of aspirin on explanted gastric mucosa, *Gut* **7**:223.

Taira, N., Narimatsu, A., and Himori, N., 1975, Mode of actions of prostaglandins E$_2$, F$_{1\alpha}$ and F$_{2\alpha}$ in the dog salivary gland, *Abst. 6th Int. Congr. Pharmacol.,* p. 159.

Takeguchi, C., and Sih, C. J., 1972, A rapid spectrophotometric assay for prostaglandin synthetase: Application to the study of nonsteroidal anti-inflammatory agents, *Prostaglandins* **2**:169.

Tonnesen, M. G., Jubiz, W., Moore, J. G., and Frailey, J., 1974, Circadian variation of prostaglandin E (PGE) production in human gastric juice, *Am. J. Digest. Dis.* **19**:644.

Vanasin, B., Greenough, W., and Schuster, M. M., 1970, Effect of prostaglandin (PG) on electrical and motor activity of isolated colonic muscle, *Gastroenterology* **58**:1004.

Vane, J. R., 1971, Inhibition of prostaglandin synthesis as a mechanism of action for aspirin-like drugs, *Nature (London) New Biol.* **231**:232.

Wada, T., and Ishizawa, M., 1970, Effects of prostaglandin on the function of the gastric secretion, *Jap. J. Clin. Med.* **28**:2465.

Waitzman, M. B., and King, C. D., 1967, Prostaglandin influences on intraocular pressure and pupil size, *Am. J. Physiol.* **212**:329.

Waller, S. L., 1973, Prostaglandins and the gastrointestinal tract, *Gut* **14**:402.

Way, L., and Durbin, R. P., 1969, Inhibition of gastric acid secretion in vitro by prostaglandin E$_1$, *Nature (London)* **221**:874.

Weeks, J. R., Sekhar, N. C., and Ducharme, D. W., 1969, Relative activity of prostaglandins E$_1$, A$_1$, E$_2$ and A$_2$ on lipolysis, platelet aggregation, smooth muscle and the cardiovascular system, *J. Pharm. Pharmacol.* **21**:103.

Whittle, B. J. R., 1972, Studies on the mode of action of cyclic 3′,5′-AMP and prostaglandin E$_2$ on rat gastric acid secretion and mucosal blood flow, *Br. J. Pharmacol.* **46**:546P.

Whittle, B. J. R., 1976, Gastric anti-secretory and anti-ulcer activity of prostaglandin E$_2$ methyl analogs, in: *Advances in Prostaglandin and Thromboxane Research,* p. 948, Raven Press, New York.

Williams, E. D., Karim, S. M. M., and Sandler, M., 1968, Prostaglandin secretion by medullary carcinoma of the thyroid, *Lancet* **1**:22.

Wilson, D. E., 1974, Prostaglandins: Their actions on the gastrointestinal tract, *Arch. Intern. Med.* **133**:112.

Wilson, D. E., and Levine, R. A., 1969, Decreased canine gastric mucosal blood flow induced by prostaglandin E$_1$: A mechanism for its inhibitory effect on gastric secretion, *Gastroenterology* **56**:1268.

Wilson, D. E., and Levine, R. A., 1972, The effect of prostaglandin E$_1$ on canine gastric acid secretion and gastric mucosal blood flow, *Am. J. Digest. Dis.* **17**:527.

Wilson, D. E., Phillips, C., and Levine, R. A., 1970, Inhibition of gastric secretion in man by prostaglandin A$_1$ (PGA$_1$), *Gastroenterology* **58**:1007.

Wilson, D. E., Phillips, C., and Levine, R. A., 1971, Inhibition of gastric secretion in man by prostaglandin A$_1$, *Gastroenterology* **61**:201.

Wilson, D. E., El-Hindi, S., Tao, P., and Poppe, L., 1975a, Effects of indomethacin on intestinal secretion, prostaglandin E and cyclic AMP: Evidence against a role for prostaglandins in cholera toxin-induced secretion, *Prostaglandins* **10**:581.

Wilson, D. E., Winnan, G., Quertermus, J., and Tao, P., 1975b, Effects of an orally administered prostaglandin analogue (16,16-dimethyl prostaglandin E$_2$) on human gastric secretion, *Gastroenterology* **69**:607.

Wilson, D. E., Quertermus, J., Raiser, M., Curran, J., and Robert, A., 1976, Inhibition of stimulated gastric secretion by an orally administered prostaglandin capsule: A study in normal man, *Ann. Intern. Med.* **84**:688.

Wolfe, L. S., Coceani, F., and Pace-Asciak, C., 1967, The relationship between nerve stimulation and the formation and release of prostaglandins, *Pharmacologist* **9**:171.

Wollin, A., Code, C. F., and Dousa, T. P., 1974, Evidence for separate histamine and prostaglandin sensitive adenylate cyclases (AC) in guinea pig gastric mucosa (GM), *Clin. Res.* **22**:606A.

Yamagata, S., Masuda, H., Ishimori, A., Mita, M., Inoue, S., Arakawa, H., Sakurada, H., Nemoto, K., and Shimoyama, M., 1970, Study of the inhibitory action of prostaglandin on gastric secretion in rat and man, *Gastroenterol. Japon.* **5**:321.

Yankee, E. W., and Bundy, G. L., 1972, (15*S*)-15-Methyprostaglandins, *J. Am. Chem. Soc.* **94**:3651.

Yankee, E. W., Axen, U., and Bundy, G. L., 1974, Total synthesis of 15-methyprostaglandins, *J. Am. Chem. Soc.* **96**:5865.

Prostaglandins and Renal Function, or "A Trip Down the Rabbit Hole"

Walter Flamenbaum* and Jack G. Kleinman‡

Department of Nephrology
Walter Reed Army Institute of Research
Washington, D.C. 20012

I. INTRODUCTION

> "If there's no meaning in it," said the King, "that saves a world of trouble, as you know, as we needn't try to find any. And yet I don't know," he went on . . . "I seem to see some meaning in them after all . . ." (From *Alice in Wonderland* by Lewis Carroll)

Ten or fifteen years ago a request to review the relationship of a family of lipids known as the prostaglandins to renal function would have presented a light task. The ubiquitous nature of these compounds has led to a parallel burgeoning of scientific and literary activity (Wilson, 1974; Weeks, 1974; Higgins and Braunwald, 1972; Anderson and Ramwell, 1974). In order to maintain a modicum of organization and lucidity, this discussion will be essentially limited to the intrarenal role of prostaglandins in renal physiological homeostasis and pathophysiology (Horton, 1969). In addition, a specific attempt has been made to identify and dispell, if possible, some of the myths which have arisen concerning prostaglandins and the kidney. Areas of potentially wider interest, such as the role of prostaglandins in systemic hypertension, will not be considered. Furthermore, to adhere to the major intent of

*Current address: Veterans Administration Hospital, Boston, Massachusetts 01230.
‡Current address: Veterans Administration Hospital, Milwaukee, Wisconsin 53193.

this chapter the biochemistry of the prostaglandins will be discussed only in relationship to distinct physiological events. The reader is referred to some reviews of the biochemistry of prostaglandins (McMurray and Magee, 1972; Bergström *et al.*, 1968; Anderson and Ramwell, 1974; Änggård, 1971; Crowshaw and McGiff, 1973; van Dorp, 1971; Larsson and Änggård, 1973; Nakano, 1970).

II. PROSTAGLANDIN SYNTHESIS AND METABOLISM

One may begin a chemical consideration of prostaglandins in an almost classical manner by commenting on the work of Kurzok and Lieb (1930), von Euler (1934), and Bergström and colleagues (Bergström and Sjövall, 1960; Bergström *et al.*, 1962, 1964*a*). A delineation of the prostaglandins of known interest and importance and a consideration of the compounds of potential physiological effect would be more worthwhile. The prostaglandins are a family of 20-carbon unsaturated lipid acids, which may be subdivided and classified based on the structure of cyclopentane ring and the number of carbon–carbon double bonds present (Anderson and Ramwell, 1974). With reference to the current physiological literature, it appears that prostagandins of the A type, E type, and F type are of the greatest potential interest (Gréen *et al.*, 1973; Bergström *et al.*, 1962; Bagli *et al.*, 1966). Ten years ago, Lee *et al.* (1965, 1967) isolated three prostaglandinlike compounds from rabbit renal medulla, which have been identified as prostaglandins A_2 ("medullin"), E_2, and $F_{2\alpha}$. Although a larger number of prostaglandins, including mono-, bis-, and tris-unsaturated compounds (indicated by the arabic subscript) as well as α-steroisomers, have been described as "primary" or "naturally" occurring prostaglandins, most investigations have concerned prostaglandins A, E, and F (Hamberg and Samuelsson, 1966, 1967; Israelsson *et al.*, 1969).

Two very simple but very crucial problems must be raised at this point. Although these may be commonly investigated prostaglandins, it must be recognized that there may be more physiologically active and important precursors, metabolic products, diastereomers, or yet unidentified prostaglandins which are not being actively investigated at this point (Anderson and Ramwell, 1974; Zins, 1975). The second point concerns whether prostaglandin A is a naturally occurring compound or an artifact. Thus it has been suggested that prostaglandin A may be an artifact resulting from dehydration of the cyclopentane ring of prostaglandin E during an extraction procedure (McGiff *et al.*, 1974*a*). Support for this biochemical observation is also found in physiological investigations. Using the isolated perfused kidney, Itskovitz *et al.*

(1973) did not observe any alterations in the perfusate concentration of prostaglandin A, measured by radioimmunoassay, under circumstances which resulted in marked increases in prostaglandins E and F, suggesting that prostaglandin A does not arise intrarenally. Furthermore, only trace amounts of prostaglandin A are formed from renal medullary homogenates (Hamberg, 1969; Crowshaw, 1973; Crowshaw and McGiff, 1973), suggesting a low biosynthesis potential within the kidney. This consideration and the lesser potency of prostaglandin A compared to E suggest that its physiological role may be limited. However, there are several alternative possibilities which merit consideration. Prostaglandin A may be formed extrarenally, perhaps as a conversion product within the systemic circulation, and still have a biophysiological effect on the kidney even though it is not formed intrarenally. Prostaglandin A, when measured by radioimmunoassay methodologies, may represent an immunologically cross-reacting, but yet unidentified member of the prostaglandin family (Davis and Horton, 1972; Zusman *et al.*, 1973; Jaffe *et al.*, 1973). Alternatively, since prostaglandin A may be enzymatically or spontaneously isomerized to prostaglandin C or B (Jones and Cammock, 1973), both of which are extremely potent and escape transpulmonary degradation, its physiological importance may be underestimated by considering only its potency or rate of synthesis and release (McGiff *et al.*, 1974*a*). Therefore, the questions concerning prostaglandin A remain unsettled from both a biochemical and a physiological perspective (Attallah *et al.*, 1974*b*: Golub *et al.*, 1974), especially since a number of groups have been able to measure prostaglandin A concentration in both plasma and renal tissue (Jaffe *et al.*, 1971, 1973; Zusman *et al.*, 1972; Attallah *et al.*, 1974*b*).

A. Renal Prostaglandin Synthesis

Having indicated that we are concerned with prostaglandins of the E, A, and F types, a consideration of renal prostaglandin synthesis is in order. Polyunsaturated fatty acids, of which arachidonic acid is the only defined substrate for the kidney, are the acceptable substrates for prostaglandin synthesis (Pace-Asciak *et al.*, 1967, Änggård and Samuelsson, 1965; van Dorp, 1965, Van Dorp *et al.*, 1964; Bergström *et al.*, 1964). How these fatty acids are made available to prostaglandin synthetase remains unsettled. It has been suggested that phospholipase A (Kunze and Vogt, 1971) or triglyceride lipase (Nissen and Bojesen, 1969) is required for synthesis of prostaglandins to occur. The relationship of this requirement to the role of prostaglandins in renal physiological homeostasis has not yet been explored. The mechanism of prostaglandin synthesis by a variety of tissues has been investigated by Samuelsson and co-workers (Bergström *et al.*, 1964*b*; Klenberg and Samuelsson, 1965) and others (Crowshaw, 1971, 1973; Janszen and Nugteren, 1971; Samuelsson, 1965; Samuelsson *et al.*, 1971). In brief, intrarenal prostaglandin synthesis involves the conversion of arachidonic acid, via a cyclic endoperoxide intermediate (Hamberg *et al.*, 1974), to PGE and PGF. The conversion of this endoperoxide intermediate to PGE_2 and $PGF_{2\alpha}$ requires, respectively, an isomerase and a reductase (Nakano, 1973). It has been suggested, because of the biological potency of the endoperoxide intermediate, that it may have an important direct or regulatory role in the physiological actions of prostaglandins (Andersen and Ramwell, 1972). The exact role of PGA, as noted above, or other subtypes in the intrarenal synthesis of prostaglandins remains to be elucidated since they may be formed by the intrarenal conversion from PGE.

1. Relationship of Cortex and Medulla

Prostaglandin synthetase, which is really more a complex of enzymes rather than a single, specific enzyme, has been localized to the endoplasmic reticulum of renal parenchymal cells (Änggård *et al.*, 1972). Further anatomical localization is the subject of some controversy. It had been originally stated that the synthesis of prostaglandins was limited to the medulla and papilla of the kidney (Änggård *et al.*, 1972; Crowshaw, 1973; Janszen and Nutgeren, 1971). Crowshaw and colleagues (Crowshaw and Szlyk, 1970; Crowshaw, 1973; Crowshaw *et al.*, 1970) were unable to demonstrate the presence of prostaglandins in extracts of the renal cortex and did not observe *de novo* sythesis of prostaglandins when arachidonic acid was incubated with slices of rabbit renal cortex. Landolt *et al.* (1974), Vance *et al.* (1973), and Spector *et al.* (1974) observed prostaglandins, albeit in small amounts, within the cortex of human kidneys. One possible explanation for the absence of prostaglandins in the renal cortex, or their presence in small amounts as compared to the medulla, may be related to their active metabolism and degradation in this area. Änggård and *et al.* (1971) demonstrated that the cortical content of 15-hydroxyprostaglandin dehydrogenase was extremely high relative to medullary and papillary content, in both the rabbit and the pig. Furthermore, the prostaglandin biosynthetic activity of the cortex, using a microsomal fraction, was approximately 10% of that observed in the renal medulla and papilla (Larsson and Änggård, 1970). It has been suggested, therefore, that the inability to demonstrate cortical prostaglandin content and/or biosynthesis may have been related to active prostaglandin degradation in this area (Larsson and Änggård, 1973). In view of the potency of the prostaglandins even at this relatively low level of biosynthetic activity, they may have important physiological impact on renal function (McGiff *et al.*, 1974*a*).

Since prostaglandins are not stored but synthesized and released as required (Crowshaw, 1973; Änggård *et al.*, 1972; Jouvenaz, 1970), considerable attention has been paid to the synthesis of prostaglandins by the noncortical areas of the kidney. In investigations of the specific anatomical site or sites, histochemical evidence for prostaglandin synthesis by cells of the collecting duct has been obtained by Janszen and Nugteren (1971). Additional evidence indicating active prostaglandin synthesis by renal interstitial cells is also available (Osvaldo and Latta, 1966; Muirhead, 1973). These cells, which have a rich endoplasmic reticulum and numerous mitochondria (Muehrcke *et al.*, 1970) along with osmiophilic lipid-containing vesicles (Nissen and Bojesen, 1969), are primarily located in the interstitium of the inner medulla between the vasa recta, loops of Henle, and collecting ducts. These cells produce prostaglandins in tissue culture (Muirhead *et al.*, 1972) and the orientation of their prominent star-shaped processes along the vasa recta and loop of Henle may explain the observation of synthetic activity in these areas. That renomedullary homogenates (Hamberg, 1969: Crowshaw, 1971) and slices (Crowshaw, 1973) have greater biosynthetic activity than interstitial cells in tissue culture suggests that these cells are not the sole site of prostaglandin synthesis. In addition, a high concentration of prostaglandins has been described in the interstitial cells of juxtapapillary area of the kidney. The question of zonal differences in biosynthetic activity within the noncortical areas of the kidney remains unsettled. Evidence has been presented for either the medulla (Kalisker and Dyer, 1972*a,b;* Daniels *et al.*, 1967; Bohman, 1974; Änggård *et al.*, 1972; Prezyna *et al.*, 1973; Crowshaw *et al.*, 1970) or papilla (Tobian *et al.*, 1972) being the highest area of prostaglandin content and synthesis.

The absence of renal storage of prostaglandins also bears upon the various reports of tissue content of these compounds (van Dorp *et al.*, 1964; van Dorp, 1971; Änggård, 1971). If the appropriate fatty acid substrate is available for the synthesis of prostaglandins, then, depending on the methods used to obtain and extract the tissue, the reported content may be more a reflection of potential biosynthetic activity than of the absolute content. Thus a 5000-fold difference in prostaglandin content may be observed which appears to be related to the rapidity of limiting biosynthesis by the kidney prior to extracting the prostaglandins. The lack of intrarenal storage of prostaglandin would indicate that any stimulus associated with increased prostaglandin release is the result of increased prostaglandin synthesis. The various known stimuli for prostaglandin release, and their interaction and relationship to renal homeostasis, will be discussed later. To date, the documented stimuli are of such a diverse nature as to result in difficulty ascertaining the mechanism by which prostaglandins are released. The common denominator in these stimuli, which range from physiological to the pathological, appears to be related to some element of cellular damage or dysfunction (Änggård and Jonsson, 1971; Piper, 1973; Piper and Vane, 1971). It has also been suggested that phospholipase A, which may have a role in initiating prostaglandin synthesis, may participate in the alterations in cell membranes that occur as a result of many of the stimuli know to lead to prostaglandin release.

Using cell fractionation techniques, the synthesis of prostaglandins has been associated with a microsomal fraction, while prostaglandin compounds themselves are in the supernatant (van Dorp, 1967; Änggård *et al.,* 1972; Hamberg and Samuelsson, 1967). Thus after synthesis the prostaglandins may be released either into the cytoplasm or into the extracellular fluid. In contrast to the majority of tissues, the medullary area of the kidney, which is the major synthetic site for prostaglandin synthesis, does not contain appreciable amounts of prostaglandin dehydrogenase (Larsson and Änggård, 1973). An anatomical separation of areas for prostaglandin synthesis and metabolism thus exists within the kidney, which requires that prostaglandins be metabolized in cells other than their site of synthesis (Attallah *et al.,* 1973*a*). Therefore, renal prostaglandins must be released extracellularly to be metabolized, while in nonrenal sites of synthesis prostaglandins may leave the cells in an inactive form (Änggård, 1971; Raz, 1972; Andersen and Ramwell, 1974).

2. Synthetase Inhibition

Inhibition of prostaglandin synthetase has been widely used as a method for studying both the biochemistry of prostaglandin synthesis and the physiological effects of prostaglandins. Discussion of the latter use of prostaglandin inhibition will be presented later in relationship to the physiological measurements under investigation. Nonsteroidal anti-inflammatory agents, unsaturated fatty acids acceptable as substrates for prostaglandin synthesis, and synthetic prostaglandin analogues have been utilized in the biochemical study of prostaglandin synthesis (Gryglewski and Vane, 1972; Smith and Willis, 1971; Sanner, 1974; Ferreira *et al.,* 1973*b*). Examples of the false substrates used to study prostaglandin synthesis include 8,17-*cis*-12-*trans*-eiocosatrienoic acid and decanoic acid (Andersen and Ramwell, 1974). An oxa analogue of the prostaglandins, 5-oxaprostenoic acid, has also been demonstrated to be a substrate competitive inhibitor, rather than inhibiting the product formation of prostaglandin synthetase. As a class, the acidic anti-inflammatory agents have received wide attention as inhibitors of prostaglandin synthesis. The action of these agents, such as aspirin and indomethacin, to prevent the formation of the endoperoxide intermediates of prostaglandin synthesis has also been documented. For a more detailed discussion of the agents resulting in prostaglandin synthesis inhibition, the reader is referred to the recent reviews (Flowers, 1974; Vane, 1973). For the purpose of the present discussion, namely the relevance to renal physiological homeostasis, only a few comments are warranted. Although substrate analogues, prostaglandin analogues, and pharmacological inhibitors may be available to study prostaglandin synthesis and effect, none of these agents completely inhibits the synthesis/actions of prostaglandins. In addition, the anti-inflammatory agents are far from enzyme specific in their ability to inhibit prostaglandin synthetase since they also inhibit other enzyme systems (Flowers, 1974). Further confusion arises since indomethacin also inhibits, in part, some of the enzyme systems involved in prostaglandin inactivation and metabolism. Cheung and Cushman (cited in Flowers, 1974) demonstrated a 93% inhibition of PGE_2 metabolism by indomethacin using an *in vivo* preparation, and specific decreases in 9-prostaglandin dehydrogenase, 15-prostaglandin dehydrogenase, and 13-prostaglandin reductase were observed using indomethacin by Pace-Asciak and Cole (1975). There is also evidence suggesting that there is some "escape" from the prostaglandin inhibitory effect resulting from the chronic administration of aspirin (Vane, 1973). These comments should indicate that some caution is required before accepting or interpreting results obtained with these agents. In terms of evaluating the intrarenal role of prostaglandins, additional difficulties may arise in drawing conclusions from studies obtained after the consequence of changes resulting from the inhibition of prostaglandins in some other organ or the systemic circulation.

B. Prostaglandin Inactivation

Once having reached the circulation from renal cellular sites of synthesis, prostaglandins do not appear to be metabolized within the circulation *per se* (Ferreira and Vane, 1967; Raz, 1972; Silver *et al.*, 1972). The major enzyme systems which have been studied for metabolizing prostaglandins include (1) 15-hydroxyprostaglandin dehydrogenase, which catalyzes the oxidation of prostaglandins at C-15, is found in high concentrations in the lung, spleen, and renal cortex, and is most active in converting PGE_1 and PGE_2 to the 15-keto metabolites; (2) prostaglandin Δ^{13}-reductase, which catalyzes the reduction of Δ^{13}

double bond of prostaglandins, and is found in highest concentrations in the liver, spleen, small intestine, and kidney; (3) β oxidation of the carboxylic acid side chain, which has been demonstrated in rat liver mitochondria and in homogenates of rat lung and kidney, and is a metabolic event which must follow dehydrogenation; (4) ω oxidation of the prostaglandin side chain, which has been demonstrated by human liver microsomes; and (5) reduction of C-9 ketoprostaglandins, resulting in the metabolism of PGE_2 to at least seven metabolites, including $PGF_{2\alpha}$ (Änggård and Samuelsson, 1966; Nakano *et al.*, 1969; Larsson and Änggård, 1970; Nakano and Morsy, 1971; Piper, 1973; Hamberg and Israelsson, 1970; Granstrom, 1971; Nakano. 1970). The exact quantitative role of each of these enzyme systems in the various organs and in the different animal species has yet to be determined. More importantly, the various metabolites and their relative physiological importance is for the most part an unknown set of variables. In overall terms, it has been demonstrated in man that after the injection of tritium-labeled PGE_2, approximately 50% of the radioactivity appears in the urine within 5 hr (Granstrom, 1967; Hamberg and Samuelsson, 1969; Hamberg and Samuelsson, 1971). The major urinary metabolite is 7α-hydroxy-5,11-diketo-tetranorprosta-1,16-dioic acid (Samuelsson *et al.*, 1971). Alternative pathways are available in the guinea pig (Änggård and Samuelsson, 1964), and the metabolic pathways for prostaglandin metabolism in the rat are more complex than those demonstrated in either man or the guinea pig (Gréen, 1969, 1971). It is apparent, therefore, that the experimental subject under study may determine various responses to prostaglandins due to differences in prostaglandin metabolism, and must be included as an experimental variable.

The lung and 15-prostaglandin dehydrogenase are the site and enzyme system which appear to be most important in metabolizing prostaglandin compounds of the E and F types which reach the systemic circulation (Änggård, 1971; Piper *et al.*, 1970; Hook and Gillis, 1975; McGiff *et al.*, 1969; Pappanicolau and Meyer, 1972), although the liver appears to be an alternative secondary site for prostaglandin metabolism. The observation that E and F prostaglandin compounds are 90–95% inactivated during a single pass through the pulmonary circulation lends support to the conclusion that prostaglandins are local hormones. Furthermore, since PGA is not inactivated by this mechanism, its relative importance as a potential "circulating" hormone is underscored, as noted already. The decreased prostaglandin metabolism by the lung under conditions of hypothermia and hypotension and after the administration of endotoxin (Lindsey and Wyllie, 1970; Nakano and Prancan, 1973) suggests that this complete inactivation of prostaglandins is not inviolate. Under pathophysiological circumstances, a decrease in transpulmonary prostaglandin inactivation may substantially contribute to the observed perturbations in physiological function. Furthermore, a circulating role for prostaglandins cannot be ruled out since intravenously administered prostaglandins may exert a physiological effect, suggesting that even the small amounts of these compounds which escape metabolism may be of a sufficient quantity to have an effect; the pulmonary and hepatic uptake of prostaglandins may be limited even under physiological circumstances, and more limited under pathophysiological circumstances; and the physiological effect of the various metabolites formed during either pulmonary or hepatic inactivation has not yet been determined.

The intrarenal and renal alternatives for the metabolism of prostaglandins are multiple. The transit of prostaglandins from sites of medullary synthesis into the urine to sites of cortical inactivation or into the venous and/or lymphatic circulation has not been thoroughly defined. That the prostaglandins observed in the urine arise from *de novo* renal prostaglandin synthesis rather than extrarenally was suggested by the observation that metabolites of prostaglandin compounds and not intact prostaglandins are found in the urine after parenteral administration.

In contrast, both PGE and PGF compounds and metabolites have been recovered from human urine unchanged, and the administration of angiotensin or arachidonic acid increases urinary prostaglandin excretion (Granström and Samuelsson, 1969; Fröhlich *et al.*, 1973). It has been suggested, based on stop-flow studies in the dog (Tannenbaum, 1975), that prostaglandins enter the nephron in the ascending limb of the loop of Henle which would anatomically correspond to the medullary site of synthesis. The possible subsequent course of prostaglandins entering tubular fluid at this nephron site is unknown, although contact of the prostaglandins with the remainder of the nephron (i.e., the distal convoluted tubule and the macula densa segment, and the collecting duct) is readily apparent. It has also been suggested that reabsorption of these compounds from tubular fluid may occur (Bito, 1971; Attallah and Lee, 1973*b*). Since prostaglandins are not metabolized in the medulla and the cortex is rich in prostaglandin dehydrogenase, the intrarenal transport of prostaglandins through the ascending vasa recta and/or ascending limbs of the loop of Henle has been inferred (Zins, 1975). These observations on the "travels" of the prostaglandin compounds from the renal site of synthesis cast some shadow of doubt on studies performed to ascertain the physiological effect of prostaglandins by infusing them. As stated by Tobian *et al.* (1974), "the kidney has most of its PGE_2 synthesized in the medulla.

This PGE_2 could then act locally on blood vessels and tubular cells and could partly diffuse *up* into the inner cortex. This pattern is completely different from an infusion of PGE_2 *down* the renal artery." In quantitative terms, McGiff and Itskovitz (1973) have estimated that blood levels within arteries and veins rarely exceed 20 and 35 ng/ml, respectively. Infusions of prostaglandins in excess of this, therefore, may not yield results which in physiological terms bear any relationship to the endogenous release of prostaglandins.

III. ROLE OF PROSTAGLANDINS IN THE CONTROL OF RENAL BLOOD FLOW

A source of the considerable investigative interest in prostaglandins relates to their potent vasoactivity. It is not a surprise, therefore, that a great amount of this effort has centered upon the kidney, in particular renal hemodynamics, since the prostaglandins are renal vasodilators which are synthesized and metabolized within the kidney. To avoid alterations in renal function which may be the result of perturbations in systemic cardiovascular activity, this discussion of the importance of prostaglandins in renal function will focus only on those studies in which direct renal effects of prostaglandins can be identified. In general, prostaglandins of the E and A types result in increased renal blood flow (Johnston *et al.*, 1966; Vander 1968; Carriere *et al.*, 1971; Martinez-Maldonado *et al.*, 1972; Marchand *et al.*, 1973; Itskovitz *et al.*, 1973; Baer and Navar, 1973; Fulgraff and Brandenbusch, 1974; Arendshorst *et al.*, 1974) and decreased renal vascular resistance, while those of the F type either have no effect or are slightly vasopressor (Horton and Main, 1965; Lonigro *et al.*, 1973a,b; McGiff *et al.*, 1973; Tannenbaum *et al.*, 1975; Chang *et al.*, 1975). Descriptions of the effects of prostaglandins of the F series involve an extrarenal site of action on cardiac output or more central phenomena (Hinman, 1967; Ducharme *et al.*, 1968, Lavery *et al.*, 1970; Ferrario *et al.*, 1970). Thus no intrarenal action of F-type prostaglandin, specifically $PGF_{2\alpha}$, has yet been proven and the majority of the work has involved prostaglandins of the A or E series.

A. Studies within the "Autoregulatory" Range

The diverse stimuli resulting in increased prostaglandin synthesis and release by the kidney (Zins, 1975), the renal site of prostaglandin synthesis and metabolism (see previously), and the vasoactive potency of the prostaglandins have led to the suggestion that prostaglandins have a central role in the regulation of renal blood flow. In view of the demonstrated interactions, which will be discussed in greater detail later, of prostaglandins with renal nerve stimulation (Dunham and Zimmerman, 1970; McGiff *et al.*, 1970*a,b;* Fujimoto and Lockett, 1970; McGiff *et al.*, 1970*b*), angiotensin (McGiff *et al.*, 1970*b*; Aiken and Vane, 1973; Needleman *et al.*, 1973), bradykinin (McGiff *et al.*, 1972*b*), vasopressin (Kalisker and Dyer, 1972*a,b*), as well as ischemia (McGiff *et al.*, 1970*c*; Jaffe *et al.*, 1972; Herbaczynka-Cedro and Vane, 1974), it is not surprising that a primary role for prostaglandins in the autoregulation of renal blood flow has been inferred. Although the phenomenon of renal blood flow autoregulation, which is the ability of the kidney to maintain relatively constant blood flow over a range of perfusion pressure by adjusting renal vascular resistance, has been observed for many years, the mechanism responsible for this remarkable attribute of renal function has escaped definition. The abolition of autoregulatory response in the isolated perfused dog kidney by prostaglandin synthetase inhibition (Herbaczynka-Cedro and Vane, 1973) and the dependence of total renal blood flow/renal vascular resistance on the availability of prostaglandins (Lonigro *et al.*, 1973*a*; Kirschenbaum *et al.*, 1974; Larsson and Änggård, 1974; Itskovitz *et al.*, 1974) would appear to lend support to the thesis that increased synthesis and release of prostaglandins by the kidney in response to diminished perfusion pressure were the primary mechanism responsible for the phenomenon of renal autoregulation by control of renal vascular resistance (Carr, 1970; Lee, 1973; Lonigro *et al.*, 1973*a*).

While the evidence does suggest a relationship between prostaglandin synthesis/release and both total renal blood flow and its intrarenal distribution, especially in anesthetized animals, the sole responsibility of prostaglandins for the renal autoregulatory response is less clear. A number of reports (Anderson *et al.*, 1975; Owen *et al.*, 1974; Bell *et al.*, 1975; Venuto *et al.*, 1975) have not demonstrated a primary role for prostaglandins in renal autoregulation. Bell *et al.* (1975) were able to show, both with an isolated kidney preparation and using the *in situ* dog kidney, that inhibition of prostaglandin synthetase with indomethacin did not alter the progressive decline in renal vascular resistance with graded decreases in renal perfusion pressure. The observations of Venuto *et al.* (1975) were even more dramatic. These investigators documented the inhibition of prostaglandin synthesis by measuring prostaglandins with a radioimmunoassay technique. Using meclofenamate, they observed constancy of renal blood flow over a range of perfusion pressures in the dog (Fig. 1). Rather than precluding a role for prostaglandins in the renal autoregulatory response, the results of experiments designed to determine the role of prostaglandins

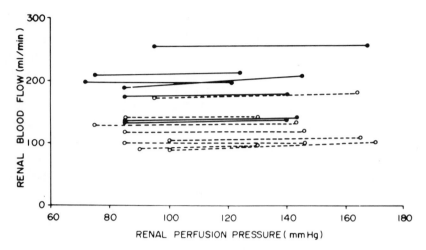

Fig. 1. Effect of prostaglandin synthesis inhibition with meclofenamate (2 mg/kg, i.v.) on the autoregulation of renal blood flow. The closed symbols represent renal blood measurements obtained in control dogs during a decrease in perfusion pressure by aortic clamping. The open symbols represent studies obtained after the administration of meclofenamate. Reproduced with permission from Venuto *et al.* (1975).

in the control of renal blood flow underline the multifactorial nature of the autoregulatory response. Thus intrarenal prostaglandins may well act in conjunction with other vasoactive mechanisms to set the tonic level of renal resistance, and consequently total blood flow, as well as its intrarenal distribution. The studies of Lonigro *et al.* (1973*a*), Larsson and Änggård (1974), Kirschenbaum *et al.* (1974), and Itskovitz *et al.* (1974) are of particular interest in this regard. The dependency of the renal circulation on prostaglandins was defined by Lonigro *et al.* (1973*a*) in both the isolated perfused dog kidney and the *in situ* dog kidney by determining the effect of prostaglandin synthetase inhibition with meclofenamate or indomethacin. In association with a decline in the renal venous efflux of PGE-like material, there was a marked decrease in total renal blood flow without significant alterations in either systemic blood pressure or cardiac output, suggesting that the resting state of renal blood flow (or the tonic state of renal vascular resistance) was dependent on prostaglandin biosynthesis. Although the intrarenal distribution of blood flow was not determined, the authors recognized that the diminution in renal blood flow of as much as 45% could not be accounted for solely by a decrease in blood flow to the medullary–papillary portion of the dog kidney (the site of prostaglandin synthesis) since this area normally accounts for no more than 10–15% of total renal blood flow. Evidence suggesting that prostaglandins might have a greater influence over inner cortical–juxtamedullary blood flow was obtained by Larsson and Änggård (1974) using radiomi-crospheres in studies of the renal hemodynamics of the rabbit kidney. The infusion of arachidonic acid increased the relative distribution of blood flow to this area. These results suggested that the tone of the vascular bed comprising the inner cortex was under control of the basal efflux of prostaglandins, which is consistent with the concept that prostaglandins act as intrarenal hormones participating in the regula-tion of the renal circulation. These observations were extended and confirmed by Kirschenbaum *et al.* (1974) in the dog using radiomicro-spheres to evaluate renal hemodynamics and both indomethacin and meclofenamate to inhibit prostaglandin synthesis. As indicated in Fig. 2, inhibition of prostaglandin synthesis was associated with a signifi-cant decrease in total renal blood flow and marked alteration in the intrarenal distribution of blood flow, characterized by more marked decreases in inner cortical–juxtamedullary blood flow. Similar obser-vations were obtained by Itskovitz *et al.* (1973) and Solez *et al.* (1974).

More specific information concerning the interrelationship of renal prostaglandins and increases in renal blood flow (decreases in renal vascular tone) and the differences between endogenous and exogenous (intrarenal and extrarenal) sources of prostaglandin is provided by the investigations of Itskovitz and co-workers (Itskovitz and McGiff, 1974) and those of Chang *et al.* (1975). Using the isolated perfused dog kidney, Itskovitz *et al.* (1974) demonstrated that the progressive vaso-dilatation observed with time during the course of renal perfusion was

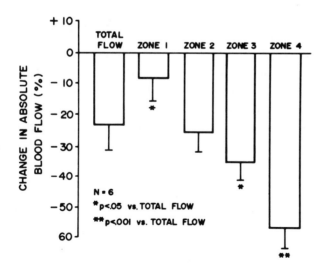

Fig. 2. Effect of indomethacin (2 mg/kg, i.v.) on the intrarenal distribution of blood flow determined using radiomicrospheres. Reproduced with permission from Kirschenbaum *et al.* (1974).

associated with an increase in perfusate concentrations of PGE-like material and a redistribution of blood flow towards inner cortex. When indomethacin was added to the perfusate, inner cortical blood flow decreased rather than increased and total blood flow decreased. When PGE_2 was added to the perfusate after prostaglandin synthetase inhibition, total renal blood flow and inner cortical blood flow still decreased, consistent with the concept that the prostaglandins had to be generated intrarenally to be effective in altering renal hemodynamics. Using the intact dog, Chang *et al.* (1975) obtained similar results concerning the differential action of intrarenal and extrarenal sources of prostaglandin. Exogenous infusion of PGE_2 resulted in a marked increase in total renal blood flow and increases in blood flow throughout the cortex, although the increase was more marked in the inner cortical area. In contrast, infusions of arachidonic acid resulted in only a modest increase in total renal blood flow and increased cortical blood flow only in the juxtamedullary region. When presented in terms of alterations in renal vascular resistance, the pattern of alteration observed with vasodilators (McNay and Abe, 1970; Stein *et al.,* 1971) suggests a relatively nonspecific effect on the resistance vessels of the inner cortex. The stimulation of endogenous prostaglandin formation by infusion of arachidonic acid, a prostaglandin precursor, more selectively increased inner cortical blood flow.

The lack of a clear-cut role for endogenous renal prostaglandin in the renal autoregulatory response and the apparent role for endogenous renal prostaglandins in setting the basal tone of the renal vascular system, especially in the inner cortical vessels, are indicative of some of the controversy concerning the role of renal prostaglandins in control of renal hemodynamics. The complex nature of this controversy is best exemplified by a recent study in unanesthetized dogs. In contrast to the marked alteration in renal hemodynamics observed in studies utilizing the isolated perfused kidney or *in situ* kidney of an anesthetized animal, Zins (1975) did not observe any alteration in total blood flow or the intrarenal distribution of blood flow after the administration of indomethacin to unanesthetized dogs. Similar results were obtained by Swain *et al.* (1975) using indomethacin. Meclofenamate, in contrast, resulted in decreased renal blood flow even in the conscious dog (Swain *et al.,* 1975). Other differences have been observed in unanesthetized vs. anesthetized animals (Higgins *et al.,* 1973). Thus, rather than interpreting these apparently contradictory results in anesthetized vs. unanesthetized dogs or the presence or absence of renal autoregulation after prostaglandin synthetase inhibition in an absolute manner, the millieu in which studies are performed should be considered (Vatner, 1974). It is well known that anesthesia activates both the sympathetic nervous system (Scott *et al.*, 1965; Brody and Kadowitz, 1974; see later), as well as the renin–angiotensin system. It is readily apparent that the increased vascular tone which may be provided by the variable influence of either increased sympathetic tone or increased angiotensin generation may severely modify any renal hemodynamic result of perturbations in prostaglandins, thus modifying the interpretation of data. Even in the isolated perfused kidney, prostaglandin action must be considered in terms of interactive effects of other vasoactive phenomena. The different renal hemodynamic results observed under different experimental conditions may, therefore, be a reflection of the variability of other influences on renal hemodynamics rather than proof for the presence or absence of a role for prostaglandins in any specific situation. The interaction of prostaglandins with other vascoactive phenomena will be discussed in greater detail below.

B. Relationship to Renal Ischemia

The observed decrease in total renal blood flow after inhibition of prostaglandin biosynthesis and the vasodilatory properties of the prostaglandins provide an additional basis for the consideration of the role of prostaglandins in alterations in renal hemodynamics aside from autoregulation. Extrapolating from an inferred antihypertensive effect of renal prostaglandins, McGiff *et al.* (1970) investigated the effects of acute renal ischemia on the release of prostaglandins in renal venous effluent. Concomitant with the decrease in renal blood flow, an increase in the amount of PGE-like material was observed. After release of the renal artery constriction, there was a rebound of both renal blood flow and PGE-like material. Furthermore, PGE-like material was released from the contralateral, nonischemic kidney which was ascribed to the simultaneous increase in renin–angiotensin system activity. Acute renal ischemia has also been demonstrated to increase the renal concentration of PGE (Jaffe *et al.*, 1972). The role of prostaglandins in renal ischemia induced by hemorrhage has been investigated by Selkurt and colleagues and Bell *et al.* (1975). The induction of hemorrhagic hypotension in dogs (Selkurt, 1974) resulted in a marked decline in renal blood flow and significant increases in both the renal release of PGE as well as an increased arterial concentration of PGE. The rise in arterial concentration was ascribed to a diminished pulmonary degradation of circulating prostaglandins as the result of a reduced pulmonary blood flow. The decline in systemic blood pressure was shown to be less marked and the decrease in renal blood flow greater when hemorrhagic hypotension was induced in dogs pretreated with indomethacin (Bell *et al.*, 1975), consistent with a role for prostaglandins in mediating changes in both systemic and renal hemodynamics in this condition. The results suggest an active role for prostaglandins in the renal response to diminutions in renal perfusion. On a mechanistic basis it may be inferred that prostaglandins, through their potent vasodilatory action, help preserve and maintain renal blood flow under conditions of renal ischemia and contribute to the decline in systemic blood pressure which accompanies hemorrhage. The conclusion is supported by the recent results demonstrating that prostaglandins mediate the renal vasodilation induced by hemorrhage (Vatner, 1974) or reactive renal hyperemia (Swain *et al.*, 1975).

The results of studies directed toward assessing the role of prostaglandins in the hemodynamic alterations which characterize the postischemic or posthemorrhagic state are also of great interest. Bell *et al.* (1975) observed that, without returning shed blood volume, blood pressure returned to approximately 70% of control values in untreated dogs and to approximately 90% of control values in dogs pretreated with indomethacin. When blood pressure was restored by transfusing shed blood, a continued increase in prostaglandin release was observed (Selkurt, 1975). If indomethacin was administered and the dogs were transfused, the release of prostaglandins was abolished and renal vascular resistance increased, suggesting that renal blood flow diminished. These results are consistent with a continued vasodilatory role, in both the systemic and renal circulations, for prostaglandins in the posthemorrhagic state. Following release of renal artery occlusion without alterations in systemic hemodynamics, increases in prostaglandin release concomitant with the reactive hyperemia characterizing these maneuvers have also been observed (McGiff *et al.*, 1970*a,c;* Vane and McGiff, 1975). Indomethacin almost completely abolishes the reactive hyperemia, suggesting that prostaglandins do mediate this response to the release of renovascular occlusion. Whether the increased release of prostaglandins after the release of the occlusion represents an accumulation of prostaglandins released at a tonic rate or an increase in prostaglandin biosynthesis as the direct consequence of ischemia has not been determined (Herbaczynka-Cedro *et al.*, 1974).

C. Relationship to Renal Hyperemia

The involvement of prostaglandins in states of increased renal blood flow has also been examined. Williamson *et al.* (1974) and Bailie *et al.* (1975) evaluated the effect of indomethacin-induced prostaglandin synthetase inhibition on the renal hemodynamic alterations associated with the administration of potent diuretic agents. The increases observed in renal blood flow after either ethacrynic acid or furosemide were abolished by treating the dogs with indomethacin, even though the ethacrynic acid-induced increases in sodium and water excretion (Williamson *et al.*, 1974) or the furosemide-induced increases in renin secretion (Bailie *et al.*, 1975) were not prevented. More detailed information is available concerning the mediation of prostaglandins in the renal hyperemia resulting from the administration of pyrogen (Gagnon *et al.*, 1975*a*). It has been known for 30 years that a variety of pyrogenic agents will induce over a short time interval a significant increase in renal blood flow. The effects of administering typhoid vaccine to anesthetized dogs on renal function were evaluated using standard techniques. A marked increase in total renal blood flow without alteration in systemic blood pressure was observed within 70 min, and then renal blood flow returned toward control values (Fig. 3). To further evaluate the renal hemodynamic effect of pyrogen-induced renal hyperemia, the intrarenal distribution of blood flow using radio-microspheres was also determined (Fig. 3). At the time of the peak renal hyperemic response to typhoid vaccine, there was a loss of the outer cortical to inner cortical blood flow distribution gradient observed during control periods. Concomitant with the return of total renal blood flow toward control values 150 min after the administration of

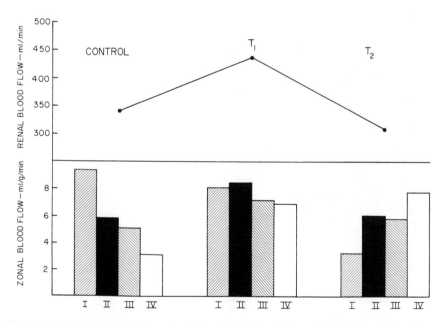

Fig. 3.   Effect of triple typhoid vaccine (0.25 ml/kg, i.v.) on total renal blood flow and zonal blood flow. The cortex was divided into four zones, from outer cortex (I) to inner cortex (IV). Measurements were obtained at 70 min (T₁) and at 150 min (T₂) after typhoid vaccine. Adapted from Gagnon *et al.* (1975*a*).

typhoid vaccine, there was a complete reversal of the cortical renal blood flow gradient as compared to control. Marked and sustained increases in renal PGE and PGF secretory rates and renal renin secretory rate, measured using radioimmunoassay techniques, occurred in conjunction with these alterations in renal blood flow. The interrelationship of these events is depicted in Fig. 4. For the convenience of comparison, the percentage distribution of cortical blood flow to the outer cortex (I + II) and the inner cortex (III + IV) has been summed. The marked increase in inner cortical blood flow earlier after the administration of typhoid vaccine (T_1, 70 min) was associated with increased secretory rates of both prostaglandins and renin. During the later time interval (T_2, 150 min), there were continued increases in prostaglandin and

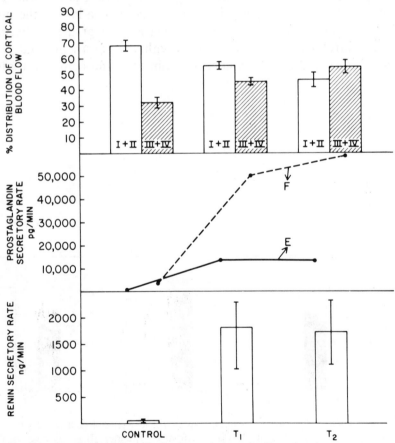

Fig. 4. Alterations in blood flow distribution and prostaglandin and renin secretory rates after typhoid toxin. See text for explanation. Adapted from Gagnon *et al.* 1975*a*).

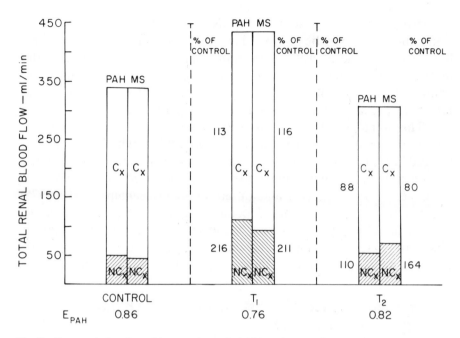

Fig. 5. Pyrogen-induced renal hyperemia cortical (C_x) and noncortical (NC_x) renal blood flows were estimated using the clearance and extraction of PAH (E_{PAH}) assuming that E_{PAH} reflected cortical blood flow. Similar calculation were obtained using microspheres (MS) assuming no entry of microspheres into the noncortical renal circulation.

renin secretory rates, a further increase in inner cortical blood flow, and a return of total blood flow toward control values. The alterations in cortical and noncortical renal blood flows, calculated using either the extraction of p-aminohippuric acid (E_{PAH}) as an estimate of noncortical blood flow or microsphere-determined blood flows as an estimate of cortical blood flow, are presented in Fig. 5. During the renal hyperemic phase of response to the administered pyrogen, there were increases in both cortical and noncortical renal blood flows, the increases in the latter being more marked. The return of total renal blood flow toward control values was associated with a decrease in cortical blood flow and a maintenance of the elevated noncortical blood flow. The administration of meclofenamate completely prevented increased prostaglandin secretory rates and the alterations in renal hemodynamics induced by typhoid toxin. These observations suggest that increased prostaglandin synthesis and release mediate the renal hemodynamic response to the administration of pyrogenic agents, such as typhoid toxin. It would appear that the increase in prostaglandins is responsible for the increase in total renal blood flow and increase in inner cortical renal blood flow initially observed. The possible interactions and net effects of increases in both vasodilator (prostaglandin) and vasoconstrictor (renin) hormones will be discussed in greater detail later. It is apparent, however, that the alterations in both total renal blood flow and the intrarenal distribution of blood flow observed with time after the administration of typhoid toxin may be the result of the differential effects of prostaglandin and renin on the renal vasculature.

The potential role for renal prostaglandins in other states characterized by renal hyperemia, other than bradykinin which will be considered later, has not been studied in detail. The association of increasing renal blood flow with time in the isolated perfused kidney observed by Itskovitz *et al.* (1974) and others (Vanherwegham *et al.*, 1975) appears to be prostaglandin dependent. Indeed, Needleman *et al.* (1974*a*) have demonstrated that adenosine triphosphate and diphosphate stimulate prostaglandin biosynthesis by many organs, suggesting that this may represent a common mode in hyperemic responses.

D. Interdependence of Prostaglandins and Renal Vasoactive Phenomena

The net effect(s) of prostaglandins on renal function is a complex issue. It should be apparent that there are a multitude of both direct and indirect effects which must be considered before interpreting any alteration in renal function which is associated with any change in the synthesis and/or release of prostaglandins. Thus any alteration in a parameter of renal function observed within the context of changes in prostaglandin activity may be ascribed to (1) direct effects of circulating or intrarenal prostaglandins on that parameter, (2) indirect effects of prostaglandin to modulate an action of another effector of change in renal function, and (3) alterations in prostaglandin activity which are causally or casually modulated by or related to other effector mechanisms. Examples of this interdependence of actions and effects are the relationships of prostaglandins to the renin–angiotensin system, the kallikrein–kinin system, and the sympathetic nervous system/catecholamines. Although the alterations in renal vascular resistance and renal blood flow are the most manifest parameters in this interdependence, it is obvious that changes in other elements of renal function may also occur independent of or in association with changes in renal hemodynamics.

1. The Renin–Angiotensin System

The effects of angiotensin on the renal vascular bed have been characterized (Carriere and Friborg, 1969; Carriere and Biron, 1970). The marked sensitivity to the vasoconstrictor effects of angiotensin has also been demonstrated (Hollenberg *et al.*, 1972). Based on the observation of concomitant alterations in renal renin and prostaglandin content in a patient with renovascular hypertension (Strong *et al.*, 1966), Vander (1968) infused varying doses of PGE into the renal artery without changing renal venous renin release rate. In contrast, Werning *et al.* (1971) demonstrated increases in plasma renin activity subsequent to the infusion of PGE_1 in dogs. The increase in plasma renin activity was prompt, maximal at 30 min, and returned to control values within 3 hr. The mechanism of this increase in renin release after the infusion of prostaglandin is unknown, and may have been the result of either a direct effect on the juxtaglomerular apparatus resulting in increased renin release or indirectly mediated by associated alterations in either renal hemodynamics or in fluid/electrolyte homeostasis. A similar variability in renin release in response to prostaglandins has been observed in humans after the infusion of either PGE_1 (Carlson *et al.*, 1969) or PGA_1 (Lee *et al.*, 1971; Fichman *et al.*, 1972). Some of this variability may be related to the direct effects of prostaglandin on the generation by renin of angiotensin from renin substrate in addition to the effect of prostaglandins on renin release. Thus PGA_1 and PGA_2 decrease the rate of angiotensin generation (Kotchen and Miller, 1974), and the renal renin response to prostaglandin infusion may be set by the level of renin–angiotensin system activity at the onset of the study. The latter effect was demonstrated by Krakoof *et al.* (1973). PGA_1 was infused intravenously, and the alterations in plasma renin activity, as influenced by prior salt balance, were determined in humans. Following sodium depletion with furosemide, a constant rise in plasma renin activity was observed with PGA_1 infusion, in contrast to the variable response observed without sodium depletion.

While alterations in renin–angiotensin system activity in response to infusion of prostaglandin may be variable, infusion of angiotensin into the renal artery has reproducibly resulted in increased release of prostaglandins by the kidney (McGiff *et al.*, 1970b; Aiken and Vane, 1973; Gagnon *et al.*, 1973; Fröhlich *et al.*, 1975) and other organs (Peskar and Hertting, 1973; Ferreira *et al.*, 1973a; Douglas *et al.*, 1973; Franklin *et al.*, 1974). This increase in prostaglandin release appears to be dose dependent (Aiken and Vane, 1973) and is readily abolished by prostaglandin synthetase inhibition (Aiken and Vane, 1973; Peskar and Hertting, 1973; Douglas *et al.*, 1973; Gagnon *et al.*, 1974; Needleman *et al.*, 1973). Furthermore, specific inhibitors of angiotensin II (Regoli *et al.*, 1973) also block the increase in prostaglandin release observed with infusions of angiotensin (Douglas *et al.*, 1973; Needleman *et al.*, 1973; Gagnon *et al.*, 1974; Needleman *et al.*, 1974b). It has also been demonstrated that both the decapeptide, angiotensin I, and the octapeptide, angiotensin II, release prostaglandins from the kidney. Using the isolated perfused rabbit kidney and a converting enzyme inhibitor (Aiken and Vane, 1972) which prevents the conversion of angiotensin I to angiotensin II, Needleman *et al.* (1973, 1974*b*) demonstrated release of prostaglandinlike substance after the infusion of the decapeptide. Although the potency of angiotensin I, as regards both as increase in renovascular resistance and the renal release of prostaglandins, is less than that of angiotensin II (Needleman *et al.*, 1974*b*), it may also function in the interdependence of the renin–angiotensin system with prostaglandins.

In addition to the quantitative interactions between prostaglandin levels and renin–angiotensin system activity, there is physiological interdependence of alterations in renal function (McGiff *et al.*, 1970a). Aiken and Vane (1973) observed that the effect of angiotensin II (3–40 ng/kg-miń) on renal vascular resistance, as judged from the percentage decrease in renal blood flow from control values, was enhanced after prostaglandin synthetase inhibition with indomethacin. This augmentation of the renal vasoconstrictor effect of angiotensin after abolition of prostaglandin synthesis and release was most marked at the lower ranges of infusion of angiotensin. Similar observations were obtained using meclofenamate to inhibit prostaglandin synthesis. Parallel effects on renal blood flow were observed after intravenous angiotensin II infusion or renal ischemia (Satoh and Zimmerman, 1975) using either indomethacin or meclofenamate. The implication is that a balance exists between the vascoconstrictor influence of renal prostaglandins which was disturbed by the inhibition of renal prostaglandin synthesis (McGiff *et al.*, 1974*b*). A similar physiological interaction between prostaglandins and angiotensin was apparent in the studies of McClatchey and Carr in dogs (1972). The infusion of angiotensin II into a renal artery in subpressor amounts resulted in a unilateral decrease in glomerular filtration rate and renal plasma flow and a decrease in urine volume and electrolyte excretion. The concomitant infusion of PGA_1 in amounts not sufficient to alter systemic blood pressure returned these parameters toward control values. Thus prostaglandin was able to modify the renal response to angiotensin. Similarly, Lonigro *et al.* (1973*b*) were able to demonstrate a reversal of the effect of infused angiotensin II on renal blood flow and urine volume with PGE_2 and PGA_2 (but not with $PGF_{2\alpha}$) as well as with acetylcholine. In addition, evidence has been obtained by Aiken (1974) suggesting that angiotensin stimulation of prostaglandin synthesis results in the tachyphylaxis observed when angiotensin is added to celiac artery spiral strips. Prostaglandins therefore may participate in both the acute and chronic (tachyphylaxis) modulation of the vascular responses to angiotensin.

The possible role of the renin–angiotensin system in the differential response to alterations in renal blood flow subsequent to inhibition of prostaglandin synthetase in the anesthetized vs. the unanesthetized animal, as well as the interaction of prostaglandins with the renin–angiotensin system in the control of the intrarenal distribution of blood flow, has been discussed above.

The inverse relationship between sodium intake and renin–angiotensin system activity is a well-known phenomenon. If there is a relationship between renin–angiotensin system activity and prostaglandin release, then sodium intake should also affect prostaglandin synthesis and release. The 49% decline in plasma PGA levels in subjects on a high sodium intake and the 34% increase in subjects on a low sodium intake, as compared to control values, observed by Zusman et al. (1973) is in agreement with this prediction. A similar inverse relationship between sodium intake and plasma PGA activity has been observed by Lee and Attallah (1974). A high sodium intake has also been demonstrated to reduce the renal content of PGE_2 by 40%, without altering the content of $PGF_{2\alpha}$ (Tobian et al., 1974). Furthermore, McGiff and Itskovitz (1973) observed that renin–angiotensin system activity as regulated by the state of sodium balance could modulate the amount of renal prostaglandin release induced by norepinephrine (McGiff et al., 1972a). An inverse relationship between plasma renin activity, as an estimate of sodium balance, and norepinephrine-induced increases in PGE levels was observed. Thus alterations in the tonic state of renovascular tone associated with changes in renin–angiotensin system activity resulting in altered prostaglandin activity may also modify the renovascular responses to other stimuli.

In addition to an interdependence between renin–angiotensin system activity and prostaglandin release, other vasoactive systems may have a direct or indirect modulating role. Activity of the renin–angiotensin system is influenced by adrenergic activity as well as the state of sodium balance. Increased adrenergic activity resulting from either increases in renal nerve activity (Vander, 1965; Loeffler et al., 1972) or increases in circulating catecholamines (Bunag et al., 1966; Gutman et al., 1973) induced an increase in renal renin release. This adrenergic activity related enhancement of renin release appears to be under control of β-adrenergic receptors (Winter et al., 1971; Pettinger, et al., 1972). The increased angiotensin generations resulting from this increased release of renin results in (1) stimulation of catecholamine release from the adrenal medulla (Peach et al., 1966) as well as from the sympathetic nerve terminals (Schumann, et al., 1970) and (2) increased prostaglandin release. The increase in catecholamines also results in increased prostaglandin release, as will be discussed later and the increase in prostaglandin release buffers the enhanced catecholamine activity both at nerve terminals and physiologically at their vascular effector sites. The interdependence of these various vasoactive systems results in a buffering of their effects and functions as a renovascular homeostatic system.

The mechanism responsible for angiotensin-induced alterations in prostaglandin synthesis and/or release has received very little attention, yet it is already the subject of some controversy. Angiotensin infusions and increases in renin–angiotensin system activity may result in systemic or renal hemodynamic alterations. That angiotensin may produce its effect on prostaglandin release via a local, intrarenal mechanism is suggested by the observation that administration of angiotensin in amounts which do not alter systemic blood pressure still results in increased prostaglandin release. Furthermore, marked increases in prostaglandin release occur when angiotensin is infused into the renal artery of isolated perfused kidney preparations. It is uncertain, however, if the intrarenal action of angiotensin on renal prostaglandin synthesis/release is a direct effect of angiotensin or is indirectly mediated by a hemodynamic or functional alteration induced by the administration of angiotensin. To evaluate this, Sirois and Gagnon (1974) determined the effect of added angiotensin I or angiotensin II on the release rate of PGE_2-like material from an isolated rabbit renal medulla preparation. The release rate of prostaglandin in control experiments was 17.0 ng/mg, not significantly different from the values of 14.7–19.4 or 20.1–24.1 ng/mg observed after the addition of angiotensin I or II, respectively. Corsini *et al.* (1974) also failed to observe increased renin release from rat renal slices *in vitro* in response to the additional PGE_1. These results would be consistent with an indirect effect of angiotensin mediated by renal hemodynamic or tubular alterations on prostaglandin activity. In contrast, Danon and Chang (1973) demonstrated that the addition of angiotensin II (50–500 ng/ml) resulted in two- to sevenfold increase in PGE_2 release from the rat renal papilla *in vitro*. These results are consistent with a direct effect of angiotensin on some element of renal prostaglandin synthesis and/or release. The observation of Aiken (1974) indicating that angiotensin-stimulated prostaglandin synthesis in artery walls may result in angiotensin tachphylaxis is also consistent with a direct local effect of angiotensin on prostaglandin synthesis.

2. The Kallikrein–Kinin System

Bradykinin formed by the enzymatic action of kallikrein on kininogen stimulates the release of prostaglandins from lung (Piper and Vane, 1969; Palmer *et al.*, 1973), spleen (Eble *et al.*, 1972; Ferreira *et al.,* 1973*b*), and kidney (McGiff *et al.,* 1972*b*, 1975), as well as from slices of bovine mesenteric vessels *in vivo* (Terragno *et al.,* 1975). McGiff *et al.* (1972*b*) infused bradykinin (20–100 ng/kg-min) into the renal artery of anesthetized dogs and measured the renal venous concentrations of prostaglandins. Concomitant with a 53% increase in renal blood flow, the concentration of PGE-like material in renal venous blood increased from a mean control value of 0.16 ng/ml to 1.05 ng/ml. With continued infusion of bradykinin, renal blood flow returned toward control values and the renal venous concentration of PGE-like material decreased. The bradykinin-associated alterations in both renal hemodynamics and prostaglandin release may not have been proportional since measurements were made on released prostaglandin concentration in renal venous effluent and measurements of local, intrarenal concentrations could not be performed. It was of note, however, that infusions of PGE_2 at a rate comparable to that observed after bradykinin infusion resulted in similar changes in total renal blood flow. Eledoisin, another vasodilatory polypeptide, did not result in increased renal venous concentrations of PGE-like material even though it resulted in a comparable increase in renal blood flow infused into the renal artery (37.5–150 ng/kg-min). Neither of these peptides resulted in significant changes in the concentration of PGF-like material.

That some vascular effects induced by bradykinin are prostaglandin mediated may also be inferred from the work of Collier (Collier *et al.,* 1971) and Brocklehurst (1971). This interrelation of bradykinin and prostaglandins as a specific interaction is strengthened by the observations that PGE_1 potentiates bradykinin-induced alterations in vascular permeability (Hanson and Williams, 1974) and by the lack of bradykinin-induced increase in renal venous levels of PGF-like material, which is not consistent with the increase in prostaglandin release being related solely to an increase in renal blood flow. The demonstration by Needleman *et al.* (1973) that angiotensin antagonists do not block bradykinin-induced prostaglandin release from the isolated perfused kidney suggests that the renin–angiotensin system does not directly mediate kinin–prostaglandin interdependence. Furthermore, bradykinin does not increase medullary prostaglandin release (Sirois and Gagnon, 1974), which is not consistent with a direct effect of kinins of prostaglandin synthesis.

To further evaluate bradykinin–prostaglandin interdependence, McGiff *et al.* (1975) performed stuides using the isolated perfused dog kidney preparation. In this preparation, the infusion of bradykinin (20–90 ng/kg-min) resulted in maintained increase in renal blood flow of 34–38% above control values and a 48% increase in inner renal cortical blood flow, as measured using radiomicrospheres. In addition, brady-kinin resulted in a decrease in urine osmolality and an increase in solute free water clearance, which was only relative since the glomerular filtration rate declined. After inhibition of prostaglandin synthetase activity with indomethacin, the administration of bradykinin resulted in a statistically insignificant increase in total renal blood flow and an 82% increase in inner cortical renal blood flow. The increase in inner cortical flow was greater after indomethacin because inner cortical blood flow had decreased by 37% subsequent to the inhibition of prostaglandin synthesis by indomethacin. Furthermore, the infusion of bradykinin after indomethacin did not result in any significant altera-tions in either urine osmolality or solute free water clearance. The interdependence of bradykinin and prostaglandins was based on the following : (1) the severe blunting of the renal vasodilatory response to bradykinin by the inhibition of prostaglandin synthetase activity, (2) the similarity in renal hemodynamic alterations induced by bradykinin and prostaglandin infusions, and (3) the abolition of changes in solvent handling induced by bradykinin, which may have been related to the antagonistic effect of prostaglandin on antidiuretic hormone (see later), by the inhibition of prostaglandin synthetase. That the renal actions of bradykinin are not totally dependent on stimulation of prostaglandin synthesis and release is suggested by the marked redistribution of intrarenal blood flow induced by bradykinin after the inhibition of prostaglandin synthesis.

The limitations of the isolated perfused kidney preparation, the interrelationship of prostaglandins and renin–angiotensin system activ-ity, and the demonstration that bradykinin stimulates prostaglandin release apart from systemic alterations (McGiff et al., 1975; Needle-man et al., 1973) led to additional studies. In these experiments (Gagnon et al., 1975b), the effects of unilateral renal artery infusion of bradykinin (80 ng/kg per min) on the renal secretory rates of PGE and PGF as well as renal renin secretory rate were determined. Intrarenal cortical blood flow distribution was determined using radiomicro-sphere and total renal blood flow using the clearance of p-aminohippur-ate corrected for extraction and packed cell volume. Bradykinin resulted in an initial rise in renal blood flow from a mean control value of 228 ml/min to a maximum value of 339 ml/min at 15 min after the onset of infusion (Fig. 6). This increase in renal blood flow was not maintained, and mean renal blood flow decreased to 273 ml/min after 117 min of bradykinin infusion, a value not statistically different from control value. In association with these alterations in total blood flow, there were marked alterations in intracortical renal blood flow charac-terized by a progressive increase in inner cortical renal blood flow (Fig.

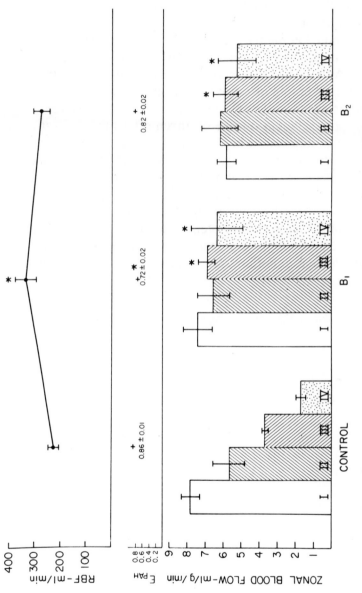

Fig. 6. Effect of renal artery bradykinin infusion (80 ng/kg per min) on total renal blood flow (RBF), PAH extraction (E_{PAH}), and intrarenal blood flow distribution (outer, I, to inner, II, cortex). Measurements were obtained 15 min (B_1) or 117 min (B_2) after onset of infusion. *Significantly different from control. Adapted from Gagnon et al. (1975b).

6). Extrapolating from the zonal blood flows and kidney weights, inner cortical renal blood flow increased from a mean control value of 40 ml/min to 92 and 77 ml/min after 15 and 117 min of bradykinin infusion, respectively. Since neither total cortical flow nor outer cortical flow changed significantly from the mean values calculated during control periods (204 and 159 ml/min, respectively), the increase in total renal blood flow was probably the result of an increase in noncortical renal blood flow. Although PGE secretory rate was determined in only one animal, and a sevenfold increase was observed, it was assumed based on this and previous studies (McGiff *et al.*, 1972*b*) that a reproducible increase in prostaglandin release occurred after the administration of bradykinin. The infusion of bradykinin also resulted in an increase in renal renin secretory rate (Fig. 7), which reached significance after continued infusion. The infusion of bradykinin also resulted in maintained increases in urine volume, solute clearance, and solute free water clearance. In a different group of dogs, bradykinin was infused after the administration of meclofenamate (10 mg/kg, intravenously). As may be seen (Fig. 8), meclofenamate reduced total renal blood flow by 29% and decreased both outer and inner cortical renal blood flows.

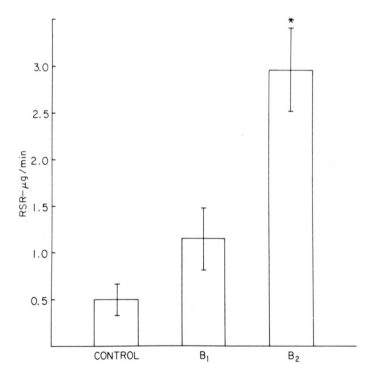

Fig. 7.   Alterations in renal renin secretory rate (RSR), estimated from the arteriovenous plasma renin activity difference, and simultaneous renal plasma flow during bradykinin administration. B_1 and B_2 as in Fig. 6. *Significantly different from control. Adapted from Gagnon *et al.* (1975*b*).

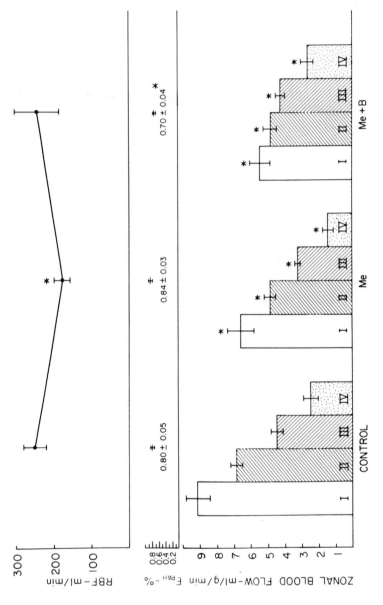

Fig. 8.   Effect of the administration of meclofenamate (Me) and subsequent administration of bradykinin (B) on renal blood flow (RBF), PAH extraction (E_{PAH}), and cortical renal blood flow distribution. *Significantly different from control. Adapted from Gagnon *et al.* (1975*b*).

The subsequent administration of bradykinin resulted in a return of total renal blood flow to control values and increase in cortical blood flow. The administration of meclofenamate (Fig. 9) resulted in a marked depression in the renal secretory rate of PGE and PGF. Concomitantly, there was also a decline in the renal renin secretory rate. The infusion of bradykinin did not result in an increase in prostaglandin secretory rate, but did return renin secretory rate to a value not different from control. Furthermore, the infusion of bradykinin after the administration of meclofenamate did not result in significant alterations in urine volume, solute clearance, and solute free water clearance.

The results of these studies indicate that bradykinin-induced alterations in renal hemodynamics are partially dependent on induced alterations in prostaglandins. That bradykinin infusion increased total renal blood flow only to control values after prostaglandin synthetase inhibition suggests that renal hyperemia after the administration of this kinin is prostaglandin dependent. The continued redistribution of intracortical renal blood flow toward the inner cortex and noncortical areas of the kidney by bradykinin after prostaglandin synthetase inhibition suggests that (1) the effects of bradykinin on juxtamedullary and medullary areas of the renal vasculature are not mediated by prostaglandins and/or (2) prostaglandin synthesis within the kidney is not

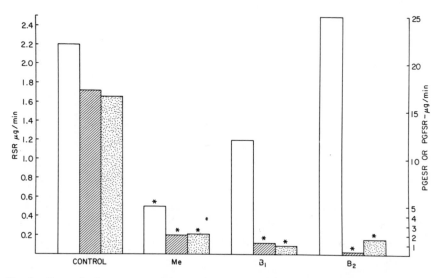

Fig. 9. Determinations of the renal secretory rates of PGE (PGESR, hatched bars), PGF (PGFSR, dotted bars), and renin (RSR, open bars) after meclofenamate (Me) or subsequent bradykinin infusion (B₁ at 15 min, B₂ at 117 min) *Significantly different from control. Adapted from Gagnon *et al.* (1975*b*).

maximally inhibited and that bradykinin still results in prostaglandin release within the intrarenal sites of prostaglandin synthesis. It would also appear that the directional changes in prostaglandin and renin secretion are similar, as would be assumed if they functioned as buffers to alteration in renal vascular resistance. In this regard, the lack of a maintained hyperemia in the control group with continued infusion of bradykinin may be the result of an increase in renin release resulting in a homeostatic return in renal blood flow toward control, as well as the lack of a sustained increase in prostaglandin release. The increase in renin secretory rate in prostaglandin synthetase inhibited animals suggests that in addition to any effect of prostaglandins there is a direct effect of bradykinin on renin–angiotensin system activity. This interrelationship of bradykinin and renin–angiotensin system activity is supported by the recent report of Wang *et al.,* (1975) demonstrating proportional alterations in plasma renin activity and the plasma concentrations of bradykinin and angiotensin II in response to the acute intravenous administration of a saline load. Similarly, the assumption of an upright posture simultaneously increased renin–angiotensin system activity and plasma bradykinin levels.

That these alterations in the response of the blood flow to an organ may also be important in the local mediation of blood flow is suggested by recent studies by Terragno *et al.* (1975). Using isolated strips of bovine mesenteric arteries and veins, Terragno *et al.* (1975) were able to demonstrate prostaglandin synthesis and release into the incubating medium, which was markedly inhibited by meclofenamate. In comparing total prostaglandin synthesis to that accounted for by the synthesis of PGE_2, the synthesis of $PGF_{2\alpha}$ was estimated to be less than half the rate of PGE_2 synthesis. The addition of bradykinin resulted in an increase in the synthesis of PGE-like material by arteries and an increase in PGF-like material synthesis by veins. In addition to being a possible explanation for the well-known variable efects of prostaglandins on various blood vessels, this differential stimulation of prostaglandin synthesis in the arterial and venous tree is of great potential importance in the local control of filtration across the intervening capillary bed of any organ, including the kidney. Messina *et al* (1975) estimated vasoconstrictor responses in rat muscle by measuring vascular lumen diameter in response to the administration of indomethacin, PGE_1 and PGE_2, bradykinin, norepinephrine, and angiotensin. Their study indicated that prostaglandin synthetase inhibition diminished arteriolar lumen diameter, diminished the increase in diameter observed after the administration of bradykinin, and potentiated the decrease in diameter observed after the addition of either angiotensin or norepinephrine. Thus within the microcirculation prostaglandins would appear to both mediate the vasodilator response to bradykinin and modulate the vasoconstrictor response to either angiotensin or norepinephrine. The interrelationship, interdependence, and interaction of prostaglandins, the kallikrein–kinin systems and the renin–angiotensin system make considerations of the influence of alterations in any one component problematic without consideration of associated changes.

3. The Sympathetic Nervous System/Catecholamines

Increased renal sympathetic nervous activity and catecholamine release, as well as circulating endogenous or exogenous catecholamines, result in increased renovascular tone/vasoconstriction. Early studies by Holmes *et al.* (1963), Bergström *et al.* (1964a), and Strong and Bohr (1967) suggested that some interaction of catecholamines and prostaglandins on vascular tone might be occurring. Using the *in situ* autoperfused dog kidney, in which renal blood flow was maintained at a constant level, Dunham and Zimmerman (1970) demonstrated a low basal efflux of prostaglandinlike material which was markedly enhanced by renal vascular constriction induced by either renal nerve stimulation (2–10 Hz, 40 V, 1 msec) or norepinephrine infusion. Similar results were obtained by McGiff *et al.* (1970a, 1972a) using norepinephrine infusion, but not with renal nerve stimulation. These investigators observed that the recovery of renal blood flow and urine flow toward normal with continued norepinephrine infusion was related to the appearance of PGE-like material in the renal venous effluent. Indeed, in one dog the absence of renal blood flow recovery was associated with a failure of increased prostaglandin release. These results, as well as the demonstration that PGE inhibits the vasoconstrictor effect of norepinephrine (Hedqvist and Brundin, 1969; Viguera and Sunahara, 1969; Lonigro *et al.*, 1973b), are consistent with prostaglandin release as being the determinant of the recovery of the renal blood flow and the antidiuresis during norepinephrine infusion. Fujimoto and Lockett (1970) have made similar observations. The lack of an increase in prostaglandin release with renal nerve stimulation in these experiments may have been related to the low frequency (2–6 Hz) or low amplitude (10 V) of stimulation (Junstad and Wennmalm, 1974). Thus Davis and Horton (1972) demonstrated a marked increase in prostaglandins A; C; or B-like substances with electrical stimulation of renal nerves in the rabbit at a frequency of 10 Hz, and this increase was reduced by the administration of indomethacin. PGE_2 has also been demonstrated to inhibit the vascular contraction and prostaglandin induced by electrical stimulation, while both indomethacin and eicosa-5,8,11,14-tetraynoic acid potentiate it (Hedqvist and Brundun, 1969; Heqvist and Wennmalm, 1971; Greenberg, 1974).

That catecholamine-induced prostaglandin release from the kidney is mediated by α-adrenergic receptor site stimulation has been demonstrated by Needleman *et al.* (1974*b,c*). Using the isolated perfused rabbit kidney, a marked increase in prostaglandin release was observed in response to the infusions of epinephrine or norepinephrine. This prostaglandin release was blocked by indomethacin and α-adrenergic blockage with phenoxybenzamine. The observations that the β-adrenergic agonist isoproterenol did not stimulate prostaglandin release were consistent with the concept that prostaglandin release was related to an α-adrenergic receptor site action. That the effect was specific was evident from the lack of an inhibition of α-adrenergic blockage on angiotensin II induced stimulation or prostaglandin release. It is of note that Ferreira and Vane (1967) observed a similar effect of α-adrenergic blockade on splenic nerve stimulation induced prostaglandin release. Parallel studies were performed using renal nerve stimulation. These experiments indicated that indomethacin blocked prostaglandin release but not catecholamine release induced by renal nerve stimulation. Phenoxybenzamine blocked both the vasoconstriction, related to catecholamine release, and prostaglandin release during nerve stimulation while propranolol was without effect. Since prostaglandin release has been observed during renal ischemia (McGiff *et al.*, 1970*c*), additional studies were performed during ischemia. Four minutes of ischemia resulted in an increase in prostaglandin release which was blocked by indomethacin but not by pretreatment with either phenoxybenzamine or propranolol.

Prostaglandins also appear to have a feedback interdependence with catecholamines which could modulate renal adrenergic tone (Brody and Kadowitz, 1974). Prostaglandins, primarily of the E type, have been demonstrated to inhibit the release of catechomanines from adrenergic nerve terminals (Stjärne, 1973*a,b;* Hedqvist, 1970, 1971, and 1974), and the inhibition of prostaglandin synthesis has been observed to result in an increase in the renal excretion of norepinephrine (Junstad and Wennmalm, 1972) as well as the facilitation of neurotransmission (Hedqvist *et al.,* 1971; Samuelsson and Wennmalm, 1971). The physiological counterpart to these alterations in neurotransmitter activity consists of the potentiation of renal nerve/ catecholamine induced vasoconstriction by prostaglandin synthetase inhibition, and the inhibition of this vasoconstriction by the administration of prostaglandin. The "feedback" response of this interdependence of prostaglandins and sympathetic nerve/catecholamines can be seen in the escape of increased vascular resistance from renal nerve stimulation. After an initial decrease in renal blood flow, there is a return toward control values which is associated with an increase concentration of prostaglandin in the renal venous effluent. Since prostaglandins decrease catecholamine release, this negative feedback inhibition would participate in this escape, as well as the physiological antagonism noted above. The observation by Pomeranz et al., (1968) that sympathetic nerve stimulation results in a decrease in outer cortical renal blood flow and an increase in outer medullary blood flow is consistent with this suggestion. There is also, however, a feedback regulation of norephinephrine release from vasoconstrictor nerves which is independent of prostaglandins (Stjärne and Gripe, 1973).

Prostaglandins of the F type appear to have a different interaction with adrenergic/catecholamine effects. There is marked facilitation of vasconstriction (Brody and Kadowitz, 1974; Ducharme et al., 1968). In this regard, Malik and McGiff (1975) noted that the concentration of $PGF_{2\alpha}$ required to affect adrenergic transmission was significantly greater than the concentrations of PGE compounds required. Furthermore, vasoconstriction has been demonstrated in response to prostaglandins of the F type in both the isolated rabbit and rat kidney (Malik and McGiff, 1975). The lack of marked or reproducible alterations in PGF-type release by renal nerve stimulation or catecholamine administration may be due to the relatively lower rate of synthesis of this prostaglandin as compared to the intrarenal synthesis of prostaglandins of the E type (Crowshaw and McGiff, 1973). Recent studies of prostaglandins of the B type have suggested that they are vasoconstrictors and induce the release of catecholamines from adrenergic nerve terminals in response to sympathetic adrenergic discharge (Greenberg et al., 1974). The demonstration that prostaglandin A compounds are converted to C compounds by an isomerase and that C-type prostaglandins are unstable and readily form B-type prostaglandins (Jones, 1972a,b) suggests that this observation is relevant to modulation of renovascular tone by prostaglandin A.

Consideration of interdependence of vasoactive systems becomes even more complex because of the interrelationships of the various components. Increased adrenergic activity, as the consequence of either exogenous catecholamines or renal nerve stimulation, results in increased renin–angiotensin system activity (Loeffler et al., 1973; Vander, 1965; Bunag, et al., 1966) which is most likely mediated by β-adrenergic receptors (Pettinger et al., 1972; Nolly et al., 1974). The induced increase in angiotensin generation, in addition to previously discussed effects on prostaglandin release, modulates adrenergic tone by stimulating catecholamine release both from the adrenal medulla (Peach et al., 1966) and from sympathetic nerve terminals (Zimmerman and Gisslen, 1968; Peach et al., 1970). The physiological consequences of this interaction were reported by Johnson et al. (1974) and Needleman et al., (1974a). Angiotensins I and II resulted in a dose-dependent enhancement of nerve stimulation induced contraction of the rat vas deferens and the renovascular response of the isolated perfused rat kidney. The augmentation of sympathetic nerve/catecholamine effect by angiotensin would appear to be the consequence of both an increase in neurotransmitter release at the presynaptic level and an increase in smooth muscle responsiveness at the postsynaptic level (Johnson *et al.*, 1974).

The mechanism of these various interactions is unclear. The observation that β-adrenergically mediated catecholamine stimulation of renin release and antidiuresis result in increased renal cyclic AMP concentrations (Beck *et al.*, 1972*b*) is consistent with a role for the adenylyl cyclase system. The potentiation of norepinephrine-induced renin release by theophylline, a phosphodiesterase inhibitor, is consistent with this view (Nolly *et al.*, 1974). Adenine nucleotides are also potent stimulators of prostaglandin release from many organs (Shio and Ramwell, 1972; Abdulla and McFarland, 1972; Needleman *et al.*, 1974*a*). Thus both adenosine monophosphate and adenosine triphosphate increase prostaglandin synthesis and release from the kidney, spleen, and liver (Needleman *et al.*, 1974*a*). It has also been observed that prostaglandin-induced norepinephrine release from tissues may be modified by the cyclic AMP content of cells, and that prostaglandins may modify the phosphodiesterase activity of cells (Al Tai and Graham, 1972). The various effects of prostaglandins on components of the adenylyl cyclase system have been recently reviewed (Horton, 1969; Hittlemen and Butcher, 1973). In view of the multitude of opposing observations and the lack of firm, experimental data with reference to the kidney, it is too early to make firm statements concerning the exact nature of this modulation.

That an interrelationship of prostaglandins and catecholamines does not extend to all species is evident from the studies of Terashima *et al.* (1974) and Malik and McGiff (1975). Using the *in situ* perfused dog kidney, Terashima *et al.* (1974) demonstrated renal vasoconstriction and prostaglandin release in response to the administration of norepinephrine (4–10 μg/kg) which was prevented by pretreatment with either acetylsalicylic acid or indomethacin. The pretreatment of this preparation with phenoxybenzamine, however, blocked neither the vascular response nor the prostaglandin release induced by norepinephrine. Analogous results were obtained using epinephrine. Malik and McGiff (1975) determined the effect of PGE_1, PGE_2, PGA_1, and PGA_2 on the vasoconstrictor response to sympathetic nerve stimulation or norepinephrine administration in isolated perfused preparations of both rabbit and rat kidneys. In the rabbit kidney, all of the prostaglandins markedly attenuated the vasoconstrictor response to nerve stimulation but not to catecholamine administration. In contrast, these prostaglandins increased the vasoconstrictor response to nerve stimulation in the rat kidney. Similarly, inhibition of prostaglandin synthesis augmented the response to nerve stimulation in the rabbit kidney and blunted the response in rat kidneys, and the converse was observed when arachidonic acid was added to these preparations. Administration of these prostaglandins alone resulted in vasodilatation of the rabbit renal vasculature and vasoconstriction of the rat renal vasculature. Furthermore β-adrenergic blockade did not alter the vasodilatory response to prostaglandins in the rabbit kidney, while α-adrenergic blockade did block the response to norepinephrine. In the rat kidney, α-adrenergic blockade did not block the vasoconstriction induced by these prostaglandins. Thus there are major species differences in the interrelation of prostaglandins and catecholamines which must be taken into consideration when interpreting the physiological response of the kidney to manipulations of either catecholamines or prostaglandins.

4. Schema of the Interdependence of Vasoactive System

Figure 10 diagrammatically represents the interactions discussed in this section and is provided solely as a basis for considering the interdependence of the various vasoactive systems and how they may interrelate. As continued investigations are undertaken with due consideration to the potential interactions of the various vasoactive components acting on the kidney, clarification of the role played by prostaglandins, the renin–angiotensin system, the kallikrein–kinin system, and sympathetic nerves/catecholamines will be forthcoming. The

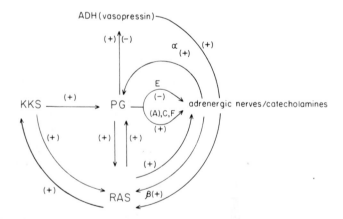

Fig. 10.   Schema depicting the interdependence of various vasoactive systems and prostaglandins on renal hemodynamics. Abbreviations: ADH, antidiuretic hormone; PG, prostaglandins of the A, C, F, or E type; KKS, kallikrein–kinin system; RAS, renin–angiotensin system; α or β, α- or β-adrenergic nerves.

schema may be used to indicate the relevent alterations induced by a change in one component on the other components. For example, it is theoretically possible that the tonic level of renin–angiotensin system activity in the rat (or its stimulation by prostaglandins) may account for the vasoconstrictor responses observed by Malik and McGiff (1975) rather than species differences in direct prostaglandin effects. That the prostaglandins do not mediate all known renal hemodynamic alterations is indicated in recent studies demonstrating that dopamine renal vasodilatation is not accompanied by prostaglandin release (Dressler *et al.*, 1975; Needleman *et al.*, 1974*b*).

IV. ROLE OF PROSTAGLANDINS IN SALT AND WATER HOMEOSTASIS

Associated with the administration of prostaglandins there are marked alterations in fluid and electrolyte excretion (Johnston *et al.*, 1967; Vander, 1968; Carriere *et al.*, 1971; Thompson *et al.*, 1971; DiScala, 1971; Martinez-Maldonado *et al.*, 1972; Gross and Bartter, 1973; Arendshorst *et al.*, 1974). These alterations consist of increases in urine flow rate and the excretion of sodium. A diuresis and natriuresis have also been observed after stimulation of endogenous prostaglandin secretion with arachidonic acid (Tannenbaum *et al.*, 1975), as well as after the infusion of exogenous prostaglandins, and the induced alterations are not prevented by α- or β-adrenergic blockade or by atropine (Fujimoto and Lockett, 1970). Although many of the observations were performed in animals, these effects have been confirmed in man (Carr, 1968, 1970; McClatchey and Carr, 1972; Fichman *et al.*, 1972; Fichman and Horton, 1973). While the prostaglandins used for the study of induced alterations in electrolyte and fluid excretion have been primarily of the E and A types, similar effects have been observed with both B type (Marchand *et al.*, 1973) and F type (Fulgraff and Brandenbusch, 1974) prostaglandins. In general, the increases in fluid and electrolyte excretion have occurred concurrent with induced renal hemodynamic alterations; the responses appear to have a threshold and are related to the dose of prostaglandin administered and are unrelated to any specific systemic effects of circulating prostaglandins.

A. Prostaglandins Are Natriuretic

The mechanism by which prostaglandins result in an altered handling of fluid and electrolytes is not readily apparent from the descriptions provided in many of the above noted studies. If one restricts the consideration of induced alterations to the major urinary cation, sodium, then one can begin to evaluate possible alterations in renal function which may be responsible for the effects of prostaglandins. Since reproducible alterations in either plasma sodium concentration or glomerular filtration rate have not been observed, one can exclude an increased load of filtered sodium as the cause of the enhanced excretion of fluid and electrolytes. Similarly, since prostaglandins increase aldosterone release (Fichman *et al.*, 1972), the natriuresis is not a result of the inhibition of sodium reabsorption in the distal nephron. In considering the mechanism of action of prostaglandins in the kidney, Gross and Bartter (1973) considered two alternative effects which might be responsible for the natriuretic and diuretic effects of these agents. Thus all of the alterations in renal function, with specific reference to the induced natriuresis or diuresis, associated with the administration of prostaglandin could be accounted for by consideration of the resultant alterations in renal hemodynamics or by direct tubular epithelial effects. Direct decreases in sodium reabsorption could, depending on the nephron segment involved, result in the changes in fluid and electrolyte excretion described. These alterations include increased urine dilution, decreased urine concentration, and enhanced sodium excretion, Alternatively, redistribution of intrarenal blood flow and glomerular filtration, or the associated alterations in the physical factors responsible for the reabsorption of tubular fluid, could also account for the observed changes in these various parameters.

Although most of the studies performed using standard clearance techniques have been interpreted as consistent with an alteration in renal hemodynamics/physical factors, the evidence is less than compelling. Attempts to clarify the effects of prostaglandins on renal function by studying the alterations induced in single nephrons using the technique of micropuncture have not clarified the issue. Fulgraff and Meiforth (1971) investigated the effect of PGE_2 on tubular function in the rat. The marked diuresis and natriuresis observed occurred without any significant change in total kidney or single nephron glomerular filtration rate. Furthermore, fluid and electrolyte absorption in proximal nephron segments was unaltered and slight decreases in distal tubular sodium reabsorption were observed. The lack of an effect of PGE_2 on proximal tubular fluid absorption was confirmed by Schneider *et al.* (1973) and Strandhoy *et al.* (1974). However, both of these investigators were able to demonstrate a significant decrease in proximal tubular sodium reabsorption after the infusion of PGE_1. Similarly, Kauker (1973) observed a diminution in proximal tubular fluid reabsorption after the infusion of PGE_2 in rats. Unfortunately, these more sophisticated methods of approach have not allowed a total separation of the possible direct effects of prostaglandins from those mediated by renal hemodynamic/physical factors. Even the observation that $PGF_{2\alpha}$ induces a natriuresis without change in renal blood flow (Fulgraff and Brandenbusch, 1974; Fulgraff et al., 1974) does not rule out renal hemodynamic factors in the genesis of altered fluid and electrolyte handling since alterations in intrarenal blood flow distribution or physical factors may occur without changes in total renal blood flow.

One of the possible direct effects of prostaglandin on fluid and electrolyte reabsorption is by altering the adenylyl cyclase system. Prostaglandins have been demonstrated to affect adenylyl cyclase activity and the generation of cyclic AMP in various tissues (Beck et al., 1972a,b, 1970; Ramwell and Shaw, 1970) and both cyclic AMP and dibutyryl cyclic AMP reproduce the natriuretic and diuretic effects of prostaglandins when administered into the renal artery (Gill and Casper, 1971; Agus et al., 1973). In order to evaluate the potential mediation of prostaglandin-induced changes in fluid and electrolyte handling by alterations in the adenylyl cyclase system, Gross and Bartter (1973) determined the effect of various prostaglandins on the urinary excretion of cyclic AMP. Direct intraarterial infusion of PGE_1 or PGA_2, in natriuretic doses, did not alter the urinary excretion of cyclic AMP. Intrarenal artery administration of $PGF_{2\alpha}$ was neither diuretic nor natriuretic, in contrast to the findings of Fulgraff et al. (1974), and also did not alter the urinary excretion of cyclic AMP. Strandhoy et al. (1974) determined the effect of PGE_1 and PGE_2 on the cyclic AMP generation of kidney slices in vitro. Neither of the prostaglandins significantly increased cyclic AMP concentration in cortical slices of dog kidney. However, both prostaglandins markedly increased cyclic AMP production by renal slices obtained from the outer medullary portion of the dog kidney. Thus, although PGE_1 was demonstrated to decrease fluid absorption in proximal nephron segments within the superficial cortex, this alteration did not correlate with any change in cyclic AMP concentration. A relationship between increased cyclic AMP concentration in the medullary portion of the kidney and increased sodium excretion was postulated. Unfortunately, these results which were obtained from in vitro experiments are in conflict with the in vivo results obtained by Gross and Bartter (1973) as well as the effects of prostaglandins on other tissues. Lafferty et al. (1972) could demonstrate no effect on cyclic AMP by either PGA_2 or PGE_2. Furthermore, both PGA_2 and PGE_2 are natriuretic, yet only PGA_2 affects renal oxygen consumption and Na^+-K^+ ATPase, suggesting that in vitro metabolic effects and in vivo natriuretic effects are separable (Lafferty et al., 1972). Thus it is not possible to establish a definitive relationship between prostaglandin-induced alterations in adenylyl cyclase and either the natriuresis or diuresis associated with the administration of prostaglandins.

Based primarily on the observations inferred from descriptive physiology, Lee and co-workers have postulated a significant role for prostaglandins, primarily PGA_2, in the renal regulation of salt and water homeostasis (Attallah and Lee, 1973b; Lafferty et al., 1965; Lee et al., 1967; Lee, 1969, 1972a,b, 1973, 1974; Lee and Attallah, 1974). According to this proposal, PGA_2 (or PGE_2) functions as a "natriuretic hormone" or "third factor" in the intrarenal regulations of the response to an increased sodium chloride load to the total organism. The postulate suggests that as a result of the intravascular volume expansion concurrent with an increase in sodium intake, there will be release of prostaglandins from the interstitial cells of the papillary area of the kidney. The released prostaglandins will, via an intrarenal circulation of prostaglandins, result in renal hemodynamic alterations characterized by a redistribution of intrarenal blood flow from medulla to cortex. The induced cortical arteriolar dilatation and increase in physical factors as well as an inhibition of Na^+-K^+ ATPase activity, result in a decrease in sodium reabsorption. Thus the ensuing natriuresis and diuresis would correct the volume expansion and an adjustment of fluid and electrolyte balance would have occurred. There are, however, some specific areas of this proposed mechanism which require further proof and understanding before a homeostatic role can be assigned to renal prostaglandins. According to the hypothesis put forward by Lee and co-workers, the initial step in this homeostatic mechanism is an increase in prostaglandin release from the interstitial cells of the renal papilla. The evidence to support this proposal consists of an alteration in renal PGA_2 content in rabbits maintained on a high sodium chloride intake (Attallah and Lee, 1973b). When the sodium content of food fed to rabbits was increased fourfold (by increasing the sodium chloride concentration of their drinking water for 5 days), the PGA_2 content of the renal cortex and outer medulla doubled while the content of the renal papilla decreased by half (Attallah and Lee, 1973b, 1974). These observations were interpreted as indicative of an increase in prostaglandin release from the papilla resulting in a diminished content, and an increase in prostaglandin content of the renal cortex and medulla due to the intrarenal circulation of prostaglandin. As previously noted, the concentration of prostaglandin dehydrogenase within the cortex makes any determination of prostaglandins in this tissue highly suspect. Furthermore, the absence of precautions to delimit the in situ generation of prostaglandin from any region of the kidney would suggest that the values are more indicative of potential prostaglandin synthesis rather than true content. If one accepts these values as indicative of potential synthetic activity, the results become more meaningful, especially when one also considers that prostaglandins are not stored but are synthesized and released as needed. The observation that papillary prostaglandin content decreases is more consistent with a decrease in prostaglandin synthetic activity in this area of the kidney, rather than the increased release proposed by Lee *et al.* (1973). Indeed, patients maintained on an increased sodium intake have lower PGA concentrations (Lee, 1974; Zusman *et al.,* 1973), and rats exposed to a high salt diet have a lower renal PGE content (Tobian *et al.,* 1974). The inverse relationship between prostaglandin concentration and salt balance is more consistent with an antinatriuretic action, or suggests that prostaglandins play no direct role in salt and water homeostasis but are indirectly modulated by some other factor(s).

Even if prostaglandin activity increases in response to increased sodium chloride intake, then the next step in the proposed mechanism, a redistribution of renal blood flow from medulla to cortex, is also open to question (Lee, 1974). This inference is based on determinations of renal blood flow distribution using inert gas washout techniques (Carriere *et al.,* 1971; Barger *et al.,* 1967) and the infusion of relatively large amounts, up to 6 μg/min, of prostaglandins into the renal artery. More recent studies (Tannenbaum *et al.,* 1975; Chang *et al.,* 1975, Kirschenbaum *et al.,* 1974), in contrast, indicate that prostaglandins result in a marked increase in inner cortical or medullary flow rate. In view of the larger, pharmacological doses used with the inert gas washout technique, as well as the inability of gas washout to accurately delineate alterations in blood flow within the renal cortex, the more current studies using radiomicrospheres for the determination of blood flow distribution seem more accurate. Furthermore, observations made on the effect of endogenous increase in prostaglandin synthesis with arachidonic acid (Chang *et al.,* 1975), or using the isolated perfused dog kidney (Itskovitz and McGiff, 1974), lend support to the conclusion that the primary effect of prostaglandins on the intrarenal distribution of blood flow is to increase flow to the innner cortical/ medullary regions of the kidney. Since inner cortical or juxtamedullary nephrons are more avid reabsorbers of sodium than superficial nephrons (Horster and Thurau, 1968), a decrease in sodium excretion should accompany this redistribution in blood flow. It seems very unlikely, therefore, that an increase in cortical blood flow would occur as the result of increased prostaglandin release *per se.*

Any alterations in enzyme activity within the renal parenchyma consistent with a decrease in sodium reabsorption might be the result of alterations in blood flow rather than direct effects of prostaglandin.

Increases in noncortical renal blood flow would result in increased peritubular capillary pressure and increased medullary blood flow with washout of medullary solute. Since decreases in medullary solute concentration may be associated with decreases in Na^+-K^+ ATPase activity (Alexander and Lee, 1970), a decrease in sodium reabsorption attributed to an alteration in the concentration of this enzyme could be the indirect result of a primary, prostaglandin-mediated alteration in renal hemodynamics. The observation that both PGE_2 and PGA_2 are natriuretic, yet only PGA_2 results in decreased Na^+-K^+ ATPase activity *in vitro* (Lafferty *et al.*, 1972; Lee, 1974), lends support to this interpretation.

As indicated by Lee (1974), "the hypothesis that PGA_2 or PGE_2 might function as intrarenal natriuretic hormone is derived from a large body of evidence that is primarily circumstantial in nature." For prostaglandins to play an important role in fluid/electrolyte homeostasis by the kidney, they must be active in physiological rather than pharmacological concentrations, and during endogenous generation rather than exogenous administration. In view of the marked alterations in the physical factors responsible for fluid reabsorption by the nephron as a result of renal vasodilatation, as well as any indirect effect of these alterations on the direct mechanisms responsible for sodium transport by renal tubular epithelium, the enhanced excretion of salt and water may be the artifact solely related to the marked changes in renal hemodynamics. This one factor makes determination of the mechanism by which prostaglandins alter fluid and electrolyte transport an important step in our understanding of the effects of these agents on renal function. In this regard, a dual action of prostaglandins seems plausible. Prostaglandins may, indirectly through renal hemodynamic alterations, result in increased fluid and electrolyte excretion which is buffered, or counterbalanced, by a direct effect of prostaglandins on the renal tubular reabsorptive mechanisms to enhance sodium reabsorption. Evidence will be presented that prostaglandins are antinatriuretic and that the observed natriuresis is a consequence of renal vasodilatation which obliterates the enhanced sodium reabsorption.

B. Prostaglandins Are Antinatriuretic

That prostaglandins, apart from any renal hemodynamic effects, increase sodium reabsorption can be inferred from "clues" available in the literature: (1) the application of prostaglandin precursors or prostaglandins *in vitro* to tissues capable of the active transport of sodium resulted in alterations consistent with a stimulation of net transport (Barry and Hall, 1969; Fassina *et al.*, 1969; Lipson and Sharp, 1971;

Hall, 1973; Lote *et al.*, 1974); (2) prostaglandins of both E and F types result in an increase in oxygen consumption by kidney slices (Rabito and Fasciolo, 1973), and there is a direct relationship between oxygen consumption and sodium transport in the kidney; (3) there appears to be an inverse, rather than a direct, relationship between prostaglandin activity and sodium intake (Zusman *et al.*, 1973; Tobian *et al.*, 1974; Lee 1974), suggesting that the increase in prostaglandins with sodium deprivation may function to promote sodium transport and conserve sodium losses; and (4) inhibition of prostaglandin synthetase activity in the isolated perfused dog kidney results in an increased fractional excretion of filtered sodium and a diminution in free water clearance (Vanherwegham *et al.*, 1975), indicative of a decrease in sodium reabsorption in association with a decrease in prostaglandin synthesis and release. If the antinatriuretic effects of prostaglandins are not apparent as the result of associated changes in renal hemodynamics, they may become manifest under conditions of prostaglandin synthetase inhibition and loss of hemodynamic effects with an associated antinatriuresis.

In recent elegant studies by Kirschenbaum and Stein (1975*a,b*), the effect of prostaglandin release on urinary sodium excretion was investigated by blocking prostaglandin synthesis. The effectiveness of prostaglandin blockade was established by the demonstration of a marked depression in renal venous PGE concentration, as measured by radioimmunoassay, after the administration of indomethacin, meclofenamate, or RO 20-5720. In order to both qualitatively and quantitatively evaluate the renal homeostatic response to inhibition of prostaglandin synthesis, all dogs were studied under conditions of Ringer's solution volume expansion, a maneuver known to markedly increase sodium excretion. It was apparent (Fig. 11A) that both control animals and dogs after prostaglandin synthetase inhibition had an equivalent natriuretic response to volume expansion, which occurred even if the synthetase inhibitor was administered after the induction of volume expansion (Fig. 11B). These results are not consistent with prostaglandins having an antinatriuretic effect, since a diminished natriuresis would have had to have been observed. Furthermore, when the effects of prostaglandin synthetase inhibition on renal blood flow were approximated by concurrent administration of angiotensin and volume expansion (Fig. 12), the natriuresis fell by 75% in the kidney receiving the angiotensin infusion! Thus angiotensin and prostaglandin synthetase inhibition had different effects on sodium reabsorption even though they have similar effects on renal blood flow. These results suggest that prostaglandin synthetase inhibition overcomes the effect of hemodynamic alteration on sodium reabsorption by some other

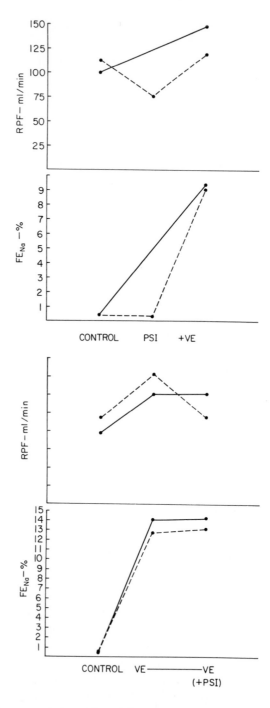

Fig. 11. (A) Effect on renal plasma flow (RPF) and fractional excretion of sodium (FE$_{Na}$) of volume expansion (VE) superimposed on prostaglandin synthetase inhibition (PSI) in anesthetized dogs. (B) Effect of prostaglandin inhibition superimposed on continuously volume-expanded dogs. Solid line, control group; interrupted line, experimented group. Adapted from Kirschenbaum *et al.* (1975).

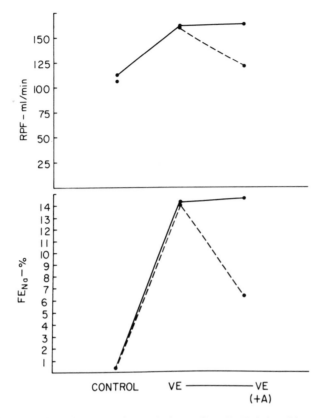

Fig. 12. Effect of unilateral decrease in renal plasma flow (RPF) induced by angiotensin (A) infusion (interrupted line) compared to the contralateral kidney (solid line) on sodium handling (FE_{Na}) in volume-expanded (VE) anesthetized dogs. Adapted from Kirschenbaum *et al.* (1975).

effect on sodium transport. In other words, the natriuresis observed in animals which are volume expanded and prostaglandin synthetase inhibited is greater than would be predicted based on the lack of the expected effect of volume expansion on renal blood flow. These observations are consistent with a direct effect of prostaglandin synthetase inhibition of sodium transport by the kidney. To confirm these observations, additional studies were performed without volume expansion on dogs in which the renal vasoconstrictor effect of diminished prostaglandin synthesis was avoided by administering either acetylcholine or bradykinin, potent vasodilators. In these studies (Fig. 13), the vasodilator was infused into one kidney after prostaglandin synthetase inhibition and the contralateral kidney was used as a control. Restoration of renal blood flow to control values was associated with a natriuresis, even in the face of a decline in glomerular filtration rate, an alteration which should be associated with a decrease in filtered sodium and diminution in sodium excretion. These findings are

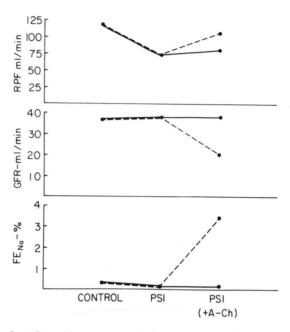

Fig. 13. Effect of a unilateral increase in renal plasma flow (RPF) induced by acetylcholine (A-Ch) infusion (interrupted line) compared to the contralateral kidney (solid line) on sodium handling (FE$_{Na}$) in prostaglandin synthetase inhibited (PSI) anesthetized dogs. GFR, glomerular filtration rate. Adapted from Kirschenbaum *et al.* (1975).

compatible with the antinatriuretic effect of prostaglandins when the concurrent depression in renal blood flow associated with decreased prostaglandin synthesis and release is avoided. The markedly different chemical structure and mechanism of action of the three prostaglandin synthetase inhibitors utilized (meclofenamate, indomethacin and RO 20-5720) suggests that the observed effects are due to the single common action of these agents, prostaglandin synthetase inhibition, rather than being direct effects of these agents on sodium transport by the renal tubules. Furthermore, when parallel studies were performed in unanesthetized animals protaglandin synthetase inhibition induced a natriuresis without any alteration in renal blood flow (Fig. 14).

It is, of course, possible that prostaglandins are both natriuretic and antinatriuretic depending on the physiology or pathophysiology of the circumstances involved. Under conditions in which there are marked and sustained increases in renal prostaglandin synthesis and release, the major effect of renal hemodynamic alterations on physical factors alone could result in a natriuresis. Conversely, the natriuresis demonstrated after inhibition of prostaglandin synthesis, especially without renal hemodynamic changes, is consistent with an alteration in sodium reabsorption which is directly related to the decrease in availa-

ble prostaglandins. On the one hand, prostaglandins mediate a natriuresis through hemodynamic mechanisms, and, on the other hand,
through direct effects on sodium reabsorption by the renal tubular
epithelium. As will be considered later, this modulating or buffering
capacity of prostaglandins may be their strongest feature.

C. Prostaglandin-Induced Alterations in Water Homeostasis

In contrast to the controversy which characterizes the effects of
prostaglandins on the renal handling of sodium, there appears to be
more agreement concerning the influence of these compounds on water
balance. Orloff *et al.* (1965) observed that prostaglandin in very low
concentrations (10^{-9} M) inhibited the stimulation of osmotically
induced water movement by antidiuretic hormone (ADH, vasopressin). This prostaglandin-mediated inhibition of ADH has been confirmed by other (Lipson *et al.*, 1968; Grantham and Orloff, 1968;

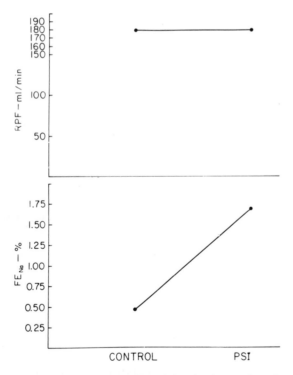

Fig. 14. Effect on renal plasma flow (RPF) and fractional excretion of sodium (FE_{Na}) of
prostaglandin synthetase inhibition (PSI) in unanesthetized conscious dogs. From Kirschenbaum
et al. (1975).

Lipson and Sharp, 1971; Ozer and Sharp, 1972, 1973; Ripoche and Bourgeut, 1973; Frith and Snart, 1973; Albert and Handler, 1974). This anti-ADH effect of prostaglandins represents the *in vitro* parallel to the observed increase in free water clearance noted after the infusion of exogenous prostaglandin into the renal artery of dogs (Johnson *et al.*, 1968; Martinez-Maldonado *et al.*, 1972). The *in vitro* antagonism of the hydro-osmotic effect of ADH noted above has been attributed to alterations in the adenylyl cyclase system since prostaglandin antagonism may be overcome by the addition of cyclic AMP (Grantham and Orloff, 1968; Lipson *et al.*, 1968); prostaglandins reduce the vasopressin-related stimulation of adenylyl cyclase activity and generation of cyclic AMP (Beck *et al.*, 1971; Kalisker and Dyer, 1972c), and perhaps prostaglandins modify the membrane coupling site of ADH (Grantham and Orloff, 1968; Beck *et al.*, 1971; Marumo and Edelman, 1971; Ozer and Sharp, 1973). In addition, antagonists of prostaglandins, such as polyphloretin phosphate and 7-oxa-13-prostenoic acid (Ozer and Sharp, 1972), or of prostaglandin synthetase, such as indomethacin (Albert and Handler, 1974), enhance the membrane response to vasopressin. Although PGE_1 enhanced net water absorption along an osmotic gradient in the isolated collecting tubule of the rabbit without the addition of vasopressin, this effect was minimal. Thus, although the finer details of the antagonism between prostaglandins and ADH-induced membrane alterations in water flow may be the subject of some controversy, the basic nature of the inhibition is understood. The observation that prostaglandins stimulate sodium transport *in vitro* (Barry and Hall, 1969; Fassina *et al.*, 1969; Lote *et al.*, 1974) but inhibit water movement is compatible with the existence of two separate adenylyl cyclase systems, one responsible for water flow and the other for sodium transport (Lipson and Sharp, 1971; Peterson and Edelman, 1964).

That the interaction of prostaglandin and ADH may serve in a balance or modulation fashion is suggested by the observation of Kalisker and Dyer (1972b) that vasopressin stimulates the release of prostaglandin from the renal medulla. The observation that PGE_1 results in a antidiuresis, as estimated from a decline in free water clearance and/or increase in urine osmolarity, in the intact dog (Murphy *et al.*, 1970; Berl and Schrier, 1973) but not in the hypophysectomized animal (Berl and Schrier 1973), suggests that such a prostaglandin–ADH interaction may also occur *in vivo*. Since similar systemic hemodynamic alterations were observed after the intravenous administration of prostaglandin to either intact or hypophysectomized animals, it was concluded that there was an increase in the release of endogenous vasopressin. In contrast, when PGE_1 was infused into the renal

artery of a dog undergoing a water diuresis which would inhibit endogenous vasopressin release, there was an increase in free water clearance. Since endogenous vasopressin was inhibited, this increase in free
water clearance could not have resulted from an antagonistic effect of
prostaglandin on vasopressin at the level of the renal tubular epithelium but could be attributed to the effects of prostaglandins on renal
hemodynamics (Berl and Schrier, 1973). Regardless of whether the
anti-ADH effect results from direct ADH antagonism or from indirect
hemodynamic alterations, there are buffering intrarenal and extrarenal
effects of prostaglandins: stimulation of endogenous vasopressin vs.
diminished intrarenal ADH action. The interrelation of ADH and
prostaglandins is included in the schema presented in Fig. 10.

V. CONCLUSION

> 'Twas brillig, and the slithy toves
> Did gyre and gimble in the wabe;
> All mimsy were the borogoves,
> And the mome raths outgrabe.

"It seems very pretty," she said when she had finished it, "but it's rather hard to
understand!" . . . "Somehow it seems to fill my head with ideas—only I don't know
exactly what they are! However, somebody killed something; that's clear, at any
rate——" (From *Through the Looking Glass* by Lewis Carroll)

The trip down the "rabbit hole" of the influence of prostaglandins
on renal function has been long and somewhat circuitous. Even at this
point it may suggested that more problems were raised than answers
obtained. In terms of the approach to this subject, it has become
apparent that there are some myths related to prostaglandins and renal
function which may be dispelled. Although specific prostaglandin
types and subtypes are frequently referred to, our specific knowledge
of which compounds and/or metabolites are important is really quite
vague. For example, the role of endoperoxides in the kidney has not
yet been evaluated in any perspective. Nor, for that matter, have the
role of thromboxanes in renal function been investigated. When we
peek through the "looking glass" of published studies of prostaglandins and renal function, it must be realized that our knowledge is more
incomplete and circumstantial than required to establish concrete relationships. The role of renal, and extrarenal, prostaglandins in the
regulation or autoregulation of renal blood flow is less than clear. That
prostaglandins do not function as the sole control mechanism for renal
blood flow is apparent; their interrelationship with other vasoactive
systems is only now becoming apparent. One concept which should be

clear in the "looking glass" is the buffer or balance effects of prostaglandins alone and with other agents. This buffer extends through the control of renal blood flow and reaches the alterations in solute and solvent handling by the kidney. Were no other firm roles for prostaglandins to be established in renal function, these buffer effects might be sufficiently important to justify the extensive work performed on prostaglandins and renal function.

In gazing into the future, the impact of the various studies on the development of clinically applicable tools or therapeutic measures is of great potential importance. The clinical implications of prostaglandins in obstetrics, gynecology, allergy-immunology, and gastroenterology can be left for others to consider. Although the role of prostaglandins and hypertension has not been discussed, the importance of agents which vasodilate and lower systemic blood pressure is obvious. The possible direct or indirect role of prostaglandins in controlling pre- and post-glomerular vascular resistance (and glomerular filtration pressure/rate) or participating in a diuresis has yet to be established. The concept of an agent which simultaneously reduces systemic blood pressure, enhances glomerular filtration rate, and promotes a diuresis in a patient with congestive heart failure is, of course, appealing. Similarly, that diseases of the renal circulation, including acute renal failure, may be treated with these potent vasoactive agents is also attractive. Conversely, the ability to treat states of (renal) prostaglandin overproduction, when they become identified, is also a possibility. It must be left to the authors of the next review of the role of prostaglandins in renal function to explore these possibilities.

VI. REFERENCES

Abdulla, Y. H., and McFarland, E., 1972, Control of prostaglandin biosynthesis in rat brain homogenates by adenine nucleotides, *Biochem. Pharmacol.* **21**:2841.

Agus, Z. S., Puschett, J. B., Senesky, D., and Goldberg, M., 1973, Mode of action of parathyroid hormone and cyclic adenosine 3′,5′-monophosphate on renal tubular phosphate reabsorption in the dog, *J. Clin. Invest.* **50**:617.

Aiken, J. W., 1974, Effects of prostaglandin synthesis inhibitors on angiotensin tachyphylaxis on the isolated coeliac and mesenteric arteries of the rabbit, *Pol. J. Pharmacol. Pharm.* **26**:217.

Aiken, J. W., and Vane, J. R., 1972, Inhibition of converting enzyme of the renin angiotensin system in kidney and hindlegs of dogs, *Circ. Res.* **30**:263.

Aiken, J. W., and Vane, J. R., 1973, Intrarenal prostaglandin release attentuates in the renal vasconstrictor activity of angiotensin, *J. Pharmacol. Exp. Ther.* **184**:678.

Albert, W. C., and Handler, J. S., 1974, Effect of PGE, indomethacin and polyphloretin phosphate on toad bladder response to ADH, *Am. J. Physiol.* **226**:1382.

Alexander, J. C., and Lee, J. B., 1970, Effect of osmolality on Na$^+$-K$^+$ATPase in outer renal medulla, *Am. J. Physiol.* **219**:1742.

Al Tai, S. A., and Graham, J. D. P., 1972, The actions of prostaglandins E and F$_2$ on the perfused vessels of the isolated rabbit ear, *Br. J. Pharmacol.* **44**:699.

Andersen, N. H., and Ramwell, P. W., 1974, Biological aspects of prostaglandins, *Arch. Intern. Med. (Chicago)* **133**:30.

Anderson, R. J., Taber, M. S., Cronin, R. E., McDonald, K. M., and Schrier, R. W., 1975, Effects of beta adredergic blockade and inhibitors of angiotensin II and prostaglandins on renal autoregulation, *Am. J. Physiol.* **229**:731.

Änggård, E., 1971, Studies on the analysis and metabolism of the protaglandins, *Ann. N.Y. Acad. Sci.* **180**:200.

Änggård, E., and Jonsson, C. E., 1971, Efflux of prostaglandins from scalded tissue, *Acta Physiol. Scand.* **81**:440.

Änggård, E., and Samuelsson, B., 1964, Prostaglandins and related factors: Metabolism of prostaglandin E_1 in guinea pig lung: The structure of two metabolites, *J. Biol. Chem.* **239**:4097.

Änggård, E., and Samuelsson, B., 1965, Biosynthesis of prostaglandin from arachidonic acid in guinea pig lung, *J. Biol. Chem.* **240**:3518.

Änggård, E., and Samuelsson B., 1966, Purification and properties of a 15-hydroxy-prostaglandin dehydrogenase from the swine lung, *Ark. Kemi* **25**:293.

Änggård, E., Larsson, C., and Samuelsson, B., 1971, Distribution of 15-hydroxy-prostaglandin dehydrogenase and prostaglandin-13-reductase in tissue of the swine, *Acta Physiol. Scand.* **81**:396.

Änggård, E., Bohman, S. O., Griffin. S. E., Larsson, C., Maunsbach, A. B., 1972, Subcellular localization of the prostaglandin system in the rabbit, *Acta Physiol. Scand.* **84**:231.

Arendshorst, W. J., Johnston, P. A., and Selkurt, E. E., 1974, Effect of prostaglandin E_1 on renal hemodymanics in nondiuretic and volume expanded dogs, *Am. J. Physiol.* **226**:218.

Attallah, A. A., and Lee, J. B., 1973*a,* Specific binding sites in the rabbit kidney for prostaglandin A_1, *Prostaglandins* **4**:703.

Attallah, A. A., and Lee, J. B., 1973*b,* Radioimmunoassay of prostaglandin A: Intrarenal PGA_2 as a factor mediating saline induced natriuresis, *Circ. Res.* **33**:696.

Attallah, A. A., Payakkapan, W., and Lee, J. B., 1974*a,* Metabolism of prostaglandin A. 1. The kidney cortex as a major site of PGA_2 degradation, *Life Sci.* **14**:1521.

Attallah, A., Payakkapan, W., Lee, J. B., Carr, A., and Brazelton, E., 1974*b,* PGA: Fact not artifact, *Prostaglandins* **5**:69.

Baer, P. G., and Navar, L. G., 1973, Renal vasodilatation and uncoupling of blood flow and filtration rate autoregulation, *Kidney Int.* **4**:12.

Bagli, J. F., Bogri, T., Denghenghi, R., and Wiener, M., 1966, ProstaglandinI—Total synthesis of 9b,15-X-dehydroxyprost-13-enoic acid, *Tetrahedron Lett.* **5**:465.

Bailie, M. D., Barbour, J. A., and Hook, J. B., 1975, Effect of indomethacin on furosemide induced changes in renal blood flow, *Proc. Soc. Exp. Biol. Med.* **148**:1173.

Barger, A. C., and Herd, J. A., 1967, *Proc. 3rd Int. Congr. Nephrol. (Washington, D.C.)*, Karger, Basel **1**:174.

Barry, E., and Hall, W. J., 1969, Stimulation of sodium movement across frog skin by prostaglandin E_1, *J. Physiol. (London)* **200**:83P.

Beck, N. P., Field, J. B., and Davis, B., 1970, Effect of prostaglandin E_1, chlorpropamide and vasopressin on cyclic AMP in renal medulla of rats, *Clin. Res.* **18**:494 (abst.).

Beck, N. P., Kaneko, T., Zor, U., Field, J. B., and Davis, B. B., 1971, Effects of vasopressin and prostaglandin E_1 on the adenyl cyclase cyclic $3',5'$-adenosine monophosphate system of the renal medulla of the rat, *J. Clin. Invest.* **50**:2461.

Beck, N. P., DeRubertis, F. R., Michelis, M. F., Fusco, R. D., Field, J. B., and Davis, B. B., 1972*a,* Effect of prostaglandin E_1 on certain renal actions of parathyroid hormones, *J. Clin. Invest.* **51**:2352.

Beck, N. P., Reid, S. W., Murdaugh, H. V., and Davis, B. B., 1972*b,* Effect of catecholamines and their interaction with other hormones on cyclic $3',5'$-adenosine monophosphate in the kidney, *J. Clin. Invest.* **51**:939.

Beck, N., Kim, K. S., and Davis, B. B., 1975, Catecholamine dependent cyclic adenosine monophosphate and renin in the dog kidney, *Circ. Res.* **36**:401.

Bell, R. D., Sinclair, R. J., and Parry, W. L., 1975, The effects of indomethacin on autoregulation and the renal response to hemorrhage, *Circ. Shock* **2**:57.

Bergström, S. and Sjövall, J., 1960, The isolation of prostaglandin F from sheep prostate glands, *Acta Chem. Scand.* **14**:1693.

Bergström, S., Ryhage, R., Samuelsson, B. and Sjövall, J., 1962, The structure of prostaglandin E_1, F_1 and F_2, *Acta Chem. Scand.* **16**:501.

Bergström, S., Carlson, L. A., and Oro, L., 1964*a*, Effect of prostaglandins on catecholamine induced changes in the free fatty acids of plasma and in blood pressure in the dog, *Acta Physiol. Scand.* **60**:170.

Bergström, S., Danielsson, H., and Samuelsson, B., 1964*b*, The enzymatic formation of prostaglandin E_2 from arachidonic acid, *Biochem. Biophys. Acta* **90**:207.

Bergström, S., Carlson, L. A., and Weeks, J. R., 1968, The prostaglandins: A family of biologically active lipids, *Pharm. Rev.* **20**:1.

Berl, T., and Schrier, R. W., 1973, Mechanism of effect of prostaglandin E_1 on renal water excretion, *J. Clin. Invest.* **52**:463.

Bito, L. Z., 1971, Accumulation and apparent active transport of prostaglandins, *Pharmacologist* **13**:293.

Bohman, S. O., 1974, The ultrastructure of the rat renal medulla as observed after improved fixation methods, *J. Ultrastruct. Res.* **47**:329.

Brocklehurst, W. E., 1971, Role of kinins and prostaglandins in inflammation, *Proc. R. Soc. Med.* **64**:4.

Brody M. J., and Kadowitz, P. J., 1974, Prostaglandins as modulators of the automatic nervous system, *Fed. Proc.* **3**:48.

Bunag, R. D., Page, I. H., and McCubbin, J. R., 1966, Neural stimulation of renin release, *Circ. Res.* **19**:851.

Carlson, L. A., Ekelund, L. G., and Oro, L., 1969, Circulatory and respiratory effects of different doses of prostaglandin E_1 in man, *Acta Physiol. Scand.* **75**:161.

Carr, A. A., 1968, Effect of prostaglandin A_1 on urinary concentration, in: *Prostaglandin Symposium of the Worchester Foundation* (P. Ramwell and J. E. Shaw, eds.), pp. 163–164, Interscience, New York.

Carr, A. A., 1970, Hemodynamic and renal effects of a prostaglandin PGA_1, in subjects with essential hypertension, *Am. J. Med. Sci.* **295**:21.

Carriere, S., and Biron, P., 1970, Effects of angiotensin I on intrarenal blood flow distribution, *Am. J. Physiol.* **219**:1642.

Carriere, S., and Friborg, J., 1969, Intrarenal blood flow and PAH extraction during angiotensin infusion, *Am. J. Physiol.* **217**:1708.

Carriere, S., Friborg, J., and Guay, J. P., 1971, Vasodilators, intrarenal blood flow, and natriuresis in the dog, *Am. J. Physiol.* **221**:92.

Carroll, L., 1954, *Alice's Adventure in Wonderland* and *Through the Looking Glass,* Grosset and Dunlap, New York.

Chang, L. C. T., Splawinski, J. A., Oates, J. A., and Nies, A. S., 1975, Enhanced renal prostaglandin production in the dog. II. Effects on intrarenal hemodynamics, *Circ. Res.* **36**:204.

Chase, L. R., 1975, Selective proteolysis of the receptor from parathyroid hormone in renal cortex, *Endocrinology* **96**:70.

Collier, H. O. J., Dinneon, L. C., Perkins, A. C., and Piper, P. F., 1967, Curtailment by aspirin and meclofenamate of hypotension unduced by bradykinin in the guinea pig, *Naunyn-Schmiedebergs Arch. Pharmacol.* **259**:159.

Collier, J. H. O., 1971, Introduction to the action of kinins and prostaglandins, *Proc. R. Soc. Med.* **64**:1.

Corsini, W. A., Crosslan, K. L., and Bailie, M. D., 1974, Renin secretion by rat kidney slices *in vitro, Proc. Soc. Exp. Biol. Med.* **145**:403.

Crowshaw, K., 1971, Prostaglandin biosynthesis from endogenous precursors in rabbit kidney, *Nature (London)* **231**:240.

Crowshaw, K., 1973, The incorporation of 1-^{14}C arachidonic acid into the lipids of rabbit renal slices and conversion to prostaglandins E_2 and $F_{2\alpha}$, *Prostaglandins* **3**:607.

Crowshaw, K., and Szlyk, J. Z., 1970, Distribution of prostaglandins in rabbit kidney, *Biochem. J.* **116**:421.

Crowshaw, K., and McGiff, J. C., 1973, Prostaglandins in the kidney: A correlative study of their biochemistry and effect on renal function, in: *Mechanisms of Hypertension* (M. P. Sambhi, ed.), pp. 254–273, Exerpta Medica Foundation, Amsterdam.

Crowshaw, K., McGiff, J. C., Strand, J. C., Lonigro, A. J., and Terragno, N. A., 1970, Prostaglandins in dog renal medulla, *J. Pharm. Pharmacol.* **22**:302.

Daniels, E. G., Hinman, J. W., Leach, B. E., and Muirhead, E. E., 1967, Identification of prostaglandin E_2 as the principal vasodepressor lipid of rabbit medulla, *Nature (London)* **215**:1298.

Danon, A., and Chang, L. C. T., 1973, Release of prostaglandins from rat renal papilla *in vitro*: Effects of arachidonic acid and angiotensin II, *Fed. Proc.* **32**:788 (abst.).

Davis, H. A., and Horton, E. W., 1972, Output of prostaglandins from the rabbit kidney, its increase on renal nerve stimulation and its inhibition by idomethacin, *Br. J. Pharmacol.* **46**:658.

Douglas, J. R., Johnson, E. M., Marshall, G. R., Jaffe, B. M., and Needleman, P., 1973, Stimulation of splenic prostaglandin release by angiotensin and specific inhibition by cysteine-AII, *Prostaglandins* **3**:67.

Dressler, W. E., Rossi, G. V., and Orzechowski, R. F., 1975, Evidence that renal vosodilatation by dopamine in dogs does not involve release of prostaglandin, *J. Pharm. Pharmacol.* **27**:203.

Ducharme, D. W., Weeks, J. R., and Montgomery, R. G., 1968, Studies on the mechanism of the hypertensive effect of prostaglandin $F_{2\alpha}$, *J. Pharmacol. Exp. Ther.* **160**:1.

Dunham, E. W., and Zimmerman, B. G., 1970, Release of prostaglandin-like material from dog kidney during nerve stimulation, *Am. J. Physiol.* **219**:1279.

Eble, N. J., Gowdey, C. W., and Vane, J. R., 1972, Blood concentration of adrenaline in dogs after intravenous administration of 5-hydroxytryptamine, *Nature (London)* **238**:254.

Ellis, E., and Hutchins, P., 1974, Cardiovascular responses to prostaglandin $F_{2\alpha}$ in spontaneously hypertensive rats, *Prostaglandins* **7**:345.

Fassina, G., Carpendo, F., and Santi, R., 1969. Effect of prostaglandin E_1 on isolated short-circuited frog skin, *Life Sci.* **8**:181.

Ferreira, S. H., and Vane, J. R., 1967, Prostaglandins: Their disappearance from and release into the circulation, *Nature (London)* **216**:868.

Ferreira, S. H., Moncada, S., and Vane, J. R., 1973*a*, Some effects of inhibiting endogenous prostaglandin formation on the responsed of the cat spleen, *Br. J. Pharmacol.* **47**:48.

Ferreira, S. H., Moncada, S., and Vane, J. R., 1973*b*, Prostaglandins and the mechanism of analgesia produced by aspirin-like drugs, *Br. J. Pharmacol.* **49**:86.

Ferreiro, C. M., Page, I. H., and McCubbin, J. W., 1970, Increased cardiac output as a contributory factor in experimental renal hypertension in dogs, *Circ. Res.* **29**:799.

Fichman, M., and Horton, R., 1973, Significance of the effects of prostaglandins on renal and adrenal function in man, *Prostaglandins* **3**:629.

Fichman, M. P., Littenburg, G., Brooker, G., and Horton, R., 1972, Effect of prostaglandin A_1 on renal and adrenal function in man, *Circ. Res.* **30–31**:19 (Suppl. II).

Fichman, M., Telfer, N., Zia, P., Speckart, P., Golub, M., and Rude, R., 1976, Role of prostaglandins in the pathogenesis of Bartter's syndrome, *Am. J. Med.* **60**:785.

Flowers, R. J., 1974, Drugs which inhibit prostaglandin biosynthesis, *Pharmacol. Rev.* **26**:33.

Franklin, G. O., Dowd, A. J., Caldwell, B. V., and Speroff, L., 1974, The effect of angiotensin II intravenous infusion on plasma renin activity and prostaglandin A, E, and F, levels in the uterine vein of the pregnant monkey, *Prostaglandins* **6**:271.

Frith, D. A., and Snart, R. S., 1973, Inhibition of vasopressin-stimulated water transport across the isolated toad bladder, *Comp. Biochem. Physiol.* **45A**:313.

Fröhlich, J. C., Sweetman, B. J., Carr, K., Splawinski, J., Watson, J. T., Änggård, E., and Oates, J. A., 1973, Occurrence of prostaglandins in human urine, *Adv. Biosci.* **9**:321.

Fröhlich, J. C., Wilson, T. W., Sweetman, B. J., Smigel, M., Nies, A. S., Carr, K., Watson, J. T., and Oates, J. A., 1975, Urinary prostaglandins: Identification and origins, *J. Clin. Invest.* **55**:763.

Fujimoto, S., and Lockett, M. F., 1970, The diuretic actions of prostaglandin E_1 and of noradrenaline, and the occurrence of a prostaglandin E_1-like substance in the renal lymph of cats, *J. Physiol. (London)* **208**:1.

Fulgraff, G., and Brandenbusch, G., 1974, Comparison of the effects of the protaglandins A_1, E_2 and $F_{2\alpha}$ on kidney function of dogs, *Pflügers Arch.* **349**:9.

Fulgraff, G., and Meiforth, A., 1971, Effects of prostaglandin E_2 on excretion and reabsorption of sodium and fluid in rat kidneys (micropuncture studies), *Pflügers Arch.* **330**:243.

Fulgraff, G., Brandenbusch, G., and Heintze, K., 1974, Dose response relation of the renal effects of PGA_1, PGE_2 in dogs, *Prostaglandins* **8**:21.

Gagnon, D. J., Gauthier, R., and Regoli, D., 1974, Release of prostaglandins from the rabbit perfused kidney: Effects of vasoconstrictors, *Br. J. Pharmacol.* **50**:553.

Gagnon, J., Rice, K., and Flamenbaum, W., 1975a, Effects of prostaglandin synthetase inhibition on pyrogen-induced renal hyperemia in the dog, *Fed. Proc.* **34**:363 (abst.).

Gagnon, J., Rice, K., Ramwell, P., and Flamenbaum, W., 1975b, Bradykinin alterations in renal blood flow: Prostaglandin and renin interrelationships, *Proc. Am. Soc. Nephrol. 8th Ann. Mtg.*, p. 79.

Gill, J. R., and Casper, A. G. T., 1971, Renal effects of adenosine 3'-5' cyclic monophosphate in the regulation of proximal tubular sodium reabsorption, *J. Clin. Endocrinol.* **50**:1231.

Gill, J. R., Fröhlich, J. C., Bowden, R. E., Taylor, A. A., Keiser, H. R., Seyberth, H. W., Oates, J. A., and Bartter, F. C., 1976, Bartter's syndrome: A disorder characterized by high urinary prostaglandins and a dependence of hyper-reninemia on prostaglandin synthesis, *Am. J. Med.* **61**:43.

Goldblatt, M. W., 1935, Properties of human seminal plasma, *J. Physiol. (London)* **84**:208.

Golub, M. S., Zia, P. K., and Horton, R., 1974, Metabolism of prostaglandin A_1 and A_2 by human whole blood, *Prostaglandins* **8**:13.

Granström, E., 1971, Metabolism of prostaglandin $F_{2\alpha}$ in swine kidney, *Biochim. Biophys. Acta* **239**:120.

Granström, E., 1973, On the metabolism of prostaglandin E_1 in man, *Prog. Biochem. Pharmacol.* **3**:89.

Granström, E., and Samuelsson, B., 1969, The structure of a urinary metabolite of prostaglandin $F_{2\alpha}$ in man, *J. Am. Chem. Soc.* **91**:3398.

Grantham, J., and Orloff, J., 1968, Effects of prostaglandin E_1 on the permeability response of the isolated collecting tubule to vasopressin, adenosine 3'-5' monophosphate, and theophylline, *J. Clin. Invest.* **47**:1154.

Gréen, K., 1969, Structure of urinary metabolites of prostaglandin $F_{2\alpha}$ in the rat, *Acta Chem. Scand.* **23**:1453.

Gréen K., 1971, Metabolism of prostaglandin $F_2\alpha$ in the rat, *Biochim. Biophys. Acta* **231**:419.

Gréen, K., Granstrom, E., and Samuelsson, B., 1973, Methods for quantitative analysis of $PGF_{2\alpha}$, PGE_2, $9\alpha 11\alpha$-dihydroxy 15-ketoprost-15-enoic acid, and $9\alpha 11\alpha$ 15-trihydroxy-prost-5-enoic acid from body fluids using deuterated carriers and gas-chromatography–mass spectrometry, *Anal. Biochem.* **54**:434.

Greenberg, R., 1974, The effects of indomethacin and eicosa-5,8,11,14-tetraynoic acid on the response of the rabbit portal vein to electrical stimulation, *Br. J. Pharmacol.* **52**:61.

Greenberg, S., Engelbrecht, J. A., and Wilson, W. R., 1974, The role of the autonomic nervous system in the responses of the perfused canine paw to prostaglandins B_1 and B_2, *J. Pharmacol Exp. Ther.* **189**:130.

Gross, J. B., and Bartter, F. C., 1973, Effect of prostaglandins E_1, A_1 and $F_{2\alpha}$ on renal handling of salt and water, *Am. J. Physiol.* **225**:218.

Gryglewski, R., and Vane, J. R., 1972, Release of prostaglandins and rabbit aorta contracting substance RCS from rabbit spleen and its antagonism by anti-inflammatory drugs, *Br. J. Pharmacol.* **45**:37.

Guttman, F. D., Tagawa, E., Haber, E., and Barger, A. C., 1973, Renal arterial pressure, renin secretion, and blood pressure control in trained dogs. *Am. J. Physiol.* **224**:66.

Hall, W. J., 1973, Seasonal changes in the sensitivity of frog skin to prostaglandin and the effect of external sodium and chloride on the response, *Irish J. Med. Sci.* **142**:230.

Hamberg, M. 1969, Biosynthesis of prostaglandins in the renal medulla of the rabbit, *FEBS Lett.* **5**:127.

Hamberg, M., and Israelsson, U., 1970, Metabolism of prostaglandin E_2 in guinea pig liver. I. Identification of seven metabolites, *J. Biol. Chem.* **245**:5107.

Hamberg, M., and Samuelsson, B., 1966, Prostaglandins in human seminal plasma, *J. Biol. Chem.* **241**:257.

Hamberg, M., and Samuelsson, B., 1967, On the mechanism of the biosynthesis of prostaglandin E_1 and $F_{1\alpha}$, *J. Biol. Chem.* **242**:5336.

Hamberg, M., and Samuelsson, B., 1969, The structure of the major urinary metabolite of prostaglandin E_2 in man, *J. Am. Chem. Soc.* **91**:2177.

Hamberg, M., and Samuelsson, B., 1971, On the metabolism of prostaglandin E_1 and E_2 in man, *J. Biol. Chem.* **246**:6713.

Hamberg, M., Svenson, J., Wakabayashi, T., and Samuelsson, B., 1974, Isolation and structure of two prostaglandin endoperoxides that cause platelet aggregation, *Proc. Natl. Acad. Sci. USA* **71**:345.

Hedqvist, P., 1970, Studies on the effect of prostaglandins E_1 and E_2 on the sympathetic neuromuscular transmission in some animal tissues, *Acta Physiol. Scand.* **79**:1(Suppl. 345).

Hedqvist, P., 1971, Prostaglandin E compounds and sympathetic neuromuscular transmission, *Ann. N.Y. Acad. Sci.* **180**:410.

Hedqvist, P., 1974, Interaction between prostaglandins and calcium ions on noradrenaline release from the stimulated guinea pig vas deferens, *Acta Physiol. Scand.* **90**:153.

Hedqvist, P., and Brundin, J., 1969, Inhibition by prostaglandin E_1 of noradrenaline release and of effector response to nerve stimulation in the cat spleen, *Life Sci.* **8**:389.

Hedqvist, P., and Wennmalm, A., 1971, Comparison of the effects of prostaglandins E_1, E_2 and $F_{2\alpha}$ on the sympathetically stimulated rabbit heart, *Acta Physiol. Scand.* **83**:156.

Hedqvist, P., Stjarne, L., and Wennmalm, A., 1970, Inhibition by prostaglandin E_2 of sympathetic neurotransmission in the rabbit heart, *Acta Physiol. Scand.* **79**:139.

Hedqvist, P., Stjärne, L., and Wennmalm, A., 1971, Facilitation of sympathetic neurotransmission in the cat spleen after inhibition of prostaglandin synthesis, *Acta Physiol. Scand.* **83**:430.

Herbaczynka-Cedro, K., and Vane, J. R., 1973, Contribution of intrarenal generation of prostaglandin to autoregulation of renal blood flow in the dog, *Circ. Res.* **33**:428.

Herbaczynka-Cedro, K., and Vane, J. R., 1974. Prostaglandins as mediators of reactive hyperaemia in the kidney, *Nature (London)* **247**:402.

Herbaczynka-Cedro, K., Staszewska-Barczak, J., and Janczewska, H., 1974, The release of prostaglandin-like substance during reactive and functional hyperemia in the hind leg of the dog, *Pol. J. Pharmacol. Pharm.* **26**:167.

Higgins, C. B., and Braunwald, E., 1972, The prostaglandins: Biochemical, physiologic and clinical considerations, *Am. J. Med.* **53**:92.

Higgins, C. H., Vatner, S. F., and Braunwald, E., 1973, Regional hemodynamic effects of prostaglandin A_1 in the conscious dog, *Am. Heart J.* **85**:349.

Hinman, J. W., 1967, The prostaglandins, *Bio. Sci.* **17**:779.

Hittleman, K. J., and Butcher, R. W., 1973, Cyclic AMP and the mechanism of action of the prostaglandins in: *The Prostaglandins: Pharmacological and Therapeutic Advances* (M. F. Cuthbert, ed.), pp. 151–166, Heinemann, London.

Hollenberg, N. K., Solomon, H. S., Adams, D. F., Abrams, H. L., and Merrill, J. P., 1972, Renal vascular responses to angiotensin and norepinephrine in normal man, *Circ. Res.* **31**:750.

Holmes, S. W., Horton, E. W., and Main, I. H. M., 1963, The effect of prostaglandin E_1 on responses of smooth muscle to catecholamines, angiotensin and vasopressin, *Br. J. Pharmacol.* **21**:538.

Hook, R., and Gillis, C. N., 1975, The removal and metabolism of prostaglandin E_1 by rabbit lung, *Prostaglandins* **9**:193.

Horster, M., and Thurau, K., 1968, Micropuncture studies on the filtration rate of single superficial and juxtamedullary glomeruli in the rat kidney, *Pflügers Arch. Gesamte Physiol.* **301**:162.

Horton, E. W., 1969, Hypothesis on physiological roles of prostaglandins, *Physiol. Rev.* **49**:122.

Horton, E. W., and Jones, R. L., 1969, Prostaglandins A_1, A_2 and 19-hydroxy A_1: Their actions on smooth muscle and their inactivation on passage through the pulmonary and hepatic vascular beds, *Br. J. Pharmacol.* **37**:705.

Horton, E. W., and Main, I. H. M., 1965, A comparison of the actions of prostaglandin $F_{2\alpha}$ and E_1 on smooth muscle, *Br. J. Pharmacol. Chemother.* **24**:470.

Israelsson, U., Hamberg, M., and Samuelsson, B., 1969, Biosynthesis of 19-hydroxy-prostaglandin A_1, *Eur. J. Biochem.* **11**:390.

Ito, T., Hadaka, H., and Kato, T., 1973, Response of plasma aldosterone to postural change, diuretics, angiotensin II, sodium restriction or loading and prostaglandin, *Jpn. Heart J.* **14**:518.

Itskovitz, H. D., and McGiff, J. C., 1974, Hormonal regulation of the renal circulation, *Circ. Res.* **34–35**:5 (Suppl. 1).

Itskovitz, H. D., Stemper, J., Pacholczyk, D., and McGiff, J. C., 1973, Renal prostaglandins: Determinants of intrarenal distribution of blood flow in the dog, *Clin. Sci. Mol. Med.* **45**:3215 (Suppl. 1).

Itskovitz, H., Terragno, N. A., and McGiff, J. C., 1974, Effect of a renal prostaglandin on distribution of blood flow in the isolated canine kidney, *Circ. Res.* **34**:770.

Jaffe, B. M., Smith, J. W., Newton, W. T., and Parker, C. W., 1971, Radioimmunaossay for prostaglandins, *Science* **171**:494.

Jaffe, B. M., Parker, C. W., Marshall, G. R., and Needleman, P., 1972, Renal concentration of prostaglandin E in acute and chronic renal ischemia, *Biochem. Biophys. Res. Commun.* **49**:799.

Jaffe, B. M., Behrman, H. R., and Parker, C. W., 1973, Radioimmunaossay measurement of prostaglandins E, A, and F, in human plasma, *J. Clin. Invest.* **52**:398.

Janszen, F. H. A., and Nugteren, D. H., 1971, Histochemical localization of prostaglandin synthetase, *Histochemie* **27**:159.

Johnson, E. M., Marshall, G. R., and Needleman, P., 1974, Modification of responses to sympathetic nerve stimulation by the renin–angiotensin system in rats, *Br. J. Pharmacol.* **51**:541.

Johnston, H. H., Herzog, J. P., and Lauler, D. P., 1967, Effect of prostaglandin E_1 on renal hemodynamics sodium and water excretion, *Am. J. Physiol.* **213**:939.

Jones, R. L., 1972*a*, Properties of a new prostaglandin, *Br. J. Pharmacol.* **45**:144P.

Jones, R. L., 1972*b*, 15-Hydroxy-9-oxoprosta-11,13-dienoic acid as the end product of prostaglandin isomerase, *J. Lipid Res.* **13**:511.

Jones, R. L., and Cammock, S., 1973, Purification, properties, and biological significance of prostaglandin A isomerase, *Adv. Biosci.* **9**:61.

Jouvenaz, G. H., 1970, A sensitive method for the determination of prostaglandins by gas chromatography with electron capture detection, *Biochim. Biophys. Acta* **202**:231.

Junstad, M., and Wennmalm, A., 1972, Increased renal excretion of noradrenaline in rats after treatment with prostaglandin synthetase inhibitor indomethacin, *Acta Physiol. Scand.* **85**:573.

Kalisker, A., and Dyer, D. C., 1972*a*, Prostaglandin–vasopressin interactions in the medulla, *Pharmacologist* **13**:293.

Kalisker, A., and Dyer, D. C., 1972b, *In vitro* release of prostaglandin from the renal medulla, *Eur. J. Pharmacol.* **19**:305.

Kalisker, A., and Dyer, D. C., 1972*c,* Inhibition of the vasopressin activated adenyl cyclase from renal medulla by prostaglandins, *Eur. J. Pharmacol.* **20**:143.

Kataoka, K., Ramwell, P. W., and Jessup, S., 1967, Prostaglandins: Localization in subcellular particles of rat cerebral cortex, *Science* **157**:1187.

Kauker, M. L., 1973, Micropuncture study of renal action of prostaglandin A_2 (PGA_2) in rats, *J. Pharmacol. Exp. Ther.* **187**:632.

Kirschenbaum, M. A., and Stein, J. H., 1975, Prostaglandin (PG) inhibitors cause a natriuresis in conscious dogs, *Proc. Am. Soc. Nephrol. 8th Ann Mtg. (Washington, D.C.),* p. 43.

Kirschenbaum, M. A., White, N., Stein, J. H., and Ferris, T. F., 1974, Redistribution of renal cortical blood flow during inhibition of prostaglandin synthesis, *Am. J. Physiol.* **227**:801.

Kirschenbaum, M. A., Bay, W. H., Ferris, T. F., and Stein, J. H., 1975, The effect of prostaglandin inhibition of the natriures of drug induced vasodilation, *Clin. Res.* **23**:36A (abst).

Klenberg, D., and Samuelsson, B., 1965, The biosynthesis of prostaglandin E_1 studied with specifically H_3-labeled 8,11,14-eicosatrienoic acids, *Acta Chem. Scand.* **19**:534.

Kotchen, T. A., and Miller, M. C., 1974, Effect of prostaglandins on renin activity, *Am. J. Physiol.* **226**:314.

Krakoff, L. R., Deguia, D., Vlachakis, N., Stricker, J., Goldstein, M., 1973, Effect of sodium balance on arterial blood pressure and renal responses to prostaglandin A_1 in man, *Circ. Res.* **33**:539.

Kunze, H. E., and Vogt, W., 1971, Significance of phospolipase A for prostaglandin formation, *Ann. N.Y. Acad. Sci.* **180**:123.

Kurzok, R., and Lieb, C., 1930, Biochemical studies of human semen. II. The action of semen on the human uterus, *Proc. Soc. Exp. Biol. Med.* **28**:268.

Lafferty, J. J., Kannegiesser, H., Lee, J. B., and Parker, C. W., 1972, Metabolic mechanisms of action of the renal prostaglandins, *Adv. Biosci.* **9**:293.

Landolt, R., Shaw, S. M., Vetter, R. J., and Chen, S. H., 1974, An autoradiographic study of prostaglandin E_1 and/or its metabolities in mouse kidney, *Prostaglandins* **5**:305.

Larsson, C., and Änggård, E., 1970, Distribution of prostaglandin metabolizing enzymes in tissues of the swine, *Acta Pharmacol. Toxicol.* **28**:61 (suppl. I).

Larsson, C., and Änggård, E., 1973, Regional differences in the formation and metabolism of prostaglandins in the rabbit kidney, *Eur. J. Pharmacol.* **21**:30.

Larsson, C., and Änggård, E., 1974, Increased juxtamedullary blood flow on stimulation of intrarenal prostaglandin biosynthesis, *Eur. J. Pharmacol.* **25**:326.

Lavery, H. A., Lowe, R. D., and Scoop, G. C., 1970, Cardiovascular effects of prostaglandins mediated by the central nervous system of the dog, *Br. J. Pharmacol.* **39**:511.

Leary, W. P., Ledingham, J. G., and Vane, J. R., 1974, Impaired prostaglandin release from the kidneys of salt-loaded and hypertensive rats, *Prostaglandins* **7**:435.

Lee, J. B., 1969, Hypertension, natriuresis and the renal prostaglandins, *Ann. Intern. Med.* **70**:1033.

Lee, J. B., 1972*a*, Natriuretic "hormone" and the renal prostaglandins, *Prostaglandins* **1**:55.

Lee, J. B., 1972*b*, The antihypertensive and natriuretic endocrine function of the kidney: Vascular and metabolic mechanisms of the renal prostaglandins, in: *Prostaglandins in Cellular Biology* (P. W. Ramwell and B. B. Pharriss, eds.), pp. 399–499, Plenum Press, New York.

Lee, J. B., 1973, Hypertension, natriuresis and the renomeduallary prostaglandins: An overview, *Prostaglandins* **3**:551.

Lee, J. B., 1974, Cardiovascular–renal effects of prostaglandins: The antihypertensive, natriuretic renal "endocrine" function, *Arch. Intern. Med. (Chicago)* **133**:56.

Lee, J. B., and Attallah, A. A., 1974, Responses of intrarenal PGA_2 to salt loading in the rabbit, *Circ. Res.* **34–35**:75 (Suppl. I).

Lee, J. B., Covino, B. G., Takman, B. H., and Smith, E. R., 1965, Renomedullary substance, medullin: Isolation, chemical characterization and physiological properties, *Circ. Res.* **17**: 57.

Lee, J. B., Crowshaw, K., Takman, B. H., Attrep, K. A., and Gougoutas, J. Z., 1967, Identification of prostaglandin E_2, $F_{2\alpha}$ and A_2 from rabbit kidney medulla, *Biochem. J.* **105**:1251.

Lee, J. B., McGiff, J. C., Kannegiesser, H., Aykent, Y. Y., Mudd, J. G., and Frawley, T. F., 1971, Prostaglandin A_1: Antihypertensive and renal effects, *Ann. Intern. Med.* **74**:703.

Lindsey, H. E., and Wyllie, 1970, Release of prostaglandins from emobilized lung, *Br. J. Surg.* **57**:738.

Lipson, L. C., and Sharp, G. W. G., 1971, Effect of prostaglandin E_1 on sodium transport and osmotic water flow in the toad bladder, *Am. J. Physiol.* **220**:1046.

Lipson, L., Hynie, S., and Sharp, G., 1968, Effect of prostaglandin E_1 on osmotic water flow and sodium transport in the toad bladder, *Ann. N.Y. Acad. Sci.* **180**:261.

Loeffler, J. R., Stockigt, J. R., and Ganong, W. F., 1972, Effect of alpha- and beta-adrenergic blocking agents on the increased renin secretion produced by stimulation of the renal nerves, *Neuroendocrinology* **10**:129.

Lonigro, A. J., Itskovitz, H. D., Crowshaw, K., and McGiff, J. C., 1973*a*, Dependency of renal blood flow on prostaglandin synthesis in the dog, *Circ. Res.* **32**:712.

Lonigro, A. J., Terragno, N. A., Malik, K. U., and McGiff, J. C., 1973*b*, Differential inhibition by prostaglandins of the renal actions of pressor stimuli, *Prostaglandins* **3**:595.

Lote, C. J., Rider, J. B., and Thomas, S., 1974, The effect of prostaglandin E_1 on the short circuit current and sodium, potassium, chloride and calcium movements across isolated frog *(Rana temporaria)* skin, *Pflügers Arch.* **352**:145.

Malik, K. U., and McGiff, J. C., 1975, Modulation by prostaglandins of adrenergic transmission in the isolated perfused rabbit and rat kidney, *Circ. Res.* **36**:599.

Marchand, G. R., Greenberg, S., Wilson, W. R., and Williamson, H. E., 1973, Effect of prostaglandin B_2 on renal hemodynamics and excretion. *Proc. Soc. Exp. Biol. Med.* **143**:938.

Martinez-Maldonado, M., Tsaparas, N., Eknoyan, G., and Suki, W. N., 1972, Renal actions of prostaglandins: Comparison with acetylcholine and volume expansion, *Am. J. Physiol.* **222**:1147.

Marumo, F., and Edelman, I. S., 1971, Effects of Ca^{++} and prostaglandin E_1 on vasopressin activation of renal adenyl cyclase, *J. Clin. Invest.* **50**:1613.

McClatchey, W. M., and Carr, A. A., 1972, Prostaglandin (PGA_1), angiotensin and renal function, *Prostaglandins* **2**:213.

McGiff, J. C., Terragno, N. A., Strand, J. C., Lee, J. B., Lonigro, A. J., and Ng, K. K. F., 1969, Selective passage of prostaglandins across the lung, *Nature (London)* **223**:742.

McGiff, J. C., and Itskovitz, H. D., 1973, Prostaglandins and the kidney, *Circ. Res.* **33**:479.

McGiff, J. C., Crowshaw, K., Terragno, N. A., and Lonigrow, A. J., 1970*a*, Renal prostaglandins: Possible regulators of the renal actions of pressor hormones, *Nature (London)* **227**:1255.

McGiff, J. C., Crowshaw, K., Terragno, N. A., and Lonigro, A. J., 1970*b*, Release of a prostaglandin-like substance into renal venous blood in response to angiotensin II, *Circ. Res.* **27–28**:I-121 (Suppl. I).

McGiff, J. C., Crowshaw, K., Terragno, N. A., Lonigro, A. J., Strand, J. C., Williamson, M. A., Lee, J. B., and Ng, K. K. F., 1970*c*, Prostaglandin like substance appearing in canine renal venous blood during renal ischemia: Their partial characterization by pharmacologic and chromatographic procedures, *Circ. Res.* **27**:765.

McGiff, J. C., Crowshaw, K., Terragno, N. A., Malik, K. U., and Lonigro, A. J., 1972*a*, Differential effect of noradrenaline and renal nerve stimulation on vascular resistance in the dog kidney and release of prostaglandin E-like substance, *Clin. Sci.* **42**:223.

McGiff, J. C., Terragno, N. A., Malik, K. U., and Lonigro, A. J., 1972*b*, Release of prostaglandin E-like substance from canine kidney by bradykinin, *Circ. Res.* **31**:36.

McGiff, J. C., Crowshaw, K., and Itskovitz, H. E., 1974*a*, Prostaglandins and renal function, *Fed. Proc.* **33**:39.

McGiff, J. C., Terragno, N. A., and Itskovitz, H. D., 1974*b*, Role of renal prostaglandins revealed by inhibitors of prostaglandin synthetase, in: *Prostaglandin Synthetase Inhibitors* (H. J. Robinson and J. R. Vane, eds.), pp. 259–269, Raven Press, New York.

McGiff, J. C., Itskovitz, H. D., and Terragno, N. A., 1975. The actions of bradykinin and elodoisin in the canine isolated kidney: Relationships to prostaglandins, *Clin. Sci. Mol. Med.* **49**:125.

McMurray, W. C., and Magee, W. L., 1972, Phospholipid metabolism, *Annu. Rev. Biochem.* **41**:129.

McNay, J. L., and Abe, Y., 1970, Redistribution of cortical blood flow during renal vasodilatation in dogs, *Circ. Res.* **27**:1023.

Messina, E. J., Weiner, R., and Kaley, G., Inhibition of bradykinin vasodilatation and potentiation of norepinephrine and angiotensin vasconstriction by inhibitors of prostaglandin synthesis in skeletal muscle of the rat, *Circ. Res.* **37**:430.

Muehrcke, R. C., Mandal, A. K., and Volini, F. I., 1970, A pathophysiologic review of the renal medullary interstitial cells and their relationship to hypertension, *Circ. Res.* **27–28**:I-109 (Suppl. I).

Muirhead, E. E., 1973, Vasoactive and anti-hypertensive effects of prostaglandins and other renomedullary lipids, in: *The Prostaglandins: Pharmacological and Therapeutic Advances* (M. F. Cuthbert, ed.), pp. 201–251, Heinemann, London.

Muirhead, E. E., Germaine, G., Leach, B. E., Pitcock, J. A., Stephenson, P., Brooks, B., Brosius, W. L., Daniels, E. G., and Hinman, J. W., 1972, Production of renomedullary prostaglandins by renomedullary interstitial cells grown in tissue culture, *Circ. Res.* **31–32**:161 (Suppl. II).

Murphy, G. P., Hesse, V. E., Evers. J. L., Hobika, G., Mostert, J., Szolniky A., Schoonees, R., Abramczyk, J., and Grace, J., 1970, The renal and cardiodynamic effects of prostaglandins (PGE_1, PGA_1) in renal ischemia, *J. Surg. Res.* **10**:533.

Nakano, J., 1970, Metabolism of prostaglandin E_1 in dog kidney, *Br. J. Pharmacol.* **40**:317.

Nakano, J., 1973, General pharmacology of prostaglandins, in: *The Prostaglandins: Pharmacological and Therapeutic Advances* (M. F. Cuthbert, ed.), pp. 23–124, Heinemann, London.

Nakano, J., and Morsy, N. H., 1971, Beta-oxidation of prostaglandin E_1 and E_2 in rat lung and kidney homogenate, *Clin. Res.* **19**:142.

Nakano, J., and Prancan, A. V., 1973, Metabolic degradation of prostaglandin E_1 in the lung and kidney of rats in endotoxin shock, *Proc. Soc. Exp. Biol. Med.* **144**:506.

Nakano, J., Änggård, E., and Samuelsson, B., 1969, Substrate specificity and inhibition of 15-hydroxyprostaglandin dehydrogenase (PGDH), *Pharmacologist* **11**:238.

Needleman, P., Kaufman, A. H., Douglas, J. R., Johnson, E. M., and Marshall, G. R., 1973, Specific stimulation and inhibition of renal prostaglandin release by angiotensin analogs, *Am. J. Physiol.* **244**:1415.

Needleman, P., Minkes, M. S., and Douglas, J. R., 1974*a*, Stimulation of prostaglandin biosynthesis by adenine nucleotides, *Circ. Res.* **34**:455.

Needleman, P., Johnson, E. M., Jakschik, B., Douglas, J. R., Marshall, G. R., 1974*b*, Renal hormonal interactions and their pharmacological modification: Renin-angiotensin, catecholamines, and prostaglandins, in: *Recent Advances in Renal Physiology and Pharmacology* (L. G. Wesson and G. M. Fanelli, eds.), pp. 197–214, University Park Press, Baltimore.

Needleman, P., Douglas, J. R., Jr., Jakschik, B., Stoeklein, P. B., and Johnson, E. M., Jr., 1974*c*, Release of renal prostaglandin by catecholamines: Relationship to renal endocrine function, *J. Pharmacol. Exp. Ther.* **188**:453.

Nissen, H. M., 1969, On lipid droplets in renal interstitial cells. II. A histologic study on the number of droplets in salt depletion and acute salt repletion, *Z. Zellforsch. Mickrosk. Anat.* **85**:483.

Nissen, H. M., and Bojesen, I., 1969, On lipid droplets in renal interstitial cells. IV. Isolation and identification, *Z. Zellforsch. Mickrosk. Anat.* **97**:274.

Nolly, H. L., Reid, I. A., and Ganong, W. F., 1974, Effect of theophylline and adrenergic blocking drugs on the renin response to norepinephrine *in vitro*, *Circ. Res.* **35**:575.

Norby, L., Lentz, R., Flamenbaum, W., and Ramwell, P., 1976, Prostaglandins and aspirin therapy in Bartter's syndrome, *Lancet* **2**:604.

Orloff, J., Handler, J. S., and Bergström, S., 1965, Effect of prostaglandin (PGE₁) on the permeability response to vasopressin, theophylline and aldenosine 3′,5′-monophosphate, *Nature (London)* **205**:397.

Osvaldo, L., and Latta, H., 1966, Interstitial cells of the renal medulla, *J. Ultrastruct. Res.* **15**:589.

Owen, T. L., Ehrhart, J., Weidner, J., Haddy, F., and Scott, J., 1974, Effects of indomethacin on blood flow, reactive hyperemia and autoregulation in the dog kidney, *Fed. Proc.* **33**:348.

Ozer, A., and Sharp, G. W. G., 1972, Effect of prostaglandins and their inhibitors on osmotic water flow in the toad bladder, *Am. J. Physiol.* **222**:674.

Ozer, A., and Sharp, G. W. G., 1973, Modulation of adenyl cyclase action in the toad bladder by chlorpropamide antagonism to prostaglandin E₁, *Eur. J. Pharmacol.* **22**:227.

Pace-Asciak, C., and Cole, S., 1975, Inhibitors of prostaglandin catabolism. I. Differential sensitivity of 9-PGDH and 13-PRG and 15-PGDH to low concentrations of indomethacin, *Experientia* **31**:143.

Pace-Asciak, C., Coceani, F., and Wolfe, L. S., 1967, Comparison of the conversion of arachidonic acid into prostaglandins by some tissues, *Proc. Can. Fed. Biol. Sci.* **10**:80.

Pace-Asciak, C., Morawka, K., Coceani, F., and Wolfe, L. S., 1968, The biosynthesis of prostaglandin E₂ and F₂α in homogenates of the rat stomach, in: *Prostaglandin Symposium of the Worchester Foundation for Experimental Biology,* (P. W. Ramwell and J. E. Shaw, eds.), pp. 371–378, Interscience, New York.

Palmer, M. A., Piper, P. J., and Vane, J. R., 1973, Release of rabbit aorta contracting substance (RCS) and prostaglandins induced by chemical or mechanical stimulation of guinea-pig lung, *Br. J. Pharmacol.* **49**:226.

Papanicolaou, N., and Meyer, P., 1972, Inactivation of prostaglandins E₁ and A₂ on their single passage through the pulmonary vascular bed in anesthetized rats, *Rev. Can. Biol.* **31**:313.

Peach, M. J., Cline, W. H., and Watts, D. T., 1966, Release of adrenal catecholamines by angiotensin, *Circ. Res.* **19**:571.

Peach, M. J., Cline, W. H. Davila, D., and Khairallah, P., 1970, Angiotensin–catecholamine interactions in the rabbit, *Eur. J. Pharmacol.* **11**:286.

Peskar, B., and Hertting, G., 1973, Release of prostaglandins from isolated cat spleen by angiotensin and vasopressin, *Naunyn-Schmiedebergs Arch. Pharmacol.* **279**:227.

Peterson, J. M., and Edelman, I. S., 1964, Calcium inhibition of the action of vasopressin on the urinary bladder of the toad, *J. Clin. Invest.* **43**:583.

Pettinger, W. A., Augusto, L., and Leon, A. S. 1972, Alteration of renin release by stress and adrenergic receptor and related drugs in anesthetized rats, *Adv. Exp. Biol. Med.* **22**:105.

Piper, P. J., 1973, Distribution and metabolism, in: *Prostaglandins: Pharmacological and Thera-peutic Advances* (M. F. Cuthbert, ed.), pp. 125–150, Heinemann, London.

Piper, P. J., and Vane, J. R., 1969, Release of additional factors in anaphylaxis and its antagonism by anti-inflammatory drugs, *Nature (London)* **223**:29.

Piper, P. J., and Vane, J. R., 1971, The release of prostaglandins from lung and other tissues, *Ann. N.Y. Acad. Sci.* **180**:363.

Piper, P. J., Vane, J. R., and Wyllie, J. H., 1970, Inactivation of prostaglandins by the lung, *Nature (London)* **225**:600.

Pomeranz, B. H., Birtch, A. G., and Barger, A. C., 1968, Neural control of intrarenal blood flow, *Am. J. Physiol.* **215**:1067.

Prezyna, A., Attallah, A., Vance, A., Schoolman, M., and Lee, J., 1973, The renomedullary body, a newly recognized structure of renomedullary interstitial cell origin associated with high prostaglandin content, *Prostaglandin* **3**:699.

Rabito, C. A., and Fasciolo, J. C., 1973, Effects of prostaglandins PGE₁ and PGF₂ on oxygen consumption, sodium and potassium content of renal tissue, *Experientia* **29**:673.

Ramwell, P. W., and Shaw, J. E., 1970, Biological significance of the prostaglandins, *Rec. Progr. Hormone Res.* **26**:139.

Raz, A., 1972, Interactions of prostaglandins A₂, F₂ and E₂ to the human plasma proteins, *Biochem. J.* **130**:631.

Regoli, D., Parks, W. K., and Rioux, F., 1973, II. Pharmacology of angiotensin antagonists, *Can. J. Pharmacol.* **51**:114.

Ripoche, P., Bourguet, J., and Parisi, M., 1973, The effect of hypertonic media on water permeability of frog urinary bladder: Inhibition of catecholamines and prostaglandin E₁, *J. Gen. Physiol.* **61**:110.

Samuelsson, B., 1965, On the incorporation of oxygen in the conversion of 8,11,14-eicosatreinoic acid to prostaglandin E₁, *J. Am. Chem. Soc.* **87**:3011.

Samuelsson, B., and Wennmalm, A., 1971, Increased nerve stimulation induced release of noradrenaline from the rabbit heart after inhibition of prostaglandin synthesis, *Acta Physiol. Scand.* **83**:163.

Samuelsson, B., Granstrom, E., Gréen, K., and Hamberg, M., 1971, Metabolism of prostaglan-dins, *Ann. N.Y. Acad. Sci.* **180**:138.

Sanner, J. H., 1974, Substances that inhibit the action of prostaglandins, *Arch. Intern. Med. (Chicago)* **133**:133.

Satoh, S., and Zimmerman, G. G., 1975, Influence of the renin–angiotensin system on the effect of prostaglandin synthesis inhibitors in the renal vasculature, *Circ. Res.* **36–37**(Suppl. 1):89.

Schneider, E. G., Strandhoy, J. W., Willis, L. R., and Knox, F. G., 1973, Relationship between proximal sodium reabsorption and excretion of calcium, magnesium and potassium, *Kidney Int.* **4**:369.

Schumann, H. J., Starke, K., and Werner, U., 1970, Interaction of inhibition of noradrenaline uptake and angiotensin on the sympathetic nerves of the isolated rabbit heart, *Br. J. Pharmacol.* **39**:390.

Scott, J. B., Daugherty, R. M., Dabney, J. M., and Haddy, F. J., 1965, Role of chemical factors in the regulation of flow through the kidney, hindlimb and heart, *Am. J. Physiol.* **208**:813.

Selkurt, E. E., 1974, Current status of renal circulation and related nephron function in hemor-rhage and experimental hemorrhagic shock. II. Neurohumoral and tubular mechanisms, *Circ. Shock* **1**:89.

Shio, H., and Ramwell, P. W., 1972, Effect of prostaglandin E₂ and aspirin on the secondary aggregation of human platelets, *Nature (London) New Biol.* **236**:45.

Silver, M. J., Smith, J. B., Ingerman, C., and Kocsis, J. J., 1972, Human blood prostaglandins: Formation during clotting, *Prostaglandins* **1**:429.

Sirois, P., and Gagnon, D. J., 1974, Release of prostaglandins from the rabbit renal medulla, *Eur. J. Pharmacol.* **28**:18.

Smith, J. B., and Willis, A. L., 1971, Aspirin selectively inhibits prostaglandin production in platelets, *Nature (London) New Biol.* **231**:235.

Solez, K., Fox, J. A., Miller, M., and Heptinstall, R. H., 1974, Effects of indomethacin on renal inner medullary plasma flow, *Prostaglandins* **7**:91.

Spector, D., Zusman, R. M., Caldwell, B. V., and Speroff, 1974, The distribution of prostaglandins A, E, and F in the human kidney, *Prostaglandins* 6:263.

Stein, J. H., Ferris, T. F., Hutch, J. E. Smith, T. C., and Osgood, R. N., 1971, The effect of renal vasodilatation on the distribution of cortical blood flow in the kidney of the dog, *J. Clin. Invest.* 50:1429.

Stjärne, L., 1973a, Alpha-adrenoceptor mediated feedback control of sympathetic neurotransmitter secretion in guinea pig vas deferens, *Nature (London) New Biol.* 241:190.

Stjärne, L., 1973b, Prostaglandin verses alpha adrenoceptor mediated control of sympathetic neurotransmitter secretion in guinea pig isolated vas deferens, *Eur. J. Pharmacol.* 22:233.

Stjärne, L., and Gipe, K., 1973, Prostaglandin-dependent and -independent feedback control of noradrenaline secretion in vasconstrictor nerves of normotensive human subjects, *Naunyn-Schmiedebergs Arch. Pharmacol.* 280:441.

Strandhoy, J. W., Ott, C. E., Schneider, E. G., Willis, L. R., Beck, N. P., Davis, B. B., and Knox, F. G., 1974, Effects of prostaglandins E_1 and E_2 on renal sodium reabsorption and forces, *Am. J. Physiol.* 226:1015.

Strong, C. G., and Bohr, D. F., 1967, Effects of prostaglandins E_1, E_2, A and $F_{1\alpha}$ on isolated vascular smooth muscle, *Am. J. Physiol.* 213:725.

Strong, C. G., Boucher, R., Nowaczynka, W., and Genest, J., 1966, Renal vasodepressor lipid *Proc. Mayo Clinic* 41:433.

Swain, J. A., Heyndrick, G. R., Boettcher, D. H., and Vatner, 1975, Prostaglandin control of renal circulation in the unanesthetized dog and baboon, *Am. J. Physiol.* 229:826.

Tannenbaum, J., Slawinski, J. A., Oates, J. A., and Nies, A. S., 1975, Enhanced renal prostaglandin production in the dog, I. Effects on renal function, *Circ. Res.* 36:197.

Terashima, R., Anderson, F. L., Jubiz, W., Tsagaris, T. J., and Kuida, H., 1974, Influence of adrenergic drugs on prostaglandin E release from the dog kidney. *Proc. Soc. Exp. Biol. Med.* 147:449.

Terragno, D. A., Crowshaw, K., Terragno, N. A., and McGiff, J. C., 1975, Prostaglandin synthesis by bovine mesenteric arteries and veins, *Circ. Res.* 36–37:76 (Suppl. I).

Thompson, R. B., Kaufman, C. E., and DiScala, V. A., 1971, Effect of renal vasodilatation on divalent ion excretion and Tm_{PAH} in anethetized dogs, *Am. J. Physiol.* 221:1097.

Tobian, L., Ishii, M., and Duke, M., 1972, Relationship of cytoplasmic granules in renal papillary interstitial cells to "post-salt" hypertension, *J. Lab. Clin. Med.* 73:309.

Tobian, L., O'Donnell, M., and Smith, P., 1974, Intrarenal prostaglandin levels during normal and high sodium intake, *Circ. Res.* 34–35:I-83 (Suppl. I.)

Vance, V. K., Attallah, A., Prezyna, A., and Lee, J. B., 1973, Human renal prostaglandins, *Prostaglandins* 3:647.

Vander, A. J., 1965, Effect of catcholamines and the renal nerves on renin secretion in anethetized dogs, *Am. J. Physiol.* 209:659.

Vander, A. J., 1968, Direct effects of prostaglandin on renal function and renin release in anesthetized dog, *Am. J. Physiol.* 214:218.

van Dorp, D., 1971, Recent developments in the biosynthesis and analysis of prostaglandins, *Ann. N.Y. Acad. Sci.* 180:181.

van Dorp, D. A., 1965, The biosynthesis of prostaglandins, *Mem. Soc. Endocrinol.* 14:39.

van Dorp, D. A., Berthuis, R. K., Nugteren, D. U., and Vonkerman, H., 1964, The biosynthesis of prostaglandin, *Biochim. Biophys. Acta* 90:204.

van Dorp, A. A., Jouvenas, G. H., and Struijk, C. B., 1967, The biosynthesis of prostaglandin in pig eye iris, *Biochim. Biophys. Acta* 137:396.

Vane, J. R., 1971, Inhibition of prostaglandin synthesis as a mechanism of action for aspirin-like drugs, *Nature (London)* 231:232.

Vane, J. R., 1973, Prostaglandins and aspirin-like drug, in: *Pharmacology and the Future of Man* (Proc. 5th Int. Congr. Pharmacol., San Francisco), Vol. 5, pp. 352–378, Karger, Basel.

Vane, J. R., and McGiff, J. C., 1975, Possible contributions of endogenous prostaglandins to the control of blood pressure, *Circ. Res.* 36–37:68 (Suppl. I).

Vanherwegham, J. L., Ducobu, J., and Hollander, A. D., 1975, Effects of indomethacin on renal hemodynamics and on water and sodium excretion by the isolated dog kidney, *Pflügers Arch.* 357:243.

Vatner, S. F., 1974, Effects of hemorrhage on regional blood flow distribution in dogs and primate, *J. Clin. Invest.* **54**:225.

Venuto, R. C., O'Dorision, T., Ferris, T. F., and Stein, J. H., 1975, Prostaglandins and renal function II. The effect of prostaglandin inhibition on autoregulation of blood flow in the intact kidney of the dog, *Prostaglandins* **9**:817.

Viguera, M. G., and Sunahara, F. A., 1969, Microcirculatory effects of prostaglandins, *Can. J. Physiol. Pharmacol.* **47**:627.

von Euler, U. S., 1935, A depressor substance in the vesicular gland, *J. Physiol. (London)* **84**:21P.

Weeks, J. R., 1974, Prostaglandins, *Fed. Proc.* **33**:37.

Werning, C., Vetter, W., Weidmann, P., Schweikert, H. W., Stiel, D., and Siegenthaler, W., 1971, Effect of prostaglandin E₁ on renin in the dog, *Am. J. Physiol.* **220**:852.

Williamson, H. E., Bourland, W. A., and Marchand, G. R., 1974, Inhibition of ethacrynic acid induced increase in renal blood flow by indomethacin, *Prostaglandins* **8**:297.

Wilson, D. E., 1974, Medical significance of prostaglandins, *Arch. Intern. Med. (Chicago)* **133**:29.

Winer, N. D., Chokshi, D. S., and Walkerhorst, W. G., 1971, Effects of cyclic AMP, catecholamines, and adrenergic receptor antagonist on renin secretion, *Circ. Res.* **29**:239.

Wong, P. Y., Talamo, R. C., Williams, G. H. W., and Colman, R. W., 1975, Responses of the kallikrein–kinin and renin–angiotensin systems to saline infusion and upright posture, *J. Clin. Invest.* **55**:691.

Zimmerman, B. G., and Gisslen, J., 1968, Pattern of renal vasoconstrictor and transmitter release during sympathetic stimulation in presence of angiotensin and cocaine, *J. Pharmacol. Exp. Ther.* **163**:320.

Zins, G. R., 1975, Renal prostaglandins, *Am. J. Med.* **58**:14.

Zusman, R. M., Caldwell, B. V., and Speroff, L., 1972, Radioimmunoassay of the A prostaglandins, *Prostaglandins* **2**:41.

Zusman, R. M., Spector, D., Caldwell, B. V., Speroff, L., Schneider, G., and Mulrow, P. J., 1973, The effect of chronic sodium loading and sodium restriction on plasma prostaglandin A, E, and F concentrations in normal humans, *J. Clin. Invest.* **52**:1093.

NOTE ADDED IN PROOF

The interval between submission of this chapter and its publication has been one of continued research into prostaglandins and the kidney. One could comment on significant additions to our knowledge of prostaglandins and renal function in the areas of biochemistry, physiology, and pharmacology. I believe, however, that the most important contribution concerns the role of renal prostaglandins in Bartter's syndrome. This uncommon disorder is characterized by hypokalemia, hyperreninemia, hyperaldosteronism, juxtaglomerular hyperplasia, and resistance to the pressor effects of angiotensin in normotensive individuals. There have been several reports indicating that urinary prostaglandins are elevated in patients with this syndrome (Gill *et al.*, 1976; Norby *et al.*, 1976). When patients having Bartter's syndrome are treated with inhibitors of prostaglandin synthetase normalization of plasma potassium concentration, renin activity, and aldosterone excretion were observed (Verberckmoes *et al.*, 1976; Fichman *et al.*, 1976; Gill *et al.*, 1976; Norby *et al.*, 1976). These reports demonstrate a significant role for prostaglandins in the renal dysfunction characteristic of this syndrome. It is anticipated that additional renal disease processes resulting from an excess or deficiency of prostaglandins will be observed in the future.

Prostaglandins and Cancer

Bernard M. Jaffe and M. Gabriella Santoro

Department of Surgery
Washington University School of Medicine
St. Louis, Missouri 63110

I. INTRODUCTION

Since prostaglandins have been shown to play a regulatory role in a number of biological systems, it is not surprising that their role in oncology has been fairly carefully studied. The recent explosion of data on the relationship between prostaglandins and tumor cell growth and function has been extraordinary. Based on earlier *in vitro* observations, newer studies have implicated prostaglandins in abnormalities in humans. This chapter will attempt to review and interpret the pertinent observations, both *in vitro* and *in vivo*. Since the general interactions of prostaglandins with the adenylyl cyclase–cyclic AMP system have been extensively reviewed, they will be alluded to only briefly. Two major aspects of the interactions between prostaglandins and tumor cells will be emphasized, prostaglandin biosynthesis by tumor cells and its effect on replication and the bone-resorbing effects of prostaglandins and their role in the hypercalcemia of neoplastic diseases.

II. PROSTAGLANDIN SYNTHESIS BY TUMORS AND ITS EFFECTS ON TUMOR CELL REPLICATION

A. *In Vitro* Studies

The studies of Apps and Cater (1973) implicated the inflammatory response in the release of PGE by human and animal tumors. Rat hepatoma 223 and BP8 mouse ascites tumor were incubated in diluted serum at 37°C. The amount of bioassayable PGE released into the incubation mixture was increased with increasing amounts of tumor and of serum, and decreased by incubating with xenogeneic serum (but not altered by allogeneic serum). PGE release was abolished by inclusion in the medium of aspirin (40 μg/ml) and indomethacin (8 μg/ml). The interaction of complement and tumor cells was essential for the release of PGE; PGE release was abolished by inhibitors of C'1 and C'3 activity (EDTA, EACA, heating to 56°C for 30 min, rabbit anti-rat C'3 serum, and treatment of serum with ammonia). These observations were verified in similar experiments with canine tumors (melanoma, hemangiosarcoma, and fibroma) and human adenocarcinomas.

Hammarström *et al.* (1973) used mass spectroscopy to measure levels of PGE_2 and $PGF_{2\alpha}$ in a nontransformed Syrian hamster fibroblast, BHK Cl 13, and in two polyoma virus-transformed fibroblasts. Cultures of BHK Wt Cl 9A, a wild-type polyoma virus-transformed fibroblast, contained almost sixtyfold more PGE_2 (1989 ng/100 μg cellular DNA), and BHK *ts-3* Cl 7C cells, a temperature-sensitive mutant, contained 10 times more PGE_2 (285.2 ng/100 μg cellular DNA) than nontransformed control cells (33.5 μg/100 μg DNA). Since 95–97% of the PGE_2 was found in the medium, intracellular PGE concentrations were quite low. Concentrations of $PGF_{2\alpha}$ were only 1/20 as high as those of PGE_2. These investigators were unable to detect any differences in prostaglandin concentrations in high- and low-density cultures.

Ritzi *et al.* (1975) demonstrated that PGF levels were similarly 1.5–2.5 times higher in simian virus-transformed fibroblasts, SV3T3, than in nontransformed control 3T3 cells. Although PGF concentrations were quite low (less than 5 ng/ml/per 24 hr), concentrations of metabolites in the media were only 5–17% as high as those of the native PGF compounds. Since rates of metabolism were not significantly different, the observed differences between transformed and nontransformed cells were found to be due to differences in prostaglandin production or release.

Although the studies of Levine, Tashjian, and associates will be described in detail in the subsequent section, it is appropriate to mention that these investigators identified five clones of a mouse fibrosarcoma, $HSDM_1$, which synthesized and secreted into the media large quantities of PGE_2 (0.7–2.0 μg/mg cell protein-day) (Levine *et al.*, 1972). Five control lines did not synthesize significant amounts of prostaglandins. PGE_2 synthesis was observed during both logarithmic and stationary phases of the cell growth cycle. Indomethacin (3×10^{-9} M) inhibited prostaglandin biosynthesis by 50%. Similar observations were made using the rabbit VX_2 carcinoma *in vitro,* which produced

0.5–3.0 μg PGE$_2$/mg cell protein per day (Voelkel *et al.*, 1975). As described below, both of these tumors caused hypercalcemia *in vivo*.

Levine and his group have studied the effects of a number of stimulatory and inhibitory compounds on radioimmunoassayable prostaglandin biosynthesis by a methylcholanthrene-transformed fibroblast line, MC5-5, *in vitro* (Hong *et al.*, 1976). During the logarithmic growth phase, the addition of precursor arachidonic acid (50–100 μg/ ml) to the culture media increased synthesis of PGE$_2$ and PGE$_{2\alpha}$ several hundredfold. Similar observations were made with the addition of fetal bovine serum, presumably because it also increased the concentrations of precursor fatty acids. For reasons which are more difficult to interpret, thrombin (0.01–1 μg/ml) stimulated PGE$_2$ biosynthesis, reflected by an increase from 0.6 to 6.0 ng/ml of medium; this effect was significantly inhibited by Lys CH$_2$Cl, a thrombin inhibitor, as well as by indomethacin. Ninefold stimulation of prostaglandin biosynthesis was also observed following addition of bradykinin (but not angiotensin II or substance P) to the medium at a final concentration of 500 ng/ml. Like the effect of thrombin, bradykinin had a very rapid action which was maximal at 3 min and reversible by indomethacin.

In further studies using both HSDM$_1$ (Tashjian *et al.*, 1975) and MC5-5 cells (Hong and Levine, 1976), these investigators demonstrated that glucocorticosteroids inhibited PGE biosynthesis by tumors *in vitro*. By labeling intracellular lipids with [^{3}H]arachidonic acid, the mechanism of action of this effect was shown to be at the level of release of arachidonic acid from membrane phospholipids. Steroid-treated cells converted exogenous arachidonic acid to PGE$_2$ and PGF$_{2\alpha}$ normally. Dexamethasone was the most potent inhibitory corticosteroid (maximal 72% inhibition of PGE biosynthesis), while hydrocortisone and corticosterone produced significant inhibition; non-anti-inflammatory steroids, including cholesterol, androgens, and estrogens, were virtually ineffective. The mechanism of action, interference with arachidonic acid release, contrasts sharply with the action of indomethacin, which does not interfere with the pre-arachidonic acid steps but totally inhibits the conversion of arachidonic acid to active prostaglandins by the cyclooxygenase system.

Jaffe *et al.* (1971) demonstrated that organ cultures of human adenocarcinomas contained large amounts of PGE$_2$ (more than 25 ng/ ml). In an experiment in which colon adenocarcinoma and adjacent normal colonic mucosa were incubated simultaneously in short-term organ culture, the carcinoma synthesized 8 times more PGE than did the normal mucosa.

Jaffe and his group have also carefully studied endogenous prostaglandin biosynthesis by cloned cells in tissue culture and correlated prostaglandin levels with rates of cell replication. During log phase growth *in vitro,* N4 (a mouse neuroblastoma clone), B 82 (a mutant L-cell line), and C6 (a rat glioma clone) synthesized radioimmunoassayable amounts of PGE (Jaffe *et al.,* 1973*b*; Hamprecht *et al.,* 1973). The cumulative PGE production was linear with time during log phase growth. In the neuroblastoma clone which manifested contact inhibition of growth, at confluency, the rate of PGE production increased abruptly and markedly. Three-fourths of the measured PGE was extracellular (in the medium) and concentrations of PGA and PGF compounds were only 8% and 2%, respectively, as high as those of PGE. Addition of dibutyryl cyclic AMP to the medium (final concentration 1 mM) significantly stimulated PGE production (2.5- to 6-fold) and inhibited cell proliferation in all three lines. In further studies, these investigators evaluated the rates of cell proliferation and PGE production by cells *in vitro* under basal conditions as well as under conditions of stimulation and inhibition of PGE biosynthesis (Thomas *et al.,* 1974*a,b*). In four tumor cell lines, HeLa, HEp-2, L, and HT-29 (a monolayer derivative of human adenocarcinoma), a clear inverse relationship was demonstrated under all experimental conditions examined. The more slowly growing cells synthesized significantly more PGE, which also correlated well with an apparently increased enzymatic ability to convert known quantities of arachidonic acid to PGE_2 (Cohen and Jaffe, 1973) (Table I). Exogenous PGE_1 (1 μg/ml) added to the medium inhibited cell proliferation by an average of 41%; endogenous PGE, stimulated by 54% by 1 mM dibutyryl cyclic AMP, resulted in a similar degree of inhibition of cell replication. Suppression of PGE synthesis (61%) by 10^{-8} M indomethacin resulted in a mean 23%

Table I. Relationship between PGE Production and Doubling Times of Cells *in Vitro*

| Cell line | PGE production | | Mean doubling time (hr) |
	Cumulative (ng/10⁶ cells)[a]	Synthesized from arachidonic acid (ng/mg cells)[b]	
HeLa	33.7	244.0	37.5
L	7.0	42.3	30.0
HEp-2	4.3	22.6	23.0
HT-29	3.0	—	11.5

[a]During growth in monolayer culture.
[b]Homogenates containing 56 μg glutathione, 0.57 μg hydroquinone, 20 mM EDTA, and 12.5 μg arachidonic acid.

stimulation of cell replication; this effect was reversed by the addition of control amounts of PGE_1 (10 ng/ml) to the media (Table II). In recent studies, Santoro *et al.* (1976) demonstrated that PGE_1 (1 μg/ ml) inhibited cell replication in mouse melanoma B-16 *in vitro* by an average of 25–30%. The glucocorticoid cortisol, at 10^{-6} M, inhibited PGE biosynthesis and stimulated cell replication by an average of 36.5%; this effect was reversed by addition to the medium of subthreshold amounts of PGE_1 (10 ng/ml).

Naseem and Hollander (1973) demonstrated dose-dependent inhibition of the growth rate of a murine plasma cell tumor (MPC-11) cultured in different concentrations of PGE_1, PGE_2, and $PGF_{2\alpha}$. As described in Table III, PGE_1 was shown to be most active.

The studies described previously all utilized counting of viable cells to demonstrate prostaglandin inhibition of cell proliferation. The studies described below used either tritiated thymidine incorporation or mitotic index.

Sonis *et al.* (personal communication) demonstrated inhibition of lymphoma (EL-4) cell proliferation in tissue culture (as measured by [^{3}H]thymidine incorporation) by an inhibitory factor produced by fibrosarcoma induced by methylcholanthrene (MCA) *in vitro*. The supernatant inhibitory substance was identified as PGE by its thin-layer chromatographic characteristics and by the fact that PGE_2 treatment of lymphoma cells mimicked the inhibitory action of the supernatant inhibitory substance. Moreover, indomethacin (0.1 μg/ml) treatment of fibrosarcoma cultures inhibited the synthesis of the supernatant inhibitory substance. The degree of inhibition depended on the amount of supernatant present, the length of time that the EL-4 cells were in contact with the supernatant, and the age of the cultures used (supernatants from established cultures that were serially passaged at least twice had more inhibitory activity than supernatants from newly extracted fibrosarcoma cells).

Adolphe *et al.* (1973) also demonstrated prostaglandin-induced inhibition of HeLa cell proliferation *in vitro,* as measured by mitotic and metaphasic index. PGE_1, PGE_2, and PGA_2 caused inhibition by interfering with the metaphase; $PGF_{1\alpha}$ and $PGF_{2\alpha}$ were shown to be inactive.

PGE_1, PGE_2, and $PGF_{2\alpha}$ were also shown to inhibit the growth of mouse leukemia lymphoblasts L51784 in culture. In these studies, Yang *et al.* (1976) used very large concentrations of prostaglandins (100, 50, and 25 μg/ml) and the inhibition of cell replication was measured by [^{3}H]thymidine, uridine, and leucine incorporation. PGE_1 and PGE_2 at 100 μg/ml inhibited the incorporation of all three compounds, while at lower concentrations leucine incorporation was not

Table II. Mean Percentage Changes in Viable Cell Count and in Cumulative PGE Production

Cell line	Cell count (%)				PGE production (%)	
	3 mM PGE$_1$	1 mM dibutyryl cyclic AMP	10^{-8} M indomethacin	10^{-8} M indomethacin + PGE$_1$ (10 ng/ml)	1 mM dibutyryl cyclic AMP	10^{-8} M indomethacin
L	−26.5	−43.3	+29.3	+1.0	+84.5	−79.5
HeLa	−44.3	−22.0	+14.3	—	+26.5	−68.0
HEp-2	−35.8	−73.3	+21.0	—	+56.8	−49.8
HT-29	−55.3	−24.7	+26.0	—	+48.1	−79.0
B-16	−30.0	—	+17.0	—	—	−35.5

Table III. Inhibition of MPC-11 Growth by PGE_1, PGE_2, and $PGF_{2\alpha}$

Concentration of PG (μg/ml)	Percent inhibition of growth[a]		
	PGE_1	PGE_2	$PGF_{2\alpha}$
0.1	14	—	—
1.0	23	5	—
3.0	39	11	—
5.0	49	25	5
10.0	58	29	5

[a]2.5×10^5 cells were cultured with three different prostaglandins at the concentrations noted for 72 hr.

inhibited. $PGF_{2\alpha}$ showed significant inhibition of thymidine and uridine incorporation but not that of leucine. PGE_1 and PGE_2 (1.8 μg/ml) also suppressed the ability of the cells to form colonies in soft agar; at this concentration, $PGF_{2\alpha}$ was found to be stimulatory, while larger amounts (56 μg/ml) were inhibitory.

Regardless of the methodologies used for measuring prostaglandins (radioimmunoassay, bioassay, gas–liquid chromatography/mass spectroscopy) and for determining rates of cell replication (counting viable cells, tritiated thymidine uptake, mitotic index), the data from the *in vitro* studies have all been quite consistent. Tumor cells, *in vitro*, synthesize significant amounts of prostaglandins, predominantly PGE_2. Synthesis which can be stimulated and inhibited by a number of compounds (Table IV) has been documented both by measuring conversion of fatty acid precursors to prostaglandins and by measuring intracellular and extracellular concentrations of prostaglandins. More than 90% of the synthesized prostaglandins are liberated into the medium. Transformed cells synthesize more prostaglandins than do

Table IV. Compounds Which Influence PGE Biosynthesis by Tumor Cells *in Vitro*

Stimulation	Inhibition
Arachidonic acid	Indomethacin
Dibutyryl cyclic AMP	Aspirin
Thrombin	Hydrocortisone
Bradykinin	Dexamethasone
Mechanical manipulation	
Serum	

normal cells. The addition of exogenous and stimulation of endogenous prostaglandins result in inhibition of the rate of cell replication. Despite the fact that these studies are in tissue culture, as will be shown in the subsequent section, similar observations have been made in *in vivo* experiments.

B. *In Vivo* Studies

1. Experimental Animal Tumors

A number of investigators have demonstrated prostaglandin biosynthesis by experimental tumors *in vivo*. In this section, we will review the experimental data, concentrating on the possible effects of the synthesized prostaglandins on the rates of tumor cell growth and in causing symptoms.

Since mice bearing the BP8/P_1 tumor usually have diarrhea, Sykes and Maddox (1972, 1974) studied the ability of this tumor to synthesize prostaglandins. Using the superperfusion bioassay, thin-layer chromatography, and gas–liquid chromatography/mass spectroscopy, these investigators demonstrated that BP8/P_1 cells (grown in C3H/He mice) contained 2 μg PGE_2/g of cells. These cells contained an active prostaglandin synthetase system which was able to convert 10% of arachidonic acid to PGE_2 and was more than 80% inhibitable by indomethacin and other nonsteroidal anti-inflammatory drugs. Similar observations were made on 3LL (Lewis lung carcinoma) and S180 tumors grown in C57/Bl mice (0.6 and 5.9 μg/g of cells, respectively). Surprisingly, ascites fluid collected from tumor-bearing mice contained relatively little PGE_2 (less than 15 ng/ml) (Sykes, 1970). Indomethacin treatment did not alter the rates of growth of these tumors.

Mammary tumors induced in Sprague-Dawley rats by intravenous injection of 7,12-dimethylbenz[*a*]anthracene (DMBA) contained an average of 4 times as much PGE_2 (215 ng/g dry tissue) as control mammary glands (Tan *et al.*, 1974). Tissue slices *in vitro* also converted 2–3 times as much arachidonic acid into PGE_2 as normal rat mammary tissue.

Humes and Strausser (1974) noted that Moloney sarcoma tumors contained 53 times more PGE (15.9 ± 4.3 pmol/mg) and 7 times as much PGF (0.89 ± 0.17 pmol/mg) as contralateral normal thigh muscle. Tumor concentrations of cyclic AMP (but not GMP) were more than twice control levels. When grown in short-term organ culture, Moloney sarcoma tissue synthesized large quantities of radioimmunoassayable PGE (180 ng/100 mg per 3 hr), more than 90% of which was secreted into the medium. Although indomethacin (0.1 mM) significantly inhibited PGE biosynthesis, neither arachidonic acid (0.2 mM) nor dibutyryl cyclic AMP (2 mM) significantly stimulated it. In contrast to the observations of Sykes and Maddox, daily intraperitoneal administration of indomethacin (5 mg/kg) to Moloney sarcoma-injected BALB/c mice delayed the onset of subcutaneous tumors and inhibited the rate of sarcoma growth. Simultaneous administration of antilymphocyte serum and Moloney sarcoma tumor resulted in enhanced and accelerated tumor growth; this augmentation of sarcoma growth was reversed by indomethacin treatment of the mice (Humes *et al.*, 1974). In a later publication (Strausser and Humes, 1975), these investigators proposed that prostaglandins synthesized by tumors suppressed immunological responsiveness and allowed for tumor growth; the effect of indomethacin in slowing tumor growth was therefore due to restoring immunological competence.

The importance of prostaglandin-induced cellular and humoral immunological unresponsiveness was further emphasized by the studies of Plescia *et al.* (1975). Mice bearing chemical- and virus-induced syngeneic tumors (an ascites cell line MCDV-12 induced in BALB/c mice by Rauscher leukemia virus and a solid fibrosarcoma MC-16 induced by methylcholanthrene in C57Bl/6J mice) were shown to be immunologically depressed; injection of sheep red blood cells did not elicit normal antibody response as measured by hemagglutinin titer or by counting splenic plaque-forming cells. As few as one ML-16 cell per 1000 syngeneic spleen cells in an *in vitro* incubation mixture containing sheep erythrocytes was sufficient to suppress normal antibody response. In both the *in vivo* and *in vitro* systems, PGE_2 was immunosuppressive. However, the addition of indomethacin (5 μg/ml) to the *in vitro* system blocked the development of immunodepression. Administration of indomethacin (125 μg/day intraperitoneally) to C57Bl/6J mice injected with MC-16 tumor cells significantly reduced the size of subcutaneous tumors.

Stein-Werblowsky (1974) had a different interpretation of the role of prostaglandins in the immunological response to tumors. She suggested that tumors secrete a prostaglandin antagonist; the resulting prostaglandin deficiency causes immunological paralysis in the local vicinity of the tumor. Accordingly, inoculation of animals with tumor cell suspensions to which large quantities of exogenous prostaglandins have been added should restore immunological responsiveness and result in rejection of the malignant graft. In order to test this hypothesis, Stein-Werblowsky (1960) injected weanling Wistar rats with benzpyrene-induced tumor cells to which PGA_2, PGE_2, and $PGF_{2\alpha}$ had been added. At concentrations exceeding 1.0 mg/ml, PGA_2 prevented tumor take in more than 50% of injected animals, and profoundly suppressed tumor growth in the remainder. $PGF_{2\alpha}$ and PGE_2 had similar effects, but were considerably less effective. The sites of prostaglandin–tumor inoculation contained abundant immunoblasts but very few tumor cells. The high concentrations of prostaglandins were shown not to be tumoricidal, since tumor cells suspended in prostaglandins grew well in areas of "immunological privilege" including striated muscle and neonatal subcutaneous tissues.

In recent experiments, Santoro et al. (1976) noted that subcutaneous (and, after the appearance of tumors, intratumor) administration of 16,16-dimethyl-PGE$_2$-methylester significantly inhibited growth of B-16 tumors in C57Bl/6J mice. The drug inhibited both the rate of "take" (72 vs. 95%) and the mean tumor size (493 vs. 152 mg, 69% inhibition). Based on the initial injection of 10^6 cells, untreated animals had tumors with an average of 121×10^6 cells, compared to 20×10^6 cells/mouse in the PGE-treated mice (83% inhibition) (Table V).

Three experimental tumors have been shown to cause hypercalcemia, HSDM$_1$ fibrosarcoma in Swiss albino mice (Tashjian et al., 1974), VX$_2$ rabbit carcinoma (Voelkel et al., 1975), and Walker sarcoma in Wistar rats (Powles et al., 1973a). Tumor-bearing animals have been shown to have markedly elevated circulating levels of PGE. Indomethacin treatment of these animals has consistently lowered circulating and tumor concentrations of prostaglandins but has had varying effects of tumor size. Early administration of 100–125 μg of indomethacin to HSDM$_1$-bearing mice suppressed tumor size (control, 4.4 ± 0.24 g; indomethacin, 3.2 ± 0.27 g) (Tashjian et al., 1974). On the other hand, the same daily dose of indomethacin prevented the development of osseous metastases, but caused slight stimulation of Walker sarcoma size in the soft tissue (control, 12.9 ± 2.7 g; indomethacin, 16.4 ± 2.7 g; $P < 0.05$) (Powles et al., 1973a).

Table V. Effect of 16,16-Dimethyl-PGE$_2$ Methylester on Mouse Melanoma B-16 Growth in Vivo

	Controls ($n = 19$)	PGE group ($n = 18$)
Number of mice with tumors	18	13
Mean tumor weight (mg)	466.9 ± 66.6	109.0 ± 35.1
Cells/gram of tumor ($\times 10^5$)	258.4 ± 10.4	184.2 ± 13.1
Total number of cells in tumor ($\times 10^6$)	120.6 ± 17.2	20.2 ± 6.5

Table VI. Medullary Carcinoma of Thyroid

Mean PGE	1922 ± 541 pg/ml	
Calcitonin (ng/ml)		
> 20	2637 ± 975	(*n* = 9)
< 20	1403 ± 349	
Diarrhea		
Present	2372 ± 1261	(*n* = 8)
Absent	1596 ± 275	

2. Human Tumors

A number of recent human studies have provided clinical examples of the above experimental situations. Several tumors have been shown to cause either diarrhea or hypercalcemia in association with synthesis of large amounts of PGE_2. In this section, we will discuss in detail the recent observations on patients with medullary carcinomas of the thyroid and related malignancies; observations on hyper-prostaglandinemic, hypercalcemic patients will be alluded to for comparison, but discussed primarily in the subsequent section.

Medullary carcinoma of the thyroid (MCT) causes diarrhea in one-third of the affected patients (Hill *et al.*, 1973; Tashjian *et al.*, 1970). A number of humoral agents have been implicated in the etiology of the diarrhea. Using the jird colon bioassay and thin-layer chromatography, Williams *et al.* (1968) first noted that tumor tissue from four of seven patients with MCT contained large amounts of PGE_2 (36–674 ng/g) and $PGF_{2\alpha}$ (15–844 ng/g); prostaglandin levels were also elevated in the peripheral plasma and particularly high in blood from veins draining the tumors. This observation was verified in studies by Barrowman *et al.* (1975) as well as by Jaffe and his group using a radioimmunoassay for PGE (Jaffe *et al.*, 1973a; Kaplan *et al.*, 1973). In a recent study of the role of PGE and PGF in endocrine diarrheagenic syndromes, Jaffe and Condon (1976) noted that 18 of 19 patients with MCT had elevated peripheral plasma levels of PGE. The mean PGE level was 1922 ± 541 pg/ml (Table VI). The eight patients with diarrhea had higher peripheral concentrations of PGE (2372 ± 1261 pg/ml) than did the 11 who did not (1596 ± 275 pg/ml). Levels of thyrocalcitonin were measured in 17 patients and averaged 24.5 ± 6.3 ng/ml. In patients whose thyrocalcitonin levels exceeded 20 ng/ml, PGE levels averaged 2537 ± 975 pg/ml, whereas in the eight patients with calcitonin levels below 20 ng/ml, PGE levels averaged 1403 ± 349 pg/ml. Although there are no data on the effects of indomethacin or aspirin on the diarrhea associated with MCT, nutmeg, an effective inhibitor of prostaglandin biosynthesis (at least by human colon *in vitro;* Bennett *et al.*, 1974), has been reported

Table VII. PGE Levels in Patients with Diarrheagenic Syndromes

Patient syndrome	Number of patients with elevated PGE	Mean PGE (pg/ml)
Medullary carcinoma of thyroid	18/19	1922 ± 541
Carcinoid	20/22	1367 ± 245
Watery diarrhea, hypokalemia, achlorhydria	8/21	993 ± 490
Zollinger-Ellison	2/29	388 ± 32
Nonendocrine diarrhea	3/28	353 ± 25
Normals	0/21	272 ± 18

to reduce the diarrhea in MCT patients (Barrowman *et al.,* 1975; Fawell and Thompson, 1973).

These observations on diarrhea associated with endocrine tumors have been extended to include the carcinoid syndrome (Smith and Greaves, 1974; Delmont and Rampal, 1975) and the WDHA syndrome (Condon and Jaffe, 1976), as well as neuroblastoma (Williams *et al.,* 1968; Sandler *et al.,* 1968). Levels of immunoreactive and bioassayable PGE are consistently elevated in patients with these diarrheagenic syndromes (Table VII). High PGE levels have also been reported in related tumors which have associated vasoactive phenomena including pheochromocytoma (Sandler *et al.,* 1968) and Kaposi's sarcoma (Bhana *et al.,* 1971). The one feature common to all these tumors is their derivation from the embryonal neural crest.

Prostaglandin-induced hypercalcemia has been reported in a number of human subjects with renal cell adenocarcinomas (Brereton *et al.,* 1974; Robertson *et al.,* 1975). Cummings *et al.,* (1975) demonstrated that veins draining a renal cell carcinoma contained large amounts of both PGE (549 pg/ml) and PGA (465 pg/ml). In addition, liver metastases from the tumor contained significantly more prostaglandins than adjacent normal liver (PGE 13.59 vs. 1.97 ng/g; PGA 2.54 vs. 0.22 ng/g). Finally, explants of this tumor synthesized significant amounts of both prostaglandins *in vitro*. The relationship of renal cell carcinomas and hypercalcemia will be discussed in the following section.

III. PROSTAGLANDINS, BONE RESORPTION, AND HYPERCALCEMIA

A. The Bone Resorptive Effect of Prostaglandins

In 1970, Klein and Raisz first demonstrated that prostaglandins induced bone resorption *in vitro*. Using as a model fetal rat long bones prelabeled with ^{45}Ca, addition to the culture media of PGE_1 and PGE_2 increased the concentrations of labeled calcium after 24 hr incubation. The resorptive effects were, in general, dose related but reached near-maximal effects at concentrations as low as 10^{-8} M. The actions were sustained for as long as 4 days in culture. The effects of PGE_1 and PGE_2 were similar to those of PTH (10^{-7}–10^{-8} M) but developed somewhat more slowly. The bone resorptive effects were accompanied by histological changes, including patchy loss of bone in the centers of the shafts. These effects were inhibited by both thyrocalcitonin and cortisol (10^{-7}–10^{-5} M). The specificity of the effects of prostaglandins was carefully evaluated in studies by Dietrich and colleagues (Dietrich et al., 1975; Dietrich and Raisz, 1975). Prostaglandins of the E, F, A, and B groups all possessed bone resorptive activity; they were all able to stimulate release of ^{45}Ca from fetal rat long bones by 60–135% at maximally effective doses. PGE_1 and PGE_2 were the most active and possessed at least 10 times as much activity as any other major prostaglandin group. In general, the prostaglandins of the 2 series (two side-chain unsaturated bonds) were more active than those of the 1 series. PGE_2 stimulated resorption at concentrations of 10^{-7}–10^{-5} M; at higher concentrations, the effect on bone resorption decreased. On a weight basis, PGE_2 was more potent than parathyroid hormone. After 6 days in culture, treatment with PGE_2 resulted in virtually complete loss of ^{45}Ca from the bones. The morphological changes were similar to those of other resorbers; the prostaglandins caused an increase in the size and number of osteoclasts as well as an increase in the ruffled border area of the osteoclasts. In contrast to these observations, Tashjian et al. (1972) were unable to demonstrate an effect of either PGF or PGB compounds; however, these investigators used calvaria from 5-day-old mice in their assay rather fetal long bones. Using a similar bioassay, we have demonstrated that $PGF_{2\alpha}$ has a resorptive effect one-third as potent as that of PGE_2 but have found PGA_2 almost inactive.

It has been recognized for some time that the addition of serum to fetal bones in culture results in increased bone resorption (Stern and Raisz, 1967). Raisz et al. (1974) demonstrated the importance of complement in this system. In a bone bioassay, unheated rabbit serum stimulated bone resorption by 60%; this effect was not observed in bones treated with either heated or C'6-deficient rabbit serum, but in the latter case could be restored by adding purified C'6. Prostaglandin biosynthesis was stimulated by untreated rabbit serum but not by complement-deficient sera. Indomethacin significantly inhibited the bone resorptive effect of complement-sufficient serum. These observations implicated complement in the resorption of bone and attributed this effect, at least partially, to prostaglandin biosynthesis.

The *in vitro* osteolytic effects of PTH were inhibited by aspirin and corticosteroids, both inhibitors of prostaglandin biosynthesis. Using prelabeled neonatal mouse calvaria, Powles *et al.* (1973*b*) noted that inclusion of aspirin (15 μg/ml) in the culture medium reduced the PTH (0.5 μg/ml) induced release of ^{45}Ca from 51 to 42%. On the other hand, aspirin did not affect PGE$_1$-stimulated bone resorption. In concentrations of 10^{-5} and 10^{-6} M, cortisol inhibited control and PTH (1 ng/ml) stimulated bone resorption *in vitro* (Caputo *et al.*, 1976), but did not influence the effect of 10^{-6} M PGE$_2$ (in contrast to prior reports; Raisz *et al.*, 1972). Other noninflammatory steroids, including estrogens, androgens, and cholesterol, were essentially inactive.

Since the observations that exogenous prostaglandins cause bone resorption *in vitro*, similar effects of endogenous PGE have been reported in a number of nontumor and tumor systems. Harris and Goldhaber (1973) noted that benign dental cysts *in vitro* caused bone resorption; recent experiments have shown that they also synthesize significant amounts of bioassayable PGE$_2$ (Harris *et al.*, 1973). Similar observations have been made for monkey gingiva (Gomes *et al.*, 1976), rheumatoid synovia (Robinson *et al.*, 1975), and fetal thyroid (Feinblatt *et al.*, 1976). In 1960, Goldhaber reported that fragments of mouse fibrosarcoma HSDM$_1$ enhanced bone resorption. As described in the following section, several experimental tumors have been identified which synthesize large amounts of PGE$_2$, which in turn produce bone resorption and hypercalcemia. In order to emphasize the importance of these *in vitro* effects, *in vivo* counterparts have recently been described. Goodson *et al.* (1974) injected PGE$_1$ (0.1 mg) daily into the tissues overlying rat calvaria. Within 7 days, a resorptive lesion was consistently noted in which the bone was replaced by vascular connective tissue. Although Klein and Raisz (1970) and Beliel *et al.* (1973) were unable to induce hypercalcemia with prostaglandins, Franklin and Tashjian (1975) have recently demonstrated that continuous intravenous infusion (260 ng/min) of PGE$_2$ (but not of PGF$_{2\alpha}$) results in hypercalcemia.

B. PGE as the Mediator of Hypercalcemia *in Vivo*

1. Experimental Animal Tumors

Based on Goldhaber's (1960) observation that extracts of HSDM$_1$ cause bone resorption *in vitro*, Levine, Tashjian, and co-workers have investigated the possibility that the humoral mediator is PGE$_2$. As mentioned above, five cloned strains of HSDM$_1$ were found to synthesize large amounts of PGE$_2$ (0.7–2 μg/ml cell protein per 24 hr). The biosynthesis of PGE was, not unexpectedly, inhibited by the addition to the medium of indomethacin (1 ng/ml) and aspirin (10 μg/ml) but not stimulated by pulsing with precursor arachidonic acid (Levine *et al.*, 1972). Tashjian *et al.* (1972) demonstrated that the bone-resorbing factor produced by $HSDM_1$ tumor cells was, in fact, PGE_2. Their evidence included the following: (1) The fact it could be extracted from tumor tissue as well as from media of cells in tissue culture. (2) The characteristics of the factor, i.e., low molecular weight (as determined by gel filtration, dialysis, and ultracentrifugation), stability at 100°C, resistance to digestion by trypsin and pepsin, solubility in organic solvents (ether, chloroform, and ethyl acetate), and high biological activity. (3) The radioimmunoassay data (4) The fact that indomethacin (0.1–100 ng/ml) inhibited the *in vitro* synthesis of both the bone-resorbing factor and PGE_2. Mice bearing $HSDM_1$ tumors were shown to have significantly elevated circulating levels of PGE_2 (627 vs. control 235 pg/ml) as well as of calcium (10.8 vs. control 9.7 mg/100 ml). Oral administration of indomethacin (100–125 μg/day) to these mice lowered serum levels of PGE_2 to 140 pg/ml, normalized serum calcium levels to 10.2 mg/100 ml, and virtually abolished the tumor content of both bone-resorbing activity and PGE_2 (2.3 vs. 76 ng/g fresh weight; Tashjian *et al.*, 1973). These experiments are nicely summarized in a review article (Tashjian *et al.*, 1974).

Another animal model in which PGE_2 has been implicated in the mediation of hypercalcemia is the rabbit VX_2 carcinoma (Voelkel *et al.*, 1975). VX_2 carcinoma *in vitro* synthesized an organic-solvent-extractable bone-resorbing factor as well as large amounts of PGE_2 (0.5–3.0 μg/mg cell protein/24 hr) during growth *in vitro;* synthesis of both of these compounds was profoundly inhibited by inclusion in the culture medium of indomethacin, 100 ng/ml. VX_2 tumors contained large amounts of radioimmunoassayable PGE_2 (294 $\pm$ 51 ng/g fresh weight). Orally administered indomethacin, 40 mg/day, virtually depleted the tumors of bone-resorbing activity and PGE_2. Rabbits became profoundly hypercalcemic (17–22 mg/100 ml), about 3–4 weeks after inoculation with VX_2 carcinoma. Indomethacin treatment from the time of the inoculation prevented the onset of hypercalcemia and, when started at 3–4 weeks after inoculation, normalized circulating calcium concentrations; cessation of indomethacin therapy was rapidly followed by the development of hypercalcemia. Although the authors were able to demonstrate that the venous drainage of VX_2 tumors contained larger amounts of PGE_2 (211 $\pm$ 15 pg/ml) than that of the contralateral side (137 $\pm$ 15 pg/ml), they did not note significant elevations of peripheral plasma PGE_2 concentrations until 5 weeks after tumor inoculation, long after the rabbits had become hypercalcemic. However, Seybirth *et al.* (1976a) noted parallel elevations in serum calcium and mass spectrometrically analyzed levels of a circulating metabolite of PGE_2 (15-keto-dihydro-PGE_2: 164 ± 49 to 4300 ± 800 pg/ml) as well as a urinary metabolic end product (tetranor derivative; PGE-M 4.9–14.3 μg/g creatinine). These data solidified the relationships between serum calcium and tumor-synthesized PGE_2.

Minne *et al.* (1975) recently characterized the hypercalcemic syndrome observed in rats bearing the Walker carcinosarcoma 256. Tumor-bearing animals were shown to have hypercalcemia, hyperphosphatemia, hyperuremia, hypercalcuria, hyperphosphaturia, and ectopic calcification. These abnormalities in calcium metabolism were reversed following resection of the tumors. These observations suggested that the tumor produced a bone-resorbing factor, but no possible candidates were discussed. By demonstrating that oral aspirin (30 mg) and indomethacin (100 μg) prevented hypercalcemia as well as osteolytic bone tumor deposits, Powles *et al.* (1973*a*) inadvertently implicated a prostaglandin as the possible mediator.

2. Human Tumors

a. Renal Cell Carcinomas. Powell *et al.* (1973) defined a group of patients with non-PTH-producing tumors which cause hypercalcemia by stimulating bone resorption and suggested that PGE_2 might be the humoral mediator. Shortly thereafter, Brereton *et al.* (1974) described the first patient with renal cell carcinoma who responded to indomethacin (25 mg twice a day) by normalizing the serum calcium level. The serum calcium level rose and fell with cessation and administration of this drug. Although tissue from a hepatic metastasis contained more immunoreactive PGE (29.1 vs. 3.8 ng/g) and PGF (46.4 vs. 8.7 ng/g) than the adjacent liver, this finding was not substantiated in tissue from a pulmonary metastasis. No elevation of circulating PGE was demonstrated. The first patient with hypercalcemia and documented hyperprostaglandinemia was reported by Robertson *et al.* (1975). This patient, also a victim of renal cell carcinoma, had higher levels of PGE, PGA, and PGF in metastatic tumor tissue than in normal lung or liver. He too was responsive to indomethacin therapy. Since these reports, a number of isolated cases of indomethacin-sensitive hypercalcemia have been reported (Ito *et al.*, 1975; Robertson and Baylink, 1975; Dindogru *et al.*, 1975) in patients with malignancies. Blum (1975) raised some question as to the specificity of the calcium response to indomethacin by demonstrating that one patient with a documented parathyroid adenoma had a fall in serum calcium following administration of this drug. There have been two recent surveys to determine the frequency of this anomaly. Demers *et al.* (1976) measured calcium levels in 72 patients with neoplastic diseases. The ten hypercalcemic patients (mean 12.5 ± 0.5 mg/100 ml), nine of whom had bone metastases, had significantly elevated levels of PGE (347 ± 83 pg/ml) compared to those of 62 normocalcemic controls (96 ± 8 pg/ml). In an informal survey taken at the 1976 Vail Prostaglandin Meeting, a total of six investigators reported that in their combined clinical experience 31 of 133 patients had hyperprostaglandinemia. Furthermore, a total of nine investigators reported that 16 of 48 hypercalcemic cancer patients responded to treatment with indomethacin. In general, the feeling at that time was that approximately one-third of the patients with hypercalcemia associated with malignancy have hyperprostaglandinemia and that indomethacin treatment of these patients is almost invariably successful.

Seyberth *et al.* (1975) measured the daily urinary excretion of PGE-M (the tetranor end product of PGE metabolism) in normoparathyroid patients with neoplastic diseases. Fourteen hypercalcemic patients secreted significantly more PGE-M than ten hospitalized controls (58.4 vs. 7.1 μg/mg creatinine). Normocalcemic patients with solid tumors had slightly elevated urinary excretion of PGE-M (14.3 μg/mg calcium). In contrast, patients with hypercalcemia associated with either primary hyperparathyroidism or hematological neoplasia had PGE-M excretion rates indistinguishable from those of the controls. In six hypercalcemic patients, aspirin and indomethacin suppressed both PGE-M excretion and serum calcium levels. In more recent studies designed to predict which patients would respond to indomethacin, Seyberth *et al.* (1976*b*) treated tumor patients with either indomethacin, 75–150 mg/day, or aspirin, 1.8–4.8 g/day. In patients with markedly elevated PGE-M excretion (52.3 ± 8.3 μg/g creatinine) but no bone metastases, prostaglandin synthesis inhibitors normalized serum calcium levels from 13.0 ± 0.5 mg% to 10.4 ± 0.2 mg%. In patients with both elevated PGE-M excretion (71.9 ± 41.6 mg/g creatinine) and bone metastases, the same treatment decreased serum calcium levels from 13.9 ± 0.6 mg% to 12.7 ± 0.5 mg% but did not normalize them, despite normalization of PGE-M excretion rates. In patients with normal urinary PGE-M levels, indomethacin treatment was ineffective.

b. Breast Cancer. Evidence has recently been presented that prostaglandins may be at least partially responsible for the development of lytic bone lesions and hypercalcemia in women with carcinoma of the breast. Bennett *et al.* (1975) used the rat stomach strip bioassay to measure synthesis of PGE- and PGF-like material by human breast cancer tissue. Malignant tumors ($N = 23$) synthesized 57 (10–110) ng PGE_2 equivalents per gram wet weight compared to 3.1 (benign breast tissue; $N = 5$) and 1.4 (normal breast tissue; $N = 19$).

Tumors from 11 patients were partially characterized. Tumors from four patients with bone metastases synthesized 700–2500 ng prostaglandin per gram of tissue, whereas synthesis of active material did not exceed 190 ng/g in six of seven tumors from patients free of bony invasion. These observations have recently been updated and strengthened (Bennett *et al.*, 1976). Twelve of 47 patients studied had positive bone scans; their tumors synthesized 820 (range 460–1870) ng of bioassayable prostaglandin per gram of tumor. In contrast, tumors from patients with negative bone scans synthesized only 123 (80–300) ng per gram of tumor. Bennett postulated that tumor-synthesized prostaglandins might resorb bone and thus permit implantation of circulating cells.

The studies of Powles *et al.* (1976) supported the role of PGE in the hypercalcemia associated with breast cancer. Using the bone-resorbing bioassay, these investigators reported that 23 of 38 (60%) breast cancers caused significant (>25%) osteolysis *in vitro*. Among the 23 women with osteolytically active tumors, seven developed (or had) bone metastases and four became hypercalcemic; in contrast, none of the patients with inactive breast tumors developed either lytic bone metastases or hypercalcemia. In eight of nine tumors, culture in the presence of aspirin (16 μg/ml) virtually abolished the osteolytic activity. These studies implied that aspirin and other nonsteroidal anti-inflammatory agents might be useful in preventing bony metastases in patients with carcinoma of the breast. In further studies (Dowsett *et al.*, 1976), these investigators noted good correlation between the osteolytic activities of acid–ether extracts and concentrations of radioimmunoassayable prostaglandins. However, they suggested that breast cancers also produced bone-resorbing factors other than prostaglandins; a significant amount of the material with bone-resorbing activity was not extractable into acid–ether.

In an attempt to put the role of prostaglandins in bone resorption and hypercalcemia in perspective, at the 1976 Vail Prostaglandin Meeting, Powles separated hypercalcemic cancer patients into two groups based predominantly on the etiology of the hypercalcemia. In the first, which he named *direct*, bone metastes are widespread, serum phosphorus levels are normal or high, and the hypercalcemia is steroid responsive but not responsive to aspirin or indomethacin. The tumors implicated in these patients include myeloma, lymphoma, and carcinoma of the breast. The hypercalcemia is *directly* due to resorption of bone by tumor. In some of these patients, bone resorption appears to be secondary to the local osteolytic effects of tumor-synthesized prostaglandins; in these patients, aspirin and indomethacin might be useful in preventing skeletal metastases and hypercalcemia. In the other group, hypercalcemia is due to *ectopic* production of a circulating *humoral* agent. In these patients, most of whom have carcinoma of the kidney or bronchus, bone metastases are not necessarily obvious or widespread and serum phosphorus levels are normal or low. This group can be further divided into a hyperprostaglandinemic group in which the hypercalcemia should respond to aspirin and/or indomethacin and a normoprostaglandinemic group in which the serum calcium should not change in response to either of these agents. Although this is obviously an oversimplification, it serves well as a conceptual model. PGE appears to play a role in both groups. In the *direct* group, indomethacin may be useful in preventing bone metastases, whereas in the *humoral* group, blockade of PGE biosynthesis appears to be a useful therapeutic modality.

IV. CONCLUSIONS

The experiments described have conclusively implicated the prostaglandins in the control of tumor cell proliferation and function. Prostaglandin analogues have been shown to inhibit the growth of experimental tumors *in vivo*. Inhibitors of prostaglandin biosynthesis have been successfully used in clinical situations to treat hypercalcemia associated with neoplasia; they are currently being utilized in experimental trials to prevent the development of osteolytic bone metastases. The prospects are exciting. Despite the progress, an enormous number of questions have been posed. They include the role of the inflammatory response and the cyclic nucleotides in mediating the effects of the prostaglandins. There is little question but that with intensive research these sorts of questions can readily be answered and the role of prostaglandins in oncology be firmly established.

V. REFERENCES

Adolphe, M., Giroud, J. P., Timsit, J., and Lechat, P., 1973, Etude comparative des effets des PGE$_1$, E$_2$, A$_2$, F$_{1\alpha}$, F$_{2\alpha}$, sur la division des cellules HeLa en culture, *Compt. Rend. Acad. Sci. Paris* **277**:537.

Apps, M. C. P., and Cater, D. B., 1973, Production of histamine-like and prostaglandin-like substances from serum incubated with rat, dog, mouse or human tumors, *Br. J. Exp. Pathol.* **54**:203.

Barrowman, J. A., Bennett, A., Hillenbrand, P., Rolles, K., Pollock, D. J., and Wright, J. T., 1975, Diarrhoea in thyroid medullary carcinoma: Role of prostaglandins and therapeutic effect of nutmeg, *Br. Med. J.* **3**:11.

Beliel, O. M., Singer, F. R., and Coburn, J. W., 1973, Prostaglandins: Effect on plasma calcium concentration, *Prostaglandins* 3:237.

Bennett, A., Gradidge, C. F., and Stamford, I. F., 1974, Prostaglandins, nutmeg and diarrhea, *N. Eng. J. Med.* 290:110.

Bennett, A., McDonald, A. M., Simpson, J. S., and Stamford, I. F., 1975, Breast cancer, prostaglandins and bone metastases, *Lancet* 1:1218.

Bennett, A., Charlier, E. M., McDonald, A. M., Simpson, J. S., and Stamford, I. F., 1976, Bone destruction by breast tumors, *Prostaglandins* 11:461.

Bhana, D., Hillier, K., and Karim, S. M. M., 1971, Vasoactive substances in Kaposi's sarcoma, *Cancer* 27:233.

Blum, I. G., 1975, Indomethacin in hypercalcaemia, *Lancet* 1:866.

Brereton, H. D., Halushka, P. V., Alexander, R. W., Mason, D. M., Keiser, H. R., and DeVita, V. T., Jr., 1974, Indomethacin-responsive hypercalcemia in a patient with renal-cell adeno-carcinoma, *N. Eng. J. Med.* 291:83.

Caputo, C. B., Meadows, D., and Raisz, L., 1976, Failure of estrogens and androgens to inhibit bone resorption in tissue culture, *Endocrinology* 98:1065.

Cohen, F., and Jaffe, B. M., 1973, Production of prostaglandins by cells *in vitro:* Radioimmunoassay measurement of the conversion of arachidonic acid to PGE_2 and $PGF_{2\alpha}$, *Biochem. Biophys. Res. Commun.* 55:724.

Cummings, K. B., Wheehis, R. F., and Robertson, R. P., 1975, Prostaglandin: Increased production by renal cell carcinoma, *Surg. Forum* 26:572.

Delmont, J., and Rampal, P., 1975, Prostaglandin and carcinoid tumors, *Br. Med. J.* 4:165.

Demers, L., Allegra, J., Harvey, H., Lipton, A., Luderer, J., White, D., Gillin, M., and Brenner, D., 1976, Plasma prostaglandins in hypercalcemic patients with neoplastic disease, *Clin. Res.* 24:359A.

Dietrich, J. W., and Raisz, L. G., 1975, Prostaglandin in calcium and bone metabolism, *Clin. Orthop. Related Res.* 111:228.

Dietrich, J. W., Goodson, J. M., and Raisz, L. G., 1975, Stimulation of bone resorption by various prostaglandins in organ culture, *Prostaglandins* 10:231.

Dindogru, A., Gailani, S., Henderson, E. S., Wallace, H. J., Jr., and Fitzpatrick, J., 1975, Indomethacin in hypercalcaemia, *Lancet* 2:365.

Dowsett, M., Easty, G. C., Powles, T. J., Easty, D. M., and Neville, A. M., 1976, Human breast tumour-induced osteolysis and prostaglandins, *Prostaglandins* 11:447.

Fawell, W. M., and Thompson, G., 1973, Nutmeg for diarrhea of medullary carcinoma of thyroid, *N. Eng. J. Med.* 289:108.

Feinblatt, J. D., Tai, L., and Leone R. G., 1976, Secretion of a bone resorbing factor by thyroid glands: evidence for a prostaglandin, *Prog. Endocrine Soc.* 58:187.

Franklin, R. B., and Tashjian, A. H. Jr., 1975, Intravenous infusion of prostaglandin E_2 raises plasma calcium concentration in the rat, *Endocrinology* 97:240.

Goldhaber, P., 1960, Enhancement of bone resorption in tissue culture by mouse fibrosarcoma, *Proc. Am. Assoc. Cancer Res.* 3:113.

Gomes, B. C., Hansmann, E., Weinfeld, N., and DeLuca, C., 1976, Prostaglandins: Bone resorption stimulating factors released from monkey gingiva, *Calcif. Tiss. Res.* 19:285.

Goodson, J. M., McClatchy, K., and Revell, C., 1974, Prostaglandin-induced resorption of the adult rat calvarium, *J. Dent. Res.* 53:670.

Hammarström, S., Samuelsson, B., and Bjursell, G., 1973, Prostaglandin levels in normal and transformed baby-hamster-kidney fibroblasts, *Nature (London) New Biol.* 243:50.

Hamprecht, B., Jaffe, B. M., and Philpott, G. W., 1973, Prostaglandin production by neuroblastoma, glioma and fibroblast cell lines: Stimulation by N^6, O^2-dibutyryl adenosine 3':5'-cyclic monophosphate, *FEBS Lett.* 36:193.

Harris, M., and Goldhaber, P., 1973, The production of a bone resorbing factor by dental cysts *in vitro*, *Br. J. Oral Surg.* 10:334.

Harris, M., Jenkins, M. V., Bennett, A., and Wills, M. R., 1973, Prostaglandin production and bone resorption by dental cysts, *Nature (London)* 245:213.

Hill, C. S., Ibanez, M. L., Samaan, N, Ahearn, M. J., and Clark, R. L., 1973, Medullary (solid) carcinoma of the thyroid gland: An analysis of the M. D. Anderson experience with patients with the tumor, its special features and its histogenesis, *Medicine* **52**:141.

Hong, S. L., and Levine, L., 1976, Inhibition of arachidonic acid release from cells as the biochemical action of anti-inflammatory corticosteroids, *Proc. Natl. Acad. Sci. U.S.A.* **73**:1730.

Hong, S. L., Polsky-Cynkin, R., and Levine, L., 1976, Stimulation of prostaglandin biosynthesis by vasoactive substances in methylcholanthrene-transformed mouse BALB/3T3, *J. Biol. Chem.* **251**:776.

Humes, J. L., and Strausser, H. R., 1974, Prostaglandin and cyclic nucleotides in Moloney sarcoma tumors, *Prostaglandins* **5**:183.

Humes, J. L., Cupo, J. J., Jr., and Strausser, H. R., 1974, Effects of indomethacin on Moloney sarcoma virus-induced tumors, *Prostaglandins* **6**:463.

Ito, H., Sanada, T., Katayama, T., and Shimazaki, J., 1975, Indomethacin responsive hypercalcemia, *N. Eng. J. Med.* **293**:558.

Jaffe, B. M., and Condon, S., 1976, Prostaglandin E and F in endocrine diarrheagenic syndromes, *Ann. Surg.* **184**:516.

Jaffe, B. M., Parker, C. W., and Philpott, G. W., 1971, Immunochemical measurement of prostaglandin or prostaglandin-like activity from normal and neoplastic cultured tissue, *Surg. Forum* **22**:90.

Jaffe, B. M., Behrman, H. R., and Parker, C. W., 1973a, Radioimmunoassay measurement of prostaglandin E, A, F in human plasma, *J. Clin. Invest.* **52**:398.

Jaffe, B. M., Philpott, G. W., Hamprecht, B., and Parker, C. W., 1973b, Prostaglandin production by cells *in vitro*, *Adv. Biosci.* **9**:179.

Kaplan, E. L., Sizemore, G. W., Peskin, G. W., and Jaffe, B. M., 1973, Humoral similarities of carcinoid tumors and medullary carcinoma of the thyroid, *Surgery* **74**:21.

Klein, D. C., and Raisz, L. G., 1970, Prostaglandins: Stimulation of bone resorption in tissue culture, *Endocrinology* **86**:1436.

Levine, L., Hinkle, P. M., Voelkel, E. F., and Tashjian, A. H., Jr., 1972, Prostaglandin production by mouse fibrosarcoma cells in culture: Inhibition by indomethacin and aspirin, *Biochem. Biophys. Res. Commun.* **47**:888.

Minne, H., Raue, F., Bellwinkel, S., and Ziegler, R., 1975, The hypercalcaemic syndrome in rats bearing the Walker carcinosarcoma 256. *Acta Endocrinol.* **78**:613.

Naseem, S. M., and Hollander, V. P., 1973, Insulin reversal of growth inhibition of plasma cell tumor by prostaglandin or adenosine 3',5' monophosphate, *Cancer Res.* **33**:2909.

Plescia, O. J., Smith, A. H., and Grinwich, K., 1975, Subversion of immune system by tumor cells and role of prostaglandins, *Proc. Natl. Acad. Sci.* **72**:1848.

Powell, D., Singer, F. R., Murray, T. M., Minkin, C., and Potts, J. T., 1973, Nonparathyroid humoral hypercalcemia in patients with neoplastic diseases, *N. Eng. J. Med.* **289**:176.

Powles, T. J., Clark, S. A., Easty, D. M., Easty, G. C., and Neville, A. M., 1973a, The inhibition by aspirin and indomethacin of osteolytic tumour deposits and hypercalcaemia in rats with Walker tumour, and its possible application to human breast cancer, *Br. J. Cancer* **28**:316.

Powles, T. J., Easty, D. M., Easty, G. C., Boudy, P. K., and Neville, A. M., 1973b, Aspirin inhibition of *in vitro* osteolysis stimulated by parathyroid hormone and PGE$_1$, *Nature (London) New Biol.* **245**:83.

Powles, T. J., Dowsett, M., Easty, D. M., Easty, G. C., and Neville, A. M., 1976, Breast-cancer osteolysis, bone metastases and antiosteolytic effect of aspirin, *Lancet* **1**:608.

Raisz, L. G., Trummel, C. L., Wener, J. A., and Simmons, H., 1972, Effect of glucocorticoids on bone resorption in tissue culture, *Endocrinology* **90**:961.

Raisz, L. G., Sandberg, A. L., Goodson, J. M., Simmons, H. A., and Mergenbagen, S. E., 1974, Complement-dependent stimulation of prostaglandin synthesis and bone resorption, *Science* **185**:789.

Ritzi, E. M., Boto, W. O., and Stylos, W. A., 1975, Measurement of initial prostaglandin F metabolites in medium of BALB/c 3T3 and SV 3T3 mouse fibroblast cultures, *Biochem. Biophys. Res. Commun.* **63**:179.

Robertson, R. P., and Baylink, D., 1975, Elevated immunoreactive prostaglandin E, hypercalcemia and suppressed parathyroid hormone in neoplasia in man, *Clin. Res.* **23**:113A.

Robertson, R. P., Baylink, D. J., Marini, J. J., and Adkison, H. W., 1975, Elevated prostaglandins and suppressed parathyroid hormone associated with hypercalcemia and renal cell carcinoma, *J. Clin. Endocrinol. Metab.* **41**:164.

Robinson, D. R., Tashjian, A. H., Jr., and Levine, L., 1975, Prostaglandin-stimulated bone resorption by rheumatoid synovia: A possible mechanism for bone destruction in rheumatoid arthritis, *J. Clin. Invest.* **56**:1181.

Sandler, M., Karim, S. M. M., and Williams, E. D., 1968, Prostaglandins in amine-peptide-secreting tumors, *Lancet* **2**:1053.

Santoro, M. G., Philpott, G. W., and Jaffe, B. M., 1976, Inhibition of tumour-growth *in vivo* and *in vitro* by prostaglandin E, *Nature* **263**:777.

Seyberth, H. W., Segre, G. V., Morgan, J. L., Sweetman, B. J., Potts, J. T., and Oates, J. A., 1975, Prostaglandins as mediators of hypercalcemia associated with certain types of cancer, *N. Eng. J. Med.* **293**:1278.

Seyberth, H. W., Hubbard, W. C., Morgan, J. L., Sweetman, B. J., and Oates, J. A., 1976*a*, Prostaglandin mediated hypercalcemia in the VX_2 carcinoma-bearing rabbit, *Clin. Res.* **24**:370A.

Seyberth, H. W., Segre, G. V., Hamet, P., Potts, J. T., and Oates, J. A., 1976*b*, Characterization of the group of hypercalcemic cancer patients who respond to treatment with prostaglandin synthesis inhibitors, *Clin. Res.* **24**:488A.

Smith, A. G., and Greaves, M. W., 1974, Blood prostaglandin activity associated with noradrenaline-provoked flush in the carcinoid syndrome, *Br. J. Dermatol.* **90**:547.

Stein, P. H., and Raisz, L. G., 1967, An analysis of the role of serum in parathyroid hormone-induced bone resorption in tissue culture, *Exp. Cell. Res.* **46**:106.

Stein-Werblowsky, R., 1960, Induction of a chorion-epitheliomatous tumor in the rat, *Nature (London)* **186**:980.

Stein-Werblowsky, R., 1974, The effect of prostaglandin on tumor implantation, *Experientia* **30**:957.

Strausser, H. R., and Humes, J. L., 1975, Prostaglandin synthesis inhibition: Effect on bone changes and sarcoma tumor induction in BALB/c mice, *Int. J. Cancer* **15**:724.

Sykes, J. A. C., 1970, Pharmacologically active substances in malignant ascites fluid, *Br. J. Pharmacol.* **40**:595P.

Sykes, J. A., and Maddox, I. S., 1972, Prostaglandin production by experimental tumors and effects of anti-inflammatory compounds, *Nature (London) New Biol.* **237**:59.

Sykes, J. A., and Maddox, I. S., 1974, Prostaglandin production by experimental tumors and its inhibition by non-steroidal anti-inflammatory drugs, *Pol. J. Pharmacol. Pharm.* **26**:83.

Tan, C., Privett, O. S., and Goldyne, M. E., 1974, Studies of prostaglandins in rat mammary tumors induced by 12-dimethylbenz[*a*]anthracene, *Cancer Res.* **34**:3229.

Tashjian, A. H. Jr., Howland, B. G., Melvin, K. E. W., and Hill, C. S., Jr., 1970, Immunoassay of human calcitonin: Clinical measurement, relation to serum calcium, and studies in patients with medullary carcinoma, *N. Eng. J. Med.* **283**:890.

Tashjian, A. H., Jr., Voelkel, E. F., Levine, L., and Goldhaber, P., 1972, Evidence that the bone resorption-stimulating factor produced by mouse fibrosarcoma cells is prostaglandin E_2, *J. Exp. Med.* **136**:1329.

Tashjian, A. H., Jr., Voelkel, E. F., Goldhaber, P., and Levine, L., 1973, Successful treatment of hypercalcemia by indomethacin in mice bearing and prostaglandin-producing fibrosarcoma, *Prostaglandins* **3**:515.

Tashjian, A. H., Jr., Voelkel, E. F., Goldhaber, P., and Levine, L., 1974, Prostaglandins, calcium metabolism and cancer, *Fed. Proc.* **33**:81.

Tashjian, A. H., Jr., Voelkel, E. F., McDonough, J., and Levine, L., 1975, Hydrocortisone inhibits PG production by mouse fibrosarcoma cells, *Nature (London)* **258**:739.

Thomas, D. R., Philpott, G. W., and Jaffe, B. M., 1974*a*, The relationship between concentration of PGE and rates of cell replication, *Exp. Cell. Res.* **84**:40.

Thomas, D. R., Philpott, G. W., and Jaffe, B. M., 1974*b*, Prostaglandin E control of cell proliferation in vitro: characteristics of HT-29, *J. Surg. Res.* **16**:463.

Voelkel, E. F., Tashjian, A. H., Jr., Franklin, R., Wasserman, E., and Levine, L., 1975, Hypercalcemia and tumor-prostaglandins: The VX_2 carcinoma model in the rabbit, *Metabolism* **24**:973.

Williams, E. D., Karim, S. M. M., and Sandler, M., 1968, Prostaglandin secretion by medullary carcinoma of the thyroid, *Lancet* **1**:22.

Yang, T. J., Dale, J. B., and Machanoff, R., 1976, Effects of prostaglandins E_1, E_2 and $F_{2\alpha}$ on the growth of leukaemia cells in culture, *J. Cell Sci.* **20**:199.

Index

Abdominal colic, 228
Acetylcholine, 86-88, 182, 183, 311, 312
Acetylsalicylic acid, *see* Aspirin
Adenosine monophosphate, cyclic (AMP),
　　83, 92-94, 97, 100, 101, 208, 244, 314
　adrenal, 95
　dibutyryl, 90, 96, 334, 335
　and prostaglandin, 96-97, 207
Adenosine triphosphate (ATP), stimulation
　　of prostaglandin biosynthesis, 286
Adenylate cyclase in lung, 210
Adenylyl cyclase, 94, 97-101, 305, 314
　and prostaglandin, 77-108
ADH, *see* Antidiuretic hormone
Adrenal gland, 95
AH/POA, *see* Anterior hypothalamic preop-
　　tic area
Airway, changes in resistance, 121
　smooth muscle tone, 201, 204, 205
Amberlite XAD, 17, 19
p-Aminohippuric acid, 285
AMP, *see* Adenosine monophosphate
Anaphylaxis, 184, 185
　and liberation of prostaglandin, 184,
　　186
　and release of chemical mediators, 169
　slow reacting substance of, 182, 183
Angiotensin, 297
　I, 58
　II, 58, 59
　and bradykinin, 297
　infusion, 311
　and prostaglandin synthesis, 290
　and renal vascular bed, 286
　—renin system, 286-290
Animal, afebrile, and prostaglandin efflux,
　　159-160

Anterior hypothalamic preoptic area
　　(AH/POA), 146, 147, 161, 162
　and fever, 155
　lesions, 155
　presence of prostaglandin, 152-154
Antibody
　against prostaglandins, 43
　and antigen-binding constant, 43
　and antigen reaction, 184-189
　definition, 41
Antidiuretic hormone(ADH), 313
Antigen
　and antibody binding constant, 43
　reactions, 184-189
　sensitization, 184-187
Antipyretics inhibit prostaglandin synthesis,
　　151, 157-159
Aorta contracting substance, 181
Arachidonic acid, 55, 69, 70, 135-140, 173,
　　182, 228, 269, 270, 275, 279, 280,
　　283, 303
　hypotension in dog, 140
　platelet aggregation, 141
　as precursor of prostaglandin, 3, 33
　pressor effect, 139
　as substrate, 50
　vasoactive agent, 142
Ascorbic acid, 188-189
Aspirin, 55, 79, 139, 157-159, 192, 204,
　　247, 273, 301, 330, 339, 343
　chronic administration of, 273
　gastric mucosal damage, 255
Asthma
　allergen-provoked attack, 192
　ascorbic acid deficiency, 188
　attack, 187, 192
　bronchial, 191-193